Praise for
Dancing to the Precipice

"Absorbing . . . documents with stylistic élan and meticulous detail a reeling period of French history. . . . With a strong narrative voice that neither vamps nor moralizes, she also describes the profligacy of the royal court with deadpan precision. . . . Moorehead, to her credit, is no biographical busybody. Quite the opposite. Her restraint is not unlike her subject's, and for the most part she lets la Tour du Pin speak for herself."

—Brenda Wineapple, *New York Times Book Review*

"Caroline Moorehead draws heavily on this celebrated memoir for her own intensely readable *Dancing to the Precipice*, often using quotes that allow Lucie's no-nonsense personality to shine through. What she adds, besides the final thirty years of Lucie's life, is the perspective that only time can provide."

—Clea Simon, *Boston Globe*

"Brilliantly re-creates not only Lucie's life but also the culture that formed it. . . . Ms. Moorehead has written an elegy for a world where the emphasis on elegance in style and behavior is alien to our own anti-elitist age. It is a testament to her skill as a writer as well as a historian that, at the conclusion of *Dancing to the Precipice*, we can mourn what has been lost by the leveling winds of the modern age in the same measure as we can celebrate its indisputable gains."

—Andrew Roberts, *Wall Street Journal*

"Excellent. . . . The greater joy of this book is Ms. Moorehead's skill in building on Lucie's observations. She covers the interplay of the European powers and the roller-coaster complexity of France from Louis XVI to Napoleon III with deceptive ease. Her description of the Revolution and then the Terror shocks the reader; her summary of the human damage wrought by Napoleon's thirst for conquest is brilliantly succinct. . . . Meanwhile, she uses the misfortunes of Lucie's life to cast a wider light on an age where disease and death were all too commonplace."

—*The Economist* (London)

"[A] remarkable biography. . . . Moorehead deftly wields period detail—the Paris salon, for instance, in which 'prowled eighteen angora cats dressed in satin'—to tell the story of a captivating woman who kept her sense of self amid the vicissitudes of politics."

—Megan O'Grady, *Vogue*

"Moorehead's biography, drawing on a trove of previously unpublished correspondence, captures the rhythm of the radical contrasts in her subject's life."

—*The New Yorker*

"More than just a biography, *Dancing to the Precipice* is also a compelling, extensive look at French, European, and American history. . . . A prolific biographer and historian, Moorehead is a master of her craft. *Dancing* flows beautifully, with Moorehead retaining a conversational tone even while chronicling political movements. . . . *Dancing* is a singular book. . . . [It] possesses remarkably vast breadth on both a personal and international scale. By the end, you have become intimately familiar not only with Lucie through her own sharp, incisive observations but also an entire extraordinary, tumultuous swath of history."
—Kim Hedges, *Star Tribune* (Minneapolis–St. Paul)

"Sensational. . . . Moorehead deftly navigates a dizzying cast of characters, location and events, allowing Lucie's 'precise, cool eye' and discerning wit to shine through. . . . Sumptuous account of Revolutionary Europe." —*Kirkus Reviews*

"An astute, thoroughly engaging biography of a formidable woman."
—Anna Mundow, *Boston Globe*

"Using Lucie's memoir and her letters as the foundation of this fascinating biography, Moorehead has filled in all the blanks, placing her subject firmly into social and historical context." —Margaret Flanagan, *Booklist*

"An utterly captivating book." —Christopher Hart, *The Sunday Times* (London)

"Outstanding. . . . The exceptional Henriette Lucie Dillon, Marquise de la Tour du Pin Gouvernet (1770–1853) has long deserved a competent biographer, and Moorehead (*Gellhorn: A Twentieth-Century Life*) does her justice."
—Jim Doyle, *Library Journal*

"Scintillating. . . . Moorehead has set herself a difficult task in rewriting Lucie de la Tour du Pin's story. She succeeds triumphantly."
—Matthew Dennison, *The Telegraph* (London)

"Madame de la Tour du Pin's Journal d'une Femme de Cinquante Ans, with its vivid descriptions of her experiences during the French Revolution and the Napoleonic Empire is one of the most enthralling memoirs of the age: a hard act, one would think, for a biographer to follow. Caroline Moorehead succeeds in doing so triumphantly in a rich and satisfying book which not only adds to our appreciation of her story but brings the whole tumultuous period and its characters to life." —Linda Kelly, *The Spectator* (London)

"Never less than a gripping story of an extraordinary life. . . . Replete with vivid moments. . . . One of the great virtues of Caroline Moorehead's fine new book [is] its ability to particularize, in the experience and reactions of a single individual, the human consequences of an entire age of revolutionary, and correspondingly violent, historical change. Time and again, her biography takes us, with equal immediacy, into the very centre of some of the most tumultuous events during the decades of France's Revolutionary Wars."
—John Adamson, *Literary Review*

DANCING TO THE PRECIPICE

BY THE SAME AUTHOR

Fortune's Hostages

Sidney Bernstein: A Biography

Freya Stark: A Biography

Beyond the Rim of the World: The Letters of Freya Stark (ed.)

Troublesome People

Betrayed: Children in Today's World (ed.)

Bertrand Russell: A Life

The Lost Treasures of Troy

Dunant's Dream: War, Switzerland and the History of the Red Cross

Iris Origo: Marchesa of Val d'Orcia

Martha Gellhorn: A Life

Human Cargo: A Journey among Refugees

The Letters of Martha Gellhorn (ed.)

DANCING TO THE PRECIPICE

The Life of Lucie de la Tour du Pin, Eyewitness to an Era

CAROLINE MOOREHEAD

NEW YORK • LONDON • TORONTO • SYDNEY • NEW DELHI • AUCKLAND

HARPER PERENNIAL

Originally published in Great Britain in 2009 by Chatto & Windus, an imprint of Random House Group Limited.

A hardcover edition of this book was published in 2009 by HarperCollins Publishers.

FIRST HARPER PERENNIAL EDITION PUBLISHED 2010.

Library of Congress Cataloging-in-Publication Data is available upon request.

ISBN 978-0-06-168442-5

10 11 12 13 14 OFF/RRD 10 9 8 7 6 5 4 3 2 1

Contents

List of Illustrations

Text Illustrations

Plate Section One

1. Lucie as a young woman
2. Comte Arthur Dillon, Lucie's father. (Georges Martin collection)
3. Thérèse-Lucy de Rothe, Lucie's mother. (Georges Martin collection)
4. Fanny Bertrand, Lucie's half-sister. (Georges Martin collection, Chateauroux archives)
5. Arthur Richard Dillon, Archbishop of Narbonne, Lucie's great-great-uncle. (Georges Martin collection)
6. Frédéric-Séraphim, Comte de Gouvernet and later Marquis de la Tour du Pin, Lucie's husband. (Guy de Feuilhade)
7. Humbert, Lucie's first child. (Georges Martin collection)
8. The Princess d'Hénin, Frederic's aunt. (Comte Liederkerke Beaufort private collection)
9. Lady Jerningham, Lucie's aunt. (Georges Martin collection)
10. Lucie's children: Humbert, Charlotte, Cécile and Aymar. (Georges Martin collection)
11. Lucie's grandchildren: Cécile and Hadelin. (Comte Liederkerke Beaufort private collection)
12. Lucie as an older woman. (Guy de Feuilhade)

Plate Section Two

1. Marie Antoinette and her children, 1787, oil on canvas, painted by Elisabeth Vigée Le Brun. (Château de Versailles, France, Giraudon/Bridgeman Art Library)
2. Caricature of Louis XVI at the time of the French Revolution. (Bridgeman Art Library)
3. *'A Versailles, A Versailles'. March of the Women on Versailles*, Paris, 5 October 1789. Coloured engraving, French School. (Musée de la Ville de Paris, Musée Carnavalet, Paris, France/Lauros/Giraudon/Bridgeman Art Library)
4. Camille Desmoulins with his wife, Lucile, and their baby, Horace-Camille, *c.* 1792, oil on canvas, by Jacques Louis David. (Château de Versailles, France/Bridgeman Art Library)
5. Thérésia Cabarrus, *c.* 1805, oil on canvas, by Francois Gérard. (Musée Carnavalet, Histoire de Paris, Paris)
6. View of the Hudson River from Fort Knyphansen, water colour on paper. By Thomas Davies. (Royal Ontario Museum, Toronto, Canada/Bridgeman Art Library)

7. Charles-Maurice de Talleyrand-Perigord, 1807, oil on canvas, by Pierre-Paul Prud'hon. (Musée de Ville de Paris, Musée Carnavalet, Paris, France/Bridgeman Art Library)
8. Lucie and six-year-old Humbert among the Indians on her farm near Albany. (Artist unknown)
9. A Parisian tea party, *c.* 1800, colour litho, by Jean Fulchran Harriet. (Private Collection/Archives Charmet/Bridgeman Art Library)
10. Josephine at her Coronation (detail from *The Consecration of the Emperor Napoleon and the Coronation of the Empress Josephine, 2 December 1804*), oil on canvas, by Jacques Louis David. (Louvre, Paris, France/Bridgeman Art Library)
11. Napoleon after his Abdication. Oil on canvas, by Paul Hippolyte Delaroche. (Musée de l'Armée, Paris, France/Bridgeman Art Library)
12. Claire de Duras, engraving, French School. (Private collection, Roger-Viollet/Bridgeman Art Library)
13. Félicie de la Rochejacquelein.
14. Madame de Staël, painted by Elisabeth Vigee Le Brun, 1807. (Château de Copper, Switzerland/Erich Lessing/Art Resource)
15. The Château de Vêves in Belgium.
16. Fanny Bertrand and her children at Napoleon's death-bed on St Helena. (Chateauroux archives)

The author and the publishers thank the Bridgeman Art Library for assistance with picture research, and Comte Liederkerke Beaufort, Georges Martin and Guy de Feuilhade for the family pictures.

Characters in the book

Lucie's family and friends

Arthur, Comte Dillon (1750–94) (*father*). Colonel-proprietor of the Dillon regiment serving under Louis XVI. At 18 married his cousin Thérèse-Lucy de Rothe, and after her death Comtesse de la Touche, first cousin of the Empress Josephine. Arthur fought in the American Revolution, was promoted General and made governor of Tobago. He returned to Paris to represent Martinique at the Estates General, then fought on the side of the republican army. Having tried to save the King's life, he was himself arrested and guillotined on 13 April 1794.

Lucy de Rothe (?–1804) (*grandmother*). After the death of her only daughter, Lucie's mother, Mme de Rothe brought up her granddaughter with great severity. Assumed to be the mistress of her uncle, Archbishop Dillon, she presided over his household until the revolution, when they fled to Germany, and then to England.

Thérèse-Lucy de Rothe (1751–82) (*mother*). Married at 17 to her cousin Arthur whom she thought of as a brother, she had two children: Georges, who died before his second birthday, and Lucie. She became lady-in-waiting to Marie Antoinette, but died of tuberculosis at the age of 31.

Richard-Arthur Dillon, Archbishop of Narbonne (1721–1806) (*great-uncle*). A worldly administrator rather than a pious prelate, the Archbishop kept a famed hunt at Hautefontaine north of Paris. Lucie accompanied him on several occasions to his see, Montpellier, where he lived in great splendour. Forced to flee France after the attack on the clergy, he spent his last years in exile in London. He was the life-long companion and lover of Lucie's grandmother, Mme de Rothe.

Frédéric-Séraphim, Comte de Gouvernet and later Marquis de la Tour du Pin Gouvernet (1759–1837) (*husband*). A soldier by profession, Frédéric served with Lafayette in the American Revolution. Briefly a diplomat, he was forced into hiding by the revolution and fled with Lucie to America. Later he was chosen by Napoleon as Prefect of Brussels and then Amiens. He represented France at the Congress of Vienna and was appointed Ambassador to Turin.

Jean-Frédéric de la Tour du Pin Gouvernet (1727–1794) (*father-in-law*). A prominent soldier and Minister for War under Louis XVI, he was arrested during the Terror and sent to the guillotine.

Adelaïde-Félicité-Henriette d'Hénin (1750–1820?) (*Frédéric's aunt*). Married at 15 to the Prince d'Hénin, from whom she lived separated, she became the centre of a group of clever, influential women in Paris. By nature irascible and impetuous, but also generous and devoted, she played an important part in Lucie's life. The Princesse d'Hénin was lady-in-waiting to Marie Antoinette and spent much of her life – after her husband went to the guillotine – as companion to Trophime-Gérard, Marquis de Lally-Tollendal, deputy to the Estates General in 1789 and later member of the Académie Française.

Félicie de Duras, Comtesse de la Rochejacquelein (1798–1883) (*goddaughter*). Daughter of Lucie's friend Claire de Duras, Félicie became Lucie's main correspondent for the last 30 years of her life. Boyish and impetuous, she embroiled Lucie's son, Aymar, in a disastrous escapade.

Lady Jerningham (1748–1825) (*aunt*). When Lucie and Frédéric fled to London in 1798, they found a home with Lady Jerningham and her family at Cossey Hall in Norfolk. Lucie was very attached to her English aunt.

Fanny Dillon (1785–1836) (*half-sister*). The only surviving daughter of Arthur and his second wife, Fanny married General Bertrand, faithful follower of Napoleon, and had four childen. They accompanied the deposed Emperor to St Helena.

Lucie's six children:

Humbert (1790–1816) Sous-Préfet under Napoleon and lieutenant in the Black Musketeers; Humbert was killed in a duel.

Séraphine (1793–5).
Alix, known as *Charlotte* (1796–1822) who died of tuberculosis.
Edward (1798) who died aged a few months.
Cécile (1800–17) who died of tuberculosis soon after her 17th birthday.
Aymar (1806–67) who was the only one of her children to survive her.

Lucie had two grandchildren to whom she was close:

Cécile (1818–93), daughter of Charlotte and brought up by Lucie.
Hadelin (1816–90), son of Charlotte who rose to prominence in the political and social world of Brussels.

Characters in France

Angoulême, Marie-Thérèse d', (1778–1851). The only surviving daughter of Louis XVI, she accompanied her uncle, later Louis XVIII, into exile in England and married her cousin, the Duc d'Angoulême. During the Bourbon restoration she presided over a starchy court, but remained unpopular.

Beauharnais, Hortense de (1783–1837). The only daughter of the Empress Josephine by her first marriage, she was later married to Napoleon's brother, Louis, and became Queen of Holland. Her son became Napoleon III.

Berri, Marie Caroline de Bourbon-Sicile, Duchesse de (1798–1870). Married to the Duc de Berri, she followed Charles X into exile and tried to inspire the royalist insurrection in which Aymar and Félicie de la Rochejacquelein took part.

Cambacérès, Jean-Jacques, Duc de Parme (1753–1824). A lawyer and judge, who became Second Consul and worked on the Napoleonic Code.

Charles X, King of France (1757–1836). Younger son of Louis XVI's brother. While still the Comte d'Artois, he was one of the first to flee the revolution. On his return, he became head of the ultra-royalist party, and succeeded his brother Louis XVIII as King in 1824. His fall marked the end of the Bourbon reign in France.

Chateaubriand, François-René (1768–1848). Poet and writer, he spent the first years of the revolution in England, returning to France to have a troubled relationship with Napoleon. He inspired great devotion in women.

Danton, Georges (1759–94). The first president of the Committee of Public Safety in the French Revolution, he was considered a moderating influence on the Jacobins. Accused of leniency towards the enemies of the revolution, he was sent to the guillotine.

Desmoulins, Camille (1760–94). A political journalist and lawyer, Desmoulins played an important part in the revolution through his writings. He was a friend of Lucie's father Arthur, and refused to condemn him before the Tribunal. Falling out with Robespierre, he was tried with other moderates. His wife Lucile followed him to the guillotine, going to the scaffold on the same day as Arthur and leaving two small children.

Fouché, Joseph, Duc d'Otrante (1759–1820). One of the most efficient organisers of the Terror, his political skills contributed to the fall of Robespierre. Later, as Minister for Police under Napoleon, he created a formidable network of spies.

Josephine de Beauharnais, Empress, first wife of Napoleon (1763–1814). Imprisoned under the Terror, in which her husband was guillotined, she married Napoleon in 1796, but was unable to give him a child. He divorced her, and she lived at Malmaison until her death.

Louis XVI, King of France (1754–93). Married to Marie Antoinette at 15 and King at 20, Louis XVI was serious-minded and vacillating. Unable to respond to the challenges of the liberals and democrats, and arrested after the failure of his plans to escape, Louis was accused of secret dealings with foreigners. Brought to trial for treason, he was executed on 21 January 1793.

Louis XVIII, King of France (1755–1824). Brother to Louis XVI, and known by the title Monsieur, he fled France on the revolution and tried to put together an army of émigrés to challenge the French republican forces. After the death of his nephew in June 1795, he took the title of King and remained in England until returning to

Paris in 1814. He had to flee once more during Napoleon's Hundred Days. Immensely fat and suffering from gout, he found it increasingly hard to oppose the ultra-conservative members of his court.

Louis Philippe, Duc d'Orléans (1747–93). Cousin to Louis XVI, he lived in the Palais-Royal where he was thought to be plotting against Versailles. In the revolution, adopting the name Philippe-Égalité, he sided with the Third Estate. He voted for the death of the King, but was himself guillotined soon after.

Louis-Philippe, Duc d'Orléans, King of the French (1773–1850). After fighting for the revolutionary army, Louis-Philippe lived in America. During the reign of Charles X, Louis-Philippe became the centre of the liberal opposition and was proclaimed King of the French after Charles X was deposed. He reigned for 18 years and was ousted by the revolutionary movement which swept through Europe in 1848.

Marat, Jean-Paul (1743–93). A Swiss-born philosopher and political theorist whose journalism was central to the revolution. Briefly one of the most important men in revolutionary France, together with Danton and Robespierre, he was stabbed to death in his bath by Charlotte Corday.

Marie-Antoinette, Archduchess of Austria, Queen of France (1755–93). The pretty, frivolous 14-year-old bride of Louis XVI remained childless for eight years and became unpopular with the conservative court at Versailles and with the people of France. When the revolution broke out, she was perceived as a reactionary influence. Accused of secret dealings with the Austrians, she was imprisoned with her family in August 1792 and guillotined in October 1793.

Marie-Louise, Empress of the French (1791–1847). Born Archduchess Maria Louisa of Austria, she was the great-niece of Marie Antoinette. She was Napoleon's second wife, and mother of Napoleon II, King of Rome. After Napoleon's abdication, she fled to Vienna, becoming Duchess of Parma. Later she remarried and had three more children.

Napoleon, Emperor of France (1769–1821). A general with the revolutionary army, he organised the coup of 18 brumaire and set up a new government, the Consulate. He was First Consul from

1799 to 1804, then Emperor until 1814. Sent into exile in Elba, he returned for a Hundred Days in 1815 before being defeated at Waterloo. He died in exile in St Helena.

Robespierre, Maximilien (1758–94). A disciple of Jean-Jacques Rousseau and one of the main architects of the Terror, when he was known as the 'Incorruptible'. With his execution in 1794, one phase of the French Revolution came to an end.

Staël, Germaine de (1766–1817). A woman of letters who wrote novels, plays and political essays and whose salon flourished after the Terror. Banished from Paris by Napoleon who found her hostile and outspoken, she spent many years in Coppet in Switzerland.

Talleyrand-Périgord, Charles-Maurice de, Prince de Benevento (1754–1838). A statesman and diplomat renowned for his political intrigues and his capacity for survival. He held office during the French Revolution – spending the two years of the Terror in the United States – and under Napoleon, Louis XVIII, Charles X and Louis-Philippe.

Tallien, Jean-Lambert (1767–1820). An active popular leader in the storming of the Tuileries in August 1792, he was also a direct participant in the September massacres, before being sent to Bordeaux to enforce revolutionary Terror on the provinces. Influential in Robespierre's downfall, Tallien later accompanied Napoleon to Egypt. He died of leprosy, in great poverty.

Tallien, Thérésia (1773–1835). A famous beauty who fled revolutionary Paris for Bordeaux where she acted as a moderating influence on Tallien, whom she later married. She became one of the leaders of Parisian social life and set the fashion for the Directory. She married three times and had 11 children, several of them by other liaisons.

Foreword

On 1 January 1820, shortly before her 50th birthday, Lucie Dillon, Marquise de la Tour du Pin, decided that the moment had come to write her memoirs. Until that day, she had never written anything but letters 'to those I love'. 'Let me take advantage,' she wrote, 'of the warmth that is still in me to tell something of a troubled and restless life, in which the unhappinesses were caused less, perhaps, by the events known to all the world, than by secret griefs known only to God.'

So saying, Lucie sat down and began writing what would be one of the finest memoirs of the age, full of humour and shrewdness and affection. She wrote boldly and dispassionately, for there was nothing retiring or falsely modest in her character and she had much to say. It was, she had decided, to be a diary, for her son and grandchildren, for she had no plans for publication, either before or after her death. And it was as a diary that she wrote it, simply and without artifice, describing precisely what she saw and heard, not only of her own extraordinary life, but the exceptionally turbulent period of French history that she lived through. She wrote about domestic matters and affairs of state, about personal tragedies and public mayhem, with optimism and robustness – despite the secret griefs – and a mixture of innocence and knowingness, which makes her voice very much her own.

When her memoir was finally published, 50 years after her death, it was immediately recognised as a faithful testimony to a lost age. Never out of print since then, it has provided countless scholars with detailed, vivid information, made all the more remarkable by the fact that, for most of her very long life, she happened to be precisely where the transforming events of her time were taking place. But her many letters – which have never been published, and which cover the 40 years of her life that followed the events described in the memoirs – are just as remarkable. In some ways, they are even more so, for they show a woman without guile or malice yet possessed of considerable shrewdness about the workings of the world.

Born in Paris in 1770 in the dying days of the *ancien régime*, into a family of liberal aristocrats with many links to Versailles and the court of Louis XVI and Marie Antoinette, she survived the French Revolution, which saw many of her family and friends die or lose all they possessed. Escaping to America, she and her husband bought a farm and became increasingly concerned about the injustices of slavery. Later she lived through the eras of Napoleon and the restoration of the French kings, Louis XVIII, Charles X and Louis-Philippe. At the time of her death in 1853 Napoleon III had just ascended the throne. Almost nothing of the world into which she was born remained, neither the grandeur, nor the idea of absolute monarchy, nor the privileges; but she herself was singularly unchanged.

Because of her parents, she grew up at the court of Marie Antoinette, but it was a court riven by corruption, vendettas and profligacy. Because of who she became, her friends included Talleyrand, Wellington, Mme de Staël, Lafayette and Josephine Bonaparte – many of whom left descriptions of her. Because of who she married – Frédéric de la Tour du Pin was a soldier, administrator and diplomat – she saw the Terror unfold in Paris and Bordeaux, attended on Napoleon and Josephine, was in Brussels during the Battle of Waterloo and observed the early days of Italy's unification. Along with a taste for hard work, she possessed a natural curiosity, an enormous need to understand and to remember, not only the grandeur and the politics, but the ordinary everyday events, the food, the clothes, the expressions on people's faces. It made her a formidable witness.

Unremittingly tough on herself, she was extremely demanding of others; but she had a shrewd and self-mocking sense of humour and she possessed a generous and loving heart. When one personal tragedy followed another – the 'secret griefs' of her life – she did not complain. On the contrary, they made her more determined than ever to show fortitude. The memoirs are a portrait in resilience, the way that great pain can be endured and overcome. Lucie was not merely courageous: she was resourceful and imaginative.

Because Lucie's own life and character were so remarkable, her story offers a fascinating portrait of an 18th-century woman. But it is more than that. The times she lived through were indeed exceptional, and it is in that context that she has to be seen, against a constantly changing, frightening and troubled background, broken by periods of domestic happiness and public prosperity, with her life running like a thread through her times. It is impossible to understand why she was so admirable without understanding the world that she looked out on; and which she survived.

What she witnessed was not just the end of an era both of extremes of privilege and extremes of poverty and backwardness, but the birth of a recognisably modern world, a new ordering of society. She saw and recognised the changes and the need for them, and most she approved of. Given her intense self-awareness and her experiences of loss and tragedy – universal experiences she shares with women at all moments of history – it is sometimes tempting to think of her as a modern woman. But Lucie belonged firmly in her times, and she dealt with her life in the ways that her 18th-century upbringing had taught her; which is why it is so important to set her clearly in her background and the age she lived through.

What Lucie discovered, as she started writing, was that she had a natural talent for description, a canny eye for the telling detail and strong feelings about right and wrong. She had feared that her memory might be poor: on the contrary, it was precise and deep. And as she wrote, so the age that she had lived through and survived came alive under her pen. Others had endured the same hardships and recorded the turmoil that consumed France in the closing years of the 18th century and the first decades of the 19th. What gave Lucie's memoirs and her letters their edge was something quite different. It was to do with a kind of purity. In an era of licentiousness and expediency, when the world of seduction and deceit depicted by Choderlos de Laclos in *Les Liaisons dangereuses* offered a mirror to the aristocratic life around her, when Catholic prelates thought nothing of fathering children, and preferment owed more to intrigue than to natural talent, Lucie retained all her life a moral clarity and simplicity. It might have made her dull and priggish. Instead, it turned her into an impressive reporter who observed and recorded a lost age with candour and humour. It made her a loving and faithful wife and a devoted mother. And it made her brave, which was fortunate, for the events that befell her would have broken a frailer spirit.

CHAPTER ONE

This Magnificent Age

When Lucie-Henriette Dillon, who all her life would be known as Lucie, was born at 91 rue du Bac on 25 February 1770, the Faubourg Saint-Germain was one of the most fashionable quarters of Paris. It was here, behind heavy wooden doors opening on to courtyards with stables and coach houses, that France's noble families lived. Abandoning the overcrowded and unhealthy Marais on the right bank of the Seine, they had crossed the river in the middle of the 17th century and settled in great stone mansions, three and four storeys high, surrounding their properties with high walls and vying with each other in grandeur.

Of all the faubourg's streets, the narrow rue du Bac, wandering down towards the river, was considered by many the most desirable. The first house, along the embankment, belonged to the Comte de Mailly; on the same side was the Marquis de Custine and further up, not far from number 91, was the Princesse de Salm, who wrote verse. Just around the corner lived the Duc de Biron, as did the Rochechouarts, where another baby, Rosalie-Sabine, was born a little before Lucie. In these houses, women held salons and sang, for the Faubourg Saint-Germain was both scholarly and musical. It was on the Duchesse de Castries's harp in the rue de Varennes that Mozart, a few years later, composed his concerto for flute and harp.

At the far end of the rue du Bac, where the road ended and the open countryside began, a missionary order had built a clergy house, with lintels of carved griffins and cherubs; its orchards and a kitchen garden looked out to the woods behind. On all sides, Paris was surrounded by forest. In the spring and summer, when Lucie and her nurse walked towards the fields, the road smelt sweetly of lime from the pollarded trees, of roses, lavender and lilac and the rare and exotic plants grown by the Swiss gardeners employed by the nobility towards the end of the 18th century. Across the river lay the open countryside of the Champs-Elysées, where on Sundays Parisians brought their children to picnic and stroll under the avenues of chestnut trees.

Number 91 was an imposing, unadorned building, its main reception rooms on the first floor reached by a handsome exterior circular staircase. Inside, the drawing rooms were hung with crimson and yellow damask, and the gold and silver threads of the embroidered armchairs were reflected in mirrors that hung around the walls. Lucie's mother, who was 20 at the time of her daughter's birth, had a room elegantly furnished in acacia. Her singing voice was pleasant and she owned a pianoforte, one of the first to be seen in Paris and which Lucie, as a small child, was not allowed to touch.

The house was known locally as l'Hôtel de Rothe, after Lucie's maternal grandmother, an imperious and ill-tempered woman, whose husband, Charles Edward de Rothe, a French general of Irish extraction, had died some years before; and it was here that Lucie and her parents lived.

On both sides of her interwoven family, Lucie was descended from the Irish Dillons of Roscommon. Her parents were second cousins. Their mutual ancestor, Theobald, 7th Viscount Dillon, had raised an Irish regiment in 1688 and followed James II to France, entering into service with the French, and remaining after James II's Jacobite court in exile had found a home at the palace of Saint-Germain-en-Laye, west of Paris.

Lucie's father, Arthur, had been a soldier since childhood, waiting in the wings until judged old enough to inherit the family vacancy of proprietory-colonel of the Dillon regiment, caused by the death in battle of two older uncles, whose heroism was part of family lore. He was a good-looking man, tall, with receding hair and an aquiline nose, a small mouth and large black eyes; a friend once said of him that he resembled a parrot eating a cherry. Serving in his regiment since the age of 16, he was passionate about all things military. Lucie's mother, Thérèse-Lucy, was also tall, with 'a pretty complexion and a charming face' and the fair colouring of her Irish ancestors, though some considered her rather too thin. She was good-natured and light-hearted, if not always averse to using Lucie in her battles with her own mother. She was also poorly educated, and she loved everything Versailles and the court provided.

Soon after their arranged marriage in 1768, when Arthur was 18 and Thérèse-Lucy 17, a son was born. They christened him Georges, but like a great many children in the 18th century he died in early infancy. Lucie was born two years later. Neither Arthur nor his wife had any money of their own, and Mme de Rothe, who held the family purse-strings in a grip of iron, was extremely reluctant to pass on to Thérèse-Lucy, her only child, any of the fortune that should by rights have gone to her.

More significant, perhaps, in a household in which the young Dillons and their new daughter seemed merely to perch, tolerated but disapproved of by the domineering Mme de Rothe, was the presence of another member of the Dillon family. This was Arthur Richard Dillon, Archbishop of Toulouse and Narbonne, President of the Estates of the Languedoc and widely accepted to be the lover of Mme de Rothe, who was the daughter of his sister, Lady Forester. Though this was not a liaison of much matter in an era of worldly prelates, there were some at court who disapproved, and Mme de Rothe felt their disdain keenly. The Marquis de Bombelles, a celebrated chronicler of the *ancien régime,* who admired Thérèse-Lucy's grace and charm, spoke openly of her as having been raised by a 'mother without principles and an uncle, believed to be her father'. The Archbishop's 'indecency', he remarked, should certainly have excluded him from his exalted position in the Church. Like Arthur and his young wife, the Archbishop, a somewhat portly figure of medium height, with a round moonlike face and a great passion for hunting, found it best to abide by Mme de Rothe's wishes. In the Church, he was known as an administrator rather than an evangelist, though his thesis had been on the doctrine of grace.

On both counts, then, Lucie was closely related to France's powerful elites: the nobility and the clergy.

February 1770, the month of Lucie's birth, was extremely cold. The men at work on the new Salle de l'Opéra had fallen behind and the stage was still not ready, but at the Comédie-Française Beaumarchais's new play, *Les Deux Amis*, an intricate tale of love and money, opened to a good reception, and the young Talma, with his clear voice and commanding presence, was being hailed as the great coming tragedian. In the *Mercure de France*, Paris's most popular paper, there was a long article about an eclipse of the sun, and much comment on the new inoculation – still in its experimental stage – against smallpox; and there was talk of James Cook's recent discovery of Australia, as he sailed home from plotting the transit and eclipse of Venus in Tahiti. De Bachaumont's *Secret Diary*, put together by a group of freethinkers and sceptics calling themselves *les Paroissiens*, kept all Paris entertained with gossip, rumour and court scandals. Due to a series of bad harvests, the economy, all over France, was reported to be faltering. The Foreign Minister, the Duc de Choiseul, was losing his struggle to impose reform through ambitious plans to alleviate poverty and improve agriculture, while trouble was brewing between the King, Louis XV, and his parliamentarians.

Thérèse-Lucy, in her elegant but cold room in the rue du Bac, took some time to recover from her daughter's birth. The new baby was fair and it was thought likely that she would take after her mother and be tall.

* * *

On 16 May, when Lucie was three months old, Marie Antoinette, the 15th child and 8th surviving daughter of the Emperor and Empress of the Holy Roman Empire, arrived at Versailles from Vienna to be married to the 15-year-old Dauphin. She was 14, a graceful, fair-haired girl, with blue-grey eyes, a long neck, an aquiline nose and the famous Habsburg projecting lower lip, which gave her a pouting air. Her forehead was somewhat high and her hairline a bit uneven, which would pose a challenge to her dressmakers. Pretty rather than beautiful, Marie Antoinette was almost totally uneducated, though she sang charmingly and loved to dance. Both her written French – the lingua franca of all European courts – and her written German were extremely poor; her spoken French was far from perfect.

Her bridal journey, in 57 carriages, had been as splendid and luxurious as all the wealth and artistry of the Austrian court could make it, and she travelled in clothes of crimson taffeta, red velvet and gold embroidery in a gold and velvet coach. Parting from her Austrian suite on an island in the middle of the Rhine, after two and a half weeks on the road, she had been stripped, as ritual demanded, of everything belonging to her past, down to her undergarments and her much loved dog, a pug called Mops. The future queen of the French was permitted to retain nothing that belonged to a foreign power. Before Marie Antoinette left Vienna, her formidable mother, Maria Teresa, had instructed her never to display too much curiosity or to be over-familiar with those beneath her in rank; and, she had added, she should take great care to provoke no scandals.

Waiting to greet her in the forest near Compiègne, two days before her entry into Versailles, was her betrothed, Louis-Auguste, Dauphin of France since the death of his two older brothers. A heavily built youth in a long line of notoriously greedy and fat men, he was clumsy, tone-deaf and short-sighted, but also possessed of intelligence, curiosity, studiousness and a passion for hunting. With him had come his grandfather, Louis XV, now in the 55th year of his long reign, no longer the *bien aimé*, the much loved king, but the *mal aimé*, for a rule perceived as repressive and corrupt; and the Dauphin's two younger brothers, the Comte de Provence, at 14 even stouter than Louis-Auguste, and the Comte d'Artois, 12, and widely acknowledged to be extremely good-looking. Accompanying these four were several high-ranking members of the French court, as well as three of the Dauphin's aunts, Adélaïde, Victoire and Sophie, all in their late thirties, memorably described by Horace Walpole as 'clumsy, plump old wenches'. Only later would Marie Antoinette meet the two princesses, her future sisters-in-law, 9-year-old Clothilde, whose girth was said to exceed her height, and 6-year-old Elisabeth. Life at Versailles, with its thousand rooms, its squabbling courtiers and legions of liveried servants, its rituals and its dramas, would prove to be far more public than that at the Austrian court.

Not a great deal had changed at Versailles since the King's great-grandfather, Louis XIV, had moved the court from Paris in 1682 to the former hunting lodge on the main road to Normandy. Now, as then, ceremony and etiquette framed the royal day. In 1770, as in the 1680s, the King of France ruled by divine right, 'rendering account . . . to God alone', with wide powers over most temporal as well as ecclesiastical affairs. His court consisted of some 60 aristocratic dynasties and more than 200,000 nobles, split between the *Noblesse de Robe*, deriving their status from royal service, and the *Noblesse d'Epée* (sword) whose status came from military prowess. Public display, modes of address and formal ritualistic meals, rights and prerogatives, designed by the Sun King as a way of controlling his nobles, all remained in place, like an ancient and formal dance, even if, over the years, the squabbles had become more bitter and the rivalries more vicious.

With the years France's unwieldy administration, a patchwork of provinces, municipalities, judicial territories and bishoprics, many with their own laws and dialects, had grown steadily more complicated and arcane. Nothing, in fact, was more bewildering than the absurd array of taxes, both direct and indirect, shot through with anomalies, and from which the nobility and clergy were largely exempt. Like Spain, Prussia and the Austrian Empire, France remained a hereditary absolute monarchy, under a king who governed centrally, using the much hated and feared *lettres de cachet*, the right to imprison at will, through secret orders.

Some 6,000 people, of all ranks but admitted only by ticket, attended Louis's and Marie Antoinette's wedding at Versailles. As courtiers, Mme de Rothe, the Archbishop, Arthur and Thérèse-Lucy would have been among them. The nobility wore full court dress, the women in hooped skirts, boned bodices, puffed sleeves and trains, their hair dressed and powdered, the men in swords, silk coats and breeches. Men and women alike glittered with jewels. Marie Antoinette, in white brocade and looking more like a child than a young woman, was presented with diamonds and a collar of pearls that had once belonged to Anne of Austria; her gifts from the King, delivered in a crimson velvet coffer, included a fan encrusted in diamonds. The Dauphin, for his part, appeared sulky. The Archbishop of Rheims was on hand to bless the marriage bed, into which the Dauphin was handed, as custom dictated, by the King. When, some time later, Paris celebrated the royal marriage, the day was marked by disaster: trenches, left by workmen, blocked the exits from the Place Louis XV and as the crowds pressed forward to watch the fireworks, 132 people were crushed or smothered in the rue Royale.

Soon after Lucie's 4th birthday, just over four years later, smallpox took the life of Louis XV, and Marie Antoinette's portly, serious, 20-year-old husband mounted the throne as Louis XVI, determined to be a virtuous ruler, responsive to the interests of his people. Appointing Jacques Turgot as his first Controller General, the new King declared: 'I wish to be loved.'

* * *

In the 1770s Paris was noisy, smelly and the largest city in Europe after London. The narrow streets of the Marais were medieval stews of fetid, slippery filth. Vast crowds thronged damp, dark lanes down which, along central gutters, ran pungent rivers of rainwater and sewage, the mud so acid that it rotted anything it touched. Saltpetre caused the walls to ooze and form crusts. An appalling stench surrounded the tanners' workshops and the slaughterhouses where butchers carved up the carcasses in the open, leaving grease, blood and entrails, while live animals, mainly cows and pigs, wandered at will. There were no pavements, no numbers on the houses and very little street lighting. To advertise their wares, shops hung out wooden or even stone signs, which swung dangerously in the wind. Louis-Sebastien Mercier, devoted and tetchy chronicler of 18th-century Parisian life, remarked on a glove hanging outside a glove-maker's which was the size of a 3-year-old child. In their window boxes, dangling high above the streets, people grew flowers and herbs; and in their courtyards, they kept rabbits and chickens.

There was a constant wail of sound, as town criers shouted and merchants pushed through the people carrying produce brought in from the country; and among all this chaos sped carts and carriages causing frequent accidents. Typhus, typhoid and smallpox were rife. Bicêtre, the lunatic asylum and prison, was crammed with people who were simply poor or very old, as well as epileptics, cripples, the mad and those with venereal diseases. The year that Lucie was born, over 6,000 babies, lice-ridden, stinking of urine, bundled into filthy rags, were abandoned in doorways and church porches, the more fortunate left at l'Hôpital des Enfants Trouvés in the shadow of Notre-Dame. Very few of them reached their first birthday. Unwanted children in the provinces were often sent to l'Hôpital in Paris, strapped on to a man's back in a lined box with room for three babies, occasionally fed from a sponge soaked in wine or milk. On arrival, it was usual to find at least one dead.

Much of the life of the capital revolved around the Seine, which flowed through the centre in a south-westerly curve, and down which, from before daylight, came thousands of boats and barges bringing wood, flour, vegetables, wine and building materials to docks along the banks. Just as artisans were ruled over by guilds, so every movement of the river was regulated and taxed; oil, soap, coffee, herrings and blocks of marble from Dieppe and Holland were delivered to one place, wood to another. Fresh flowers were to be found on the Quai de la Mégisserie; wigs at the Quai de l'Horloge. On the bank by the Châtelet, six families had the concession to cook and sell tripe.

The Seine brought people too, passengers arriving by *coches d'eau*, water carriages run by the *Diligences et Messageries*. Smaller skiffs ferried people across from one bank to the other. Some of these passengers, like the wet-nurses who fed most of the children born in the city, were allowed to travel at reduced rates. Anchored here and there were barges where people could take baths, doing business between spring and late summer. It was forbidden to bathe in the open river in hours of daylight, and there were endless quarrels between those competing for the river, and fines for those who broke the rules.

But this was only one part of the capital, the part lived in by the poor. To the west, Paris was an enormous garden, dotted with magnificent houses and thickly covered in trees. Approaching the city in 1767 along the tree-lined road from Versailles, Benjamin Franklin marvelled at Paris's 'prodigious mixture of magnificence and negligence', and at the blinding pearly splendour of the steeples bathed in hazy light. The flour windmills on the hills of Montmartre reminded him of a majestic family of eagles taking flight. Franklin was not the only late 18th-century traveller to remark on the perfectly manicured paths of the Tuileries leading to the Louvre, nor on the size and majesty of the new Place Louis XV with Edmé Bouchardon's fine equestrian statue of the king. Others, coming from England, Germany or Italy, were overwhelmed by the grandeur in which the French nobility lived – though disapproving of the dirt – and by the opulence of the gold, silver and velvet liveries worn by their servants.

The Parisian garden was a world of perfection, of art and nature shaped into an oasis of delight, in which fountains trickled and caged birds sang. In the Faubourg Saint-Germain visitors stopped in the cafés which served coffee – introduced to Louis XIV by the Sultan Mahommet – to sit at marble-topped tables, read newspapers and observe the ladies who ordered their coachmen to pause while they sent a servant in to collect a cup. For Lucie's English relations, who often crossed the Channel to visit the Dillons, Paris, with its ebullient public street life, its street vendors selling sorbets, *fruits glacés* and fresh raspberries, was a source of endless entertainment and wonder. When, not long after Lucie's birth, an animal with the head of a leopard, large shining eyes, the teeth of a lion, long moustaches and feet webbed like those of a goose was captured in the Straits of Magellan and brought to France, it was the first seal ever seen in Paris and it caused a sensation.

During the Orléans Regency and the long reign of Louis XV, very little had been done to change the face of Paris. But by 1770 the city's economy, stagnant during the Seven Years War, had revived and Paris itself was in what Mercier called a '*fureur de bâtir*', a fury of building. Streets were being straightened, new squares built, the old wooden houses on the bridges over the Seine demolished. Pavements were being created, to lift pedestrians out of the mud. Windows were enlarged and given glass panes. The dark, dank, medieval streets were to be opened up to the light, illuminated, made cleaner. Religious orders, exorbitantly rich in property and land, perceiving the steep rise in land prices, were negotiating sales of some of their extensive grounds.

To fill and decorate their new *hôtels particuliers* in the up-and-coming Chaussée d'Antin, in the Faubourg Saint-Honoré and along the Champs-Elysées, with their intimate, ornate interiors and their Boucher ceilings of amorous shepherds, the nobility and the rich financiers needed furniture, hangings and portraits. They found their pictures in the biannual exhibitions mounted in the Salon Carré of the Louvre, densely hung from eye level to ceiling with Chardin's still lives and interiors, Greuze's moral tales and the new high-minded scenes of classical antiquity, inspired by recent archaeological finds at Herculaneum and in Greece and Asia Minor. The first griffins and sphinxes, returning with travellers from Baalbek and Palmyra, made their appearance not long before Lucie's birth. On panels in the new drawing rooms and libraries, Bacchus and Ceres cavorted among fawns. It was a time for collecting: shells, thimbles, lacquer boxes, telescopes, flowers – real, painted, artificial, embroidered, woven – stools, screens, porcelain from China, tiles from Delft, cups from Sèvres. Never before or since, it would be said, was so much effort expended on dress, fashion, luxury and comfort.

Architects now looked to Palladio for façades that would follow classical proportions. Rococo, the last swan song of the baroque, was fast falling from favour. The rich wanted their buildings majestic, but they also wanted them pleasing to the eye, ornamented with medallions and arabesques, with lyres, ribbons and roses, with painted wallpapers showing pastoral scenes. Whether in painting, statuary or stucco, there were to be allegories of nature, childhood and love, embracing the art of living and of happiness. 'A young gentleman,' observed Voltaire, 'is fortunately neither a painter, nor a musician, nor an architect, nor a sculptor, but he causes these arts to flower with his magnificence.' Salons were to be square if the intention was to hold serious conversation, oval if the purpose was voluptuous. Bedrooms were to be green, the colour of rest. Louis XIII's stiff high-backed chairs had long since given way to rounded sofas, ottomans, Turkish carpets and cushions. Unseen hands fed stoves through openings in other rooms. Skilled masons were summoned from the Limousin, carpenters from Normandy, armies of plasterers, roofers and joiners, who left trails of white footprints along the roads.

Nor was sculpture limited to buildings. Lucie's first sight of a formal dinner was of liveried servants bearing vast platters of sculpted food. By the 1770s master pastry cooks were vying with architects to construct miniature landscapes down the centre of dining tables, rococo scenes spun in coloured sugar, biscuit dough, wax and silk, amplified by artfully placed mirrors. People talked of food and cooking as a kind of chemistry, in which ever more arcane ingredients were blended into imaginative combinations. Larks, bunting, teal, herons and egrets appeared on menus, but peacock had been replaced by turkey as the preferred roast for banquets. One of the first lexicons of French food, the *Dictionnaire portatif de la cuisine,* published three years before Lucie's birth, listed 40 ways to prepare a bird. The potato was regarded as suitable only for pigs. There was a growing demand for freshness, for meat and game served pink and roasted, browned with dotted pieces of fat which burnt crisply. Food was glorious, elaborate, absurd: when the Prince de Ligne wanted to send a gift to the Prince de Conti, he sent him a beautiful young girl, buried under mounds of pigs' heads, with cheese from the Hainaut, capons from Campire, rabbits from Os, oysters from Ostend and shrimps from Antwerp.

Mercier left a picture of the typical Paris day, in the years just before the revolution. At one in the morning, came 6,000 peasants, bringing food and vegetables to Les Halles, the largest of the capital's many markets; at six, came the bakers, bringing fresh bread; at seven, gardeners, going to their plots; at nine, wig-makers, carrying freshly powdered wigs to clients; at ten, lawyers and plaintiffs on the way to their court cases at the Châtelet; at two, those dining in town, be-wigged, powdered, walking on the tips of their toes to keep the hems of their clothes clean. From five in the afternoon, chaos and confusion, as the aristocracy set out on their social rounds; at midnight, the sound of carriages and horses, carrying the revellers home.

It was a world, for a rich, pretty child of the nobility, heiress to a great fortune, of *fêtes champêtres*, of ceilings covered in nymphs at play, of picnics in the shade of fake ruined Roman temples, of blind man's buff played by men in tall silvery wigs, of black servant boys in turbans laying out food on white tablecloths. But for Lucie, who had all this and more, her first years were lonely and confusing.

Her father, Arthur, who loved her, was often away with his regiment. Her mother, Thérèse-Lucy, whom Lucie would always remember as 'beautiful and sweet-tempered as an angel', was completely in the power of Mme de Rothe. Married at 17 to a boy only a year older than herself, whom she had known and played with as a child and for whom she felt only sisterly affection, Thérèse-Lucy was far too afraid to ask for anything for herself, her husband or her daughter. On the very rare occasions when Thérèse-Lucy summoned up the courage to talk about money, Mme de Rothe 'flew into a passion and maternal affection gave way to one of those incredible hatreds so beloved of writers of romances and tragedies'.

How Lucie's grandmother came by her grim character is nowhere revealed; she remained a strangely one-dimensional figure in Lucie's memoirs. But it cannot have been easy for her, even in such licentious times, to carry off a liaison so at odds with Church and society. Archbishop Dillon was, after all, a leading figure in the French Catholic Church, and the court, often forgiving towards men who transgressed, could be merciless towards women. Whatever the reason, Mme de Rothe's dark nature seems to have cast an unremittingly bleak light over Lucie's childhood; never, in anything she wrote later, would she recall a moment of tenderness or affection. Duty, obligation, occasionally; but never, towards her or towards anyone else, love.

Lucie herself, an only child in a house at war, in which both her mother and her grandmother wished to use her as their spy, was aware, even when very young, of the powerlessness of her frail mother and the strength of her malicious grandmother, who, when crossed, would beat and lock up the small girl for the most minor misdeeds. 'The continual warring in the house,' Lucie would write, 'meant that I was perpetually on the defensive . . . If my mother wanted me to do something, my grandmother would forbid it. I was silent, and therefore accused of being surly and taciturn. I became the butt for the moods of all and sundry.' Caught between her somewhat frivolous and weak mother and her angry grandmother, Lucie, while pretending to play with her doll or read her books, observed and remembered. 'I acquired the habit of hiding my feelings and judging for myself the actions of my parents.' To escape, she took refuge in fantasy, as many small children do, imagining another world, inventing changes of fortune where her own resourcefulness would bring her freedom and happiness. Already, at an age when more fortunate children begin to comprehend the love that binds families together, Lucie was learning about duplicity, guile and power. Later, she would write that her first thoughts were all connected with this hatred, and that 'reserve and discretion' became her earliest and most useful weapons.

Forty years later, recalling her years in her grandmother's house, Lucie would write: 'I had no real childhood.'

* * *

Number 91 rue du Bac was a restless, uneasy house, but it was also an extremely cultivated one. In Lucie's grandmother's drawing room, and among her father's friends, there was constant talk of natural history, of exploration and of the possibilities offered by scientific enquiry. The house contained a large library, exceptionally well stocked for the period, and by the age of 7 Lucie was reading 'voraciously and indiscriminately'. A tutor was found to teach her the harpsichord, a young organist from Béziers called M. Combes, who, discovering his pupil to be full of curiosity, tried to share some of his other studies with her. Lucie was fortunate in that she was born not simply at the moment when the

Encyclopédistes were completing their monumental reordering of human knowledge, but into a family with roots and connections among philosophers and writers. Lucie was exposed, not only to the works of the great *Encyclopédistes*, but to those of their number still alive in the 1770s – men such as Voltaire, Rousseau and Condorcet – who were all visitors to the salons frequented by Arthur and Thérèse-Lucy. It was these ideas, enormously exciting and often very daring in Paris just before the revolution, that provided Lucie with her first taste of knowledge. M. Combes would later say that he was sometimes forced to slow down her studies in order that she should not overtake him. Curiosity and loneliness became Lucie's spurs towards a world of the mind.

Ever since Aristotle, philosophers had been arranging and rearranging the map of mental knowledge. The origins of the remarkable intellectual experience that became known as the Enlightenment lay neither in the 18th century nor in France; but in the Paris of Louis XV. The quest to sort out and clarify phenomena, to open man up to scientific scrutiny, took a particular form, shaped by men such as Diderot and Montesquieu and promoted and even paid for by the women who, for well over a century, perfected the art of the French salon. Neither Mme de Rothe nor Thérèse-Lucy held a salon of their own, but their friends and relations did, and what they discussed there was much talked about in the rue du Bac.

The *Encyclopédie*, published in 17 volumes between 1751 and 1772, was the work of 150 known and dozens of unknown contributors, who, under the driving spirit of Denis Diderot, set out to draw up a systematic account of the 'order and concatenation of human knowledge'. It was to be, said one of its founders, Jean d'Alembert, a 'kind of world map' showing not only the principal countries of the mind, but the roads leading between them, a 'history of the human spirit, not of men's vanity'. He saw it as a Lockean version of Bacon's tree of knowledge, starting from the premise that we can know nothing beyond what comes to us from sensation and reflection, and that, as sentient, cogent beings, we have no choice but to sweep away the cobwebs of superstition and darkness.

Knowledge, said Diderot, was power; by charting its contours, the *Encyclopédistes* thought that they might conquer the world.

No longer would the universe be a mystery, but a machine that could be taken to bits, examined, altered and improved. Even death, with its ritual of confession, resignation and absolution, was no longer to be feared, but accepted as a natural and gradual process. The frontispiece to the first volume showed Reason pulling a veil from the eyes of Truth, while grey clouds behind it drifted away. Hardly surprising, then, that the *Encyclopédistes* found themselves increasingly unpopular with the Church and the court, or that Diderot spent some time in prison for an essay on heresy.* Or that Lucie, an only child in a house full of adults, permitted to sit silently in her grandmother's drawing room, was intrigued by what she heard.

To delineate this new order of knowledge and draw fresh lines between the known and the unknown, d'Alembert, himself a mathematician, commissioned entries on astronomy, architecture, food, the arts, mathematics, literature, the occult, love, mechanics, optics, and these were available to Lucie as she grew up. 'I was,' she would write, 'remarkably eager to learn. I wanted to know about everything, from cooking to experiments in chemistry.' Appetite for the printed word was rising year by year. The timing for the *Encyclopédie* had been good: though the sources of many of the great rivers remained mysterious, the surface of the oceans was being explored and mapped, while exotic new species of plants and animals were being brought back from the colonies. Even so, it was inconceivable that any other part of the world would match Europe: 'All Asia is buried in the most profound darkness,' observed the Comte de Volney. 'The Chinese . . . offer to my view an abortive civilisation and a race of automatons . . . The Indian vegetates in an incurable apathy. The Tartar . . . lives in the barbarity of his ancestors.'

Among the most celebrated of the philosophers was Voltaire, whose ideas on personal and religious freedom and material progress had been refined during a visit to England in the 1720s. What Voltaire called for, before retiring to his house at Ferney on the Swiss border, was a representative government, a spirit

* When the *Encyclopédie* reached its seventh volume, it was placed on the banned Index and its authors, the *philosophes*, were labelled as sexual deviants. Sodomy was known as *le pêché philosophique*, the philosophers' sin, and philosophical texts as pornography.

of tolerance and material happiness, even, if need be, luxury, providing it was of the right kind, 'polite' and not 'frivolous or lazy'.

In these views, he was opposed by another of the great figures of the Enlightenment years, Jean-Jacques Rousseau, who argued that, on the contrary, man, in becoming modern, had lost his innocence and health and was now miserable: born free, he wrote in his widely quoted *Social Contract*, which appeared in 1762, man was now 'everywhere in chains'. The material prosperity born of progress had served only to corrupt his pristine purity. In the years immediately before Lucie's birth, Rousseau published two of the 18th century's best-selling novels, *Émile* and *La Nouvelle Héloïse*, in which he called for readers to abandon the false lures of society and retire to nature and solitude, there to ponder the unmediated word of God. In Rousseau's novels, lovers teach each other to love, and to read so deeply that literature is absorbed into life.

Both Voltaire and Rousseau feared for France's future. In 1764, Voltaire wrote that he regretted that he would not be alive to witness a 'revolution that cannot fail to happen'. Rousseau, for his part, believed that it was not only the French monarchy, but the royal families of all Europe who did not have much time left on their thrones. 'They have all shone brightly, and every state that shines so brightly is on its path to decline . . . We are approaching a moment of crisis and a century of revolutions.' But that was in the 1760s, and neither Versailles nor the court was listening, though the message did not go unheard in the Faubourg Saint-Germain, where the Dillons were not the only family drawn to the heady world of new freedoms.

The Enlightenment, as it unfolded, touched most of educated Europe, but in France, and particularly in Paris, its direction was determined early by a succession of highly intelligent, imaginative, bold women who invited into their salons '*honnêtes gens*', men of letters, scholars and socialites who were, like themselves, tolerant, reasonable, full of restraint and self-respect, hostile to the idea of a powerful and controlling Church and monarchy. The life of the Parisian salon had started in 1613 when Madame de Rambouillet opened her famous Blue Room to a collection of witty, erudite friends. And it had been carried on down the years, one hostess succeeding, and sometimes rivalling, another, handing down their guests as one died and another took over. It was into this world that Lucie was born and where she grew up, and it left an indelible print on her.

In rooms that were themselves especially charming, intimate and conducive to conversation, her mother's and her grandmother's friends presided over talk that embraced gallantry but not love; morals but not religion; philosophy, literature and the sciences but neither domestic matters (too boring) nor politics (too dangerous). Over dinners habitués debated moral dilemmas, composed maxims and satirical verses, discussed free will, geometry, economics, and read aloud to each other from their new works. In the second volume of the *Encyclopédie*, published in 1752 – and immediately viewed as subversive – conversation was described as a river of talk, flowing lightly, without affectation, moving from topic to topic, neither a game of chess nor a contest of arms. (Rousseau complained that this exquisite courtesy was nothing but a mask for sterility and sophistry.) For the philosophers of the Enlightenment and their friends, the salons were the one place where ideas of this kind could be aired in safety, where no questions were deemed too sensitive to debate, no thoughts too perilous to think. Many severed their links with their religious upbringings.

It was not all, of course, about ideas. For it was in the Parisian salons of Mme du Deffand and Mme Geoffrin – the latter was the last real patroness of the *Encyclopédie*, a woman neither clever nor educated enough to join in the conversations but shrewd enough to preside and keep control – that the idea of the '*douceur de vivre*', that elusive and untranslatable concept, was born. It was an art that consisted in living well, with courtesy, elegance and mutual pleasure, in which pleasing others was a form of pleasure in itself, and where etiquette, '*politesse*', and '*bon ton*' provided protection from confusion and the uncertainties of the outside world. Voice, gesture, self-awareness, amiability, and even silence, all possessed meanings. Even Hume, who held that English political life was greatly superior to French, agreed that the French, in their salons, had 'perfected that art, the most useful and agreeable of any, *l'art de vivre*, the art of society and conversation'. Mme du Deffand and Mme Geoffrin did not die until Lucie was nearly 10, and their influence was felt strongly throughout the Faubourg Saint-Germain.

And as the nobility, across the 18th century, became more and more alienated from the governance of France, stifled by those in power at Versailles, repelled by the licentiousness of the Orléans Regency, so in their salons these women and their friends redefined themselves through their attachment to exquisite manners, and through their wit and epigrams and word games. Flattery was tolerated, provided it did not turn into adulation; teasing was permitted, but not malicious mockery, for that would transgress against courtesy. And it was happiness, that concept dear to Voltaire, defined by Montesquieu as the perpetual satisfaction of endlessly deferred desires on one hand and a state of tranquillity on the other, that most interested them. 'It is to this noble subordination,' wrote Talleyrand, the man who would later mourn the loss of the '*douceur de vivre*', 'that we owe the art of seemliness, the elegance of custom, the exquisite good manners with which this magnificent age is imprinted.' It was in Talleyrand's world that Arthur and Thérèse-Lucy moved.

And nor was it all serious or all very kind. The elderly Princesse de Ligne, whose pale, plump, shiny face ended in three chins, was described as resembling a melting candle, while of the Duchesse de Mazarin it was said that she had the freshness not of the proverbial rose, but of meat in a butcher's shop. It was at her reception that a flock of sheep, newly washed and guarded by a shepherdess, a dancer at the opera, intended to look decorous in the garden, panicked and got loose among the guests, bleating and crashing into the wall of mirrors that ran the length of a long gallery.

For her entire childhood, until the revolution put an end to it for ever, Lucie lived in a world in which elegance of performance was a form of freedom of expression. The people who filled her life were witty, full of curiosity, eager to learn, attentive to the meanings of words and their most subtle nuances, convinced that culture could overcome prejudice, ignorance and the brutality of the instincts. And they sincerely believed that France itself was more cultured, more intellectually interesting, more attuned to manners and taste than any other country in the world. It gave her a cast of mind she never lost, a taste for conversation that went well beyond the simple imparting of ideas, and an attachment to manners and the need to give pleasure to others before oneself. Allied to her own innate intelligence, and her too early understanding of unhappiness and the importance of self-reliance, it gave her a strength remarkable in so young a girl.

* * *

Rousseau's call for a return to nature and behaviour that was natural, rather than artificial, had struck a chord with the French nobility. By the 1770s they were beginning to look to the countryside for a retreat into a welcome and healthy simplicity. The habit of swaddling newborn babies and sending them to wet-nurses was abandoned in favour of breast-feeding them at home. (Crawling, however, was discouraged as 'animal-like'.) In the wake of *Émile* and *La Nouvelle Héloïse*, people flocked to the country, to picnic, to walk, and to look for plants. Increasingly, the nobility took to spending longer periods of time in their country estates, though many of their pleasures travelled with them from Paris. By the 18th century such was the obsession with theatre that country houses had miniature halls complete with boxes built for amateur theatricals, in which house parties acted out proverbs, staged comic operas and wrote their own plays. Where there were no little theatres, they used the orangeries or outhouses.

In 1764, Lucie's grandfather, General de Rothe, had bought a château at Hautefontaine, 20 kilometres west of Compiègne, perched on the side of a hill overlooking a gorge above a hamlet. The wooded valley, of beech and oak, was surrounded by lakes, by fields of wheat and meadows, by a few vines, and by quarries which sent stone to Paris, 60 kilometres away. The château, rebuilt in 1720 round a central hall, with a large dining room giving a view down a long avenue of trees to a fortified 12th-century church, had a particularly fine staircase, rising to the first floor in stone, continuing in wood above. There were 25 separate apartments for guests, each with its own adjoining bedroom, plus a closet, a dressing room and a room for a servant, reached by an internal staircase. The fireplaces were of marble and the house, unusually for the times, had a bathroom attached to the main bedroom. Surrounding the château were walled gardens, a park, a field for archery and a dovecot, this last a status symbol reserved for the nobility. The General lived to enjoy his retreat for just two years.

Archbishop Dillon, however, loved Hautefontaine and as it now belonged to Mme de Rothe, it was his to use. Early each spring, Mme de Rothe, Thérèse-Lucy and Lucie moved from Paris, bringing with them servants, horses, carriages and books from the Archbishop's library. They were joined in late summer by Arthur, back from his annual four months' service with his regiment. Together with extra people brought in from the village there were 40 servants, from a maître d'hôtel to a *frotteur*, a man whose sole job it was to keep the floors of the château polished. The Rothes were liked locally, having brought prosperity to the valley, and the village now had among its inhabitants a tailor, a locksmith and a shoemaker to serve the château. Mindful of Rousseau's celebration of the simple life, the Archbishop and his guests attended local weddings and feast days and acted as witnesses to the marriages between their servants. For the *Fête des Roses*, the prettiest and most virtuous village girl was crowned with flowers and given a dowry. On Sundays, Lucie accompanied her mother, grandmother and great-great-uncle to the wooden pew reserved for them at the front of the church, though it would be said that the books carried by their guests were more likely to be novels of a scabrous nature than prayer books. Something of the irreverence of the household, its disdain for conventional religious observance, marked Lucie, even as a small child.

In the 1770s, the English were much in vogue for their horses and their hunt, and the Archbishop, along with his other worldly pursuits, was a keen follower, keeping a pack of hounds just outside the village, in order that their barking should not disturb his guests. The hunt servants, in their Dillon liveries, were all English, as was the gardener's wife, and with them Lucie read *Robinson Crusoe* and practised her English. The surrounding forests of Compiègne and Villers-Cotterêts were rich in stag and boar, and as soon as Lucie was able to ride she was allowed to join the hunt. Out hunting one day as a small child, she fell off her horse and broke her leg. Borne home on a stretcher made of branches, bearing the pain without complaint, she was put to bed for six weeks while it healed. During the day, her mother and her friends sat by her side, reading to her from the *Arabian Nights*. In the evenings a small puppet theatre would be wheeled into her room, Thérèse-Lucy and her Parisian visitors taking different parts, which they either sang or spoke, giving Lucie a delight in plays and 'works of romance and the imagination' that she never lost.

She would remember the time she spent in bed with her broken leg as one of the happiest moments of her childhood.

It was at Hautefontaine that she made a companion and a friend of a servant, a young woman from nearby Compiègne who could neither read nor write but who was evidently as devoted to her small charge as Lucie was to her. Marguerite would remain with Lucie until her death. She had, Lucie wrote, 'the heaven-sent gift of healthy judgement, fairness of mind and strength of soul . . . She helped me to see evil wherever it existed and . . . encouraged me in virtue.' Wary of the bickerings and jealousies of the household, and lonely, Lucie had found an ally. It was as well that she had, she noted later, for the things she witnessed 'might have been expected to warp my mind, pervert my affections, deprave my character and destroy every notion of religion and morality'.

At Hautefontaine, the Archbishop kept open house, even if some of the guests complained that Mme de Rothe was a trouble-maker with disagreeable manners. Members of Paris's *beau monde* came for long stays, and with them many Dillon relations, Irish and English and French, descendants of the soldiers who had come to France with James II, and others who had followed in their wake to become merchants and bankers. There was Édouard, '*le beau Dillon*', a famously handsome man, popular at court, and Arthur's sister Frances, married to Sir William Jerningham. François Sheldon, Lucie's cousin, celebrated his marriage in Hautefontaine Church. Not all the Dillon diaspora had prospered. Robert, a wine merchant in Bordeaux, had died some years earlier leaving a widow of 32, expecting her 13th child, and as they grew up these impecunious children looked to the Archbishop for patronage.

The Archbishop shared the considerable expenses of his excellent hunt, said to be the envy even of the King, with two younger men, both courtiers at Versailles. One was the Duc de Lauzun, a soldier and friend of Arthur's, a buccaneering figure rapidly going through the fortunes of his pale and unhappy wife, Amélie. The other was the Prince de Guéménée. Both men were said to be in thrall to Thérèse-Lucy's charm and gentle manner. On New Year's Day 1777, Lauzun presented Lucie with a doll, with a full wardrobe of exquisite clothes, of the kind that did the rounds of foreign courts, advertising French fashions, *la grande Pandora* in court clothes, *la petite* in everyday wear. He had ordered it from the Queen's dressmaker, Rose Bertin, and the doll had a 'well-made foot and a very good wig' as well as silk stockings, high-heeled shoes, a petticoat hemmed in embroidery, a bone corset, and a number of caps, hats and bouquets of flowers. The Prince de Guéménée, whose own wife was frequently detained at court by duties, became a constant visitor to the château, often bringing with him sheet music and sometimes even musicians from Paris. In Paris, Lucie's mother sang with Niccolò Piccinni, the new Italian favourite, 'a lively agreeable little man, rather grave for an Italian, so full of fire and genius', whose compositions were rapidly rivalling those of Gluck in popularity and causing bitter feuds between *Gluckinistes* and *Piccinnistes*. (Gluck's compositions, noted Lord Herbert, on a visit to Paris, were 'worse than ten thousand Cats and Dogs howling'.)

Adèle d'Osmond, a distant cousin of Lucie's, whose acerbic memoirs dwelt at some length on the Dillon family, would later write that her mother was so appalled by the libertine tone of Hautefontaine during her visits that she was frequently in tears. Unable to leave because of her need for support, she found herself mercilessly teased for her prudery, until the day when a visiting prelate, as worldly as the Archbishop, took her to one side. 'If you wish to be happy here,' he said, 'you must conceal your love for your husband. Conjugal love is the only kind that is not tolerated here.' It was all, as the habitués of the salons had perceived so clearly, a question of '*politesse*' and '*bon ton*', at both of which Thérèse-Lucy excelled. Etiquette dictated that while there could be no display of physical intimacy – for a man to place his hand on the back of a woman's chair was considered a grave breach of manners – a play on words, however risqué, was all part of the wit and art of 18th-century conversation. Later, Adèle d'Osmond would admit that when she arrived at Hautefontaine for the first time she had been convinced that Mme Dillon and the Prince de Guéménée were lovers; but after six months, she doubted it. Even Lucie, however, wondered later whether her mother was 'sufficiently distant in her relations with the men she liked'. Arthur, much disliked by his autocratic mother-in-law, seldom came to Hautefontaine, and there had been no child born after Lucie. Thérèse-Lucy's arranged marriage had not brought happiness to either one of them.

In the evenings, after meals that were more like banquets, in the large dining room with its carved wooden furniture and sumptuous hangings, the party would sit down to the gaming tables, to gamble at tric trac – a form of backgammon – or to play whist. Some evenings there were charades or short plays, in which all the guests, and some of the servants, had parts; on others, the visitors gathered round to listen to Thérèse-Lucy play.

When Lucie was seven, Thérèse-Lucy was named lady-in-waiting to Marie Antoinette; much of her time would now be spent at court, and Lucie was left to the mercy of Mme de Rothe's whims. But long after Hautefontaine had disappeared, Lucie remembered how she had grown up, the only child in this large, rich, hospitable, ungodly family, in which they rarely sat down to meals without guests, knowing that one day it would all be hers.

CHAPTER TWO

A Talent for Deception

That Lucie would receive a good education was never in doubt. The *Encyclopédistes* and the salons of Mme du Deffand and Mme Geoffrin had made certain of that. Lucie was also extremely fortunate in the timing of her childhood. With the revolution would disappear much of the equality won for women by men such as Diderot – who taught his daughter Angélique to '*raisonner juste*', think clearly, saying that knowledge would make the world a place in which 'children, becoming better instructed than us, may at the same time become more virtuous and happy'. It was Diderot who pointed out that girls needed to accept their biological condition, but that their education could provide the way for making that prison as comfortable as possible.

The only question was what form Lucie's schooling should take. By the 1770s, Saint-Cyr, the celebrated school started by Madame de Maintenon – where nobly born girls were urged never to forget that thcy descended from warriors, and that their appearance mattered, since beauty was a gift from God – had long since closed. But convent schools, many of them run by the Ursulines, survived and much of the nobility continued to send their daughters away at 7, seeing them only occasionally until they emerged to marry and even then only in a parlour and in the presence of a nun. One possibility was that Lucie might join her English cousin Charlotte, Lady Jerningham's daughter, in a convent not far from the rue du Bac. From time to time, Lucie was taken to visit her, but Charlotte herself never left the convent or its grounds. But Mme de Rothe and Thérèse-Lucy, schooled by Diderot and the salons, had no sympathy for what they regarded as the meagre offerings of the fashionable Parisian convents, where girls studied little beyond literature, dancing and mathematics and where the emphasis was on learning to please, while mastering and understanding nuances of gesture and demeanour.

To Lucie's great relief, Arthur and Thérèse-Lucy decided to educate their only child at home. Nothing could have pleased her more, for she continued to show signs of being hugely curious about the world, certain that some great adventure lay in store for her. She envied Marguerite her village life and, when the young woman returned from visits to her family, begged her to describe in detail every minute of her time away. She was already conscious, she wrote later, of longing for a world in which people were not forced, as she was, 'to hide their tastes and ideas'.

M. Combes was asked to stay on and teach her French, mathematics, history, geography and the sciences, and a maid was brought from London with whom she could practise her English. At Hautefontaine, whenever she could escape this Englishwoman who was meant, to Lucie's distress, to replace Marguerite as her daily companion, she would walk down to the village to watch the apothecary conducting experiments in his small laboratory. Learning was rapidly becoming not just a pleasure but a necessary distraction. With the shrewdness of a lonely child, she was discovering that the way to escape punishment and ridicule was to appear at all times impassive and obedient. 'How careful,' Lucie would write, almost 50 years later, 'one should be when bringing children up not to wound their affections, nor to be deceived by the apparent shallowness of their natures.' In old age, recalling the indignities to which Marguerite was subjected, she would still feel angry.

Like her mother, Lucie was musical. Round the corner from 91 rue du Bac, at 110 rue de Grenelle, lived Thérèse-Lucy's closest friend, Mme de Rochechouart, whose daughter Rosalie-Sabine had been born just before Lucie. It was here, as she grew up, that Lucie went to play the violin with other members of the family. In 1770, Paris was full of music teachers and organists, many of them embroiled in the squabbles between those who followed Piccinni's tender and intimate melodies, and those who supported the Bohemian Gluck and argued that music should take on all the grandeur and pathos of great theatre, with dignity rather than gallantry, and a minimum of unnecessary dances.

Up and down the Faubourg Saint-Germain, celebrated musicians and singers, come from Germany, Italy and England, performed at small private gatherings in special rooms painted with nymphs and Pans, clutching hautbois, lutes and tambourines. Paris, rather than Mannheim, was rapidly becoming the most important European musical centre, especially for symphonies. In 1778, when Lucie was 8, Mozart composed his Paris Symphony, the 31st. On the Quai Voltaire, the Marquise de Villette seated her guests on chairs carved in the shape of lyres. *Thé à l'anglaise*, a meal much in fashion in the 1770s, at which guests not only drank tea but consumed large quantities of food, was invariably taken to the sound of a harp or a violin. In the Rochechouarts' house, Lucie learnt the art of graceful performance.

Lucie herself, with her quick ways, was growing up to be interesting-looking rather than conventionally pretty. She lacked the perfect 18th-century oval face with small straight nose and delicate features and her grey eyes were rather small. After an attack of smallpox at the age of 4 – which had left her face remarkably unscarred – her eyelashes and eyebrows were somewhat sparse. Her nose, like her father's, was long and a little heavy at the tip, but her mouth and her lips were full, her teeth excellent, and she had thick ash-blonde hair. She also had a charming smile.

* * *

There was another strand to Lucie's life, and it came not from Paris and the world of the French nobility, but from America, where, after the colonies united against British rule and mobilised a militia, and Britain sent troops, fighting broke out in 1775. In 1776, 13 colonies voted to adopt a Declaration of Independence. As a soldier, the commander and proprietor of a regiment, Arthur had been following the rebellion closely. From her earliest childhood, Lucie had heard constant talk about this vast land, much of it of a wildly contradictory nature. The most damning picture came from a Dutchman who had never been there, Cornelius de Pauw. In his *Recherches philosophiques sur les Américains*, published in Paris in 1768, de Pauw reported that the New World was putrefying and swampy, covered in snakes, insects and lizards of monstrous size with odd numbers of toes. The men who inhabited this land of 'noxious vapours' were themselves very strange, more like orang-utans than humans, degenerate, sexless and absurdly small. The Indians, observed de Pauw, from a safe distance of 12,000 miles, were not only sexually frigid, but insensitive to pain, cowardly, indolent, and lacking in all curiosity; the same fate, he warned, would surely befall Europeans who ventured to settle in America. (Blacks who came to Europe, however, could hope to turn white.*)

Hector St John de Crèvecoeur, a Norman nobleman who had actually seen America for himself, described on the other hand an Arcadia of meadows and orchards, in which even the kingbirds guarding the cornfields were known for 'their extreme vigilance, indefatigable perseverance, and their audacity'. Arthur and his military young friends warmed to this vision of a Utopian land of plenty, settled by wise farmers at peace with themselves and the world, which they seemed to be refashioning in the very mould the 18th-century rationalists aspired to. Arthur had read Voltaire, and with him admired William Penn and the Quakers in tolerant, contented Philadelphia. And when in 1767 and again in 1769, Benjamin Franklin had visited Paris, where he became very popular with the nobility and the court, Arthur and his friends were delighted to learn about life in this country of free trade and political radicalism. Arthur's sympathies increasingly lay with the reformers and radicals whose unease about the profligacy of the French court was growing all the time.

Benjamin Franklin, for Arthur and his friends, was the perfect American emissary. Celebrated by Voltaire as a man of genius, discoverer of electricity, instrumental in bringing pavements and lighting to Philadelphia's streets, he was courteous, sweet-tempered, prudent and wily; and he spoke passable French. He also looked the part. In a Paris of men in powdered wigs, ruffled lace and silk stockings, he wore a rebel's plain brown coat when visiting Versailles, and wandered around Paris with a fur hat perched on his high domed forehead. Soon, a hairstyle *à la Franklin* was all the rage. The Comte de Ségur, contrasting the polish and magnificence of the French courtiers with the rustic simplicity

* How could a country filled, asked the Abbé Raynal, a popular commentator on political and social affairs, with feeble-minded iguana-eaters, conceivably compare with societies that had produced Locke, Newton, Leibniz and Descartes?

and directness of Franklin, said that it made him think of sages of the time of Plato or Cato, introduced into 'the midst of the effeminate and servile refinement of the 18th century'.

Franklin was also extremely shrewd. Perceiving the French fascination with the natural world, he played his homespun card to perfection. In the salons of Mme du Luxembourg and Mme du Deffand, where Arthur and the Duc de Lauzun met to debate metaphysics, he charmed his audience with his skill at mastering the rules of '*bon ton*', his subtlety and understatement. He was '*sensible*', sensitive, like the best of the salon habitués, and simple, and thus pleasing to the followers of Rousseau; but equally, he was scientific-minded and rational, which endeared him to Voltaire. The French philosophers, the liberal aristocratic soldiers, the worldly prelates and the essayists, all liked his energy and his versatility and they enjoyed listening to him talk about his glass-works and his tannery. And when, on 4 July 1776, word came that the 13 United States of America had declared independence from England, and, in Thomas Jefferson's words, proclaimed the equality of all men and their 'inalienable right' to life, liberty and the pursuit of happiness, the young French officers, longing to see action, began to think that an American campaign, among people like Benjamin Franklin, was exactly what they wanted. The American rebels themselves were desperate for French aid: they needed money, weapons, gunpowder and material for uniforms.

Franklin himself, back in Paris in 1777 to promote the cause of American independence as ambassador to the court of Louis XVI, in his spectacles and tall beaver fur hat, encouraged them, making friends with the influential philosopher and mathematician Condorcet, and becoming a member of the Académie des Sciences and of a Parisian masonic lodge. One night, at a dinner attended by the Abbé Raynal, a convinced sceptic about the charms of the New World, at which the meagre size of the American people was brought up, Franklin asked first the Americans present, and then the French, in two separate groups, to stand up. As it happened, the Americans at the dinner were tall, vigorous men, and the French rather small; afterwards, Franklin referred to the Abbé as a 'mere shrimp'.

In January 1777, the King, spurred on by visions of the economic and political rewards that might follow for France, granted the

American rebels 2 million *livres*, without interest, to be repaid only when 'the United States are settled in peace and prosperity'. The deal was to remain, for the moment, secret, the idea of insurgency terrifying to those Frenchmen who believed in absolute monarchy.

The risk was considerable. France, though politically at its most stable for some time, was financially faltering. There had been a series of bad harvests, and attacks on farmers and bakers. Turgot, the King's Controller General, the man committed to humane reform and the sway of reason, who had spoken of 'six years of despotism to establish liberty', who had sought to unfetter trade from crippling restrictions, abolished forced labour for road-making and tried to curb expenditure, had been dismissed the previous year. 'Monsieur Turgot wants to be me,' the King had declared, 'and I do not want him to be me.' It was Jacques Necker, a rich Swiss banker, who had made his money from successful speculations, who would now, as Director General of finance, steer France's fortunes through the costly coming American war, with a commitment to greater transparency in financial affairs, a leaner and more efficient tax policy and greater central control. Necker, a rather effeminate-looking man, with a severe expression and little interest in leisure, was humane and imaginative; he was also a master at floating loans.

A friend of Arthur Dillon's, the Marquis de Lafayette was the first to slip away to Philadelphia where he offered his services as soldier to Washington. He was soon followed by other young aristocratic officers, longing for military glory. Expecting to be greeted as saviours, many were disappointed by their reception; they complained that the sheets in the inns were filthy and that the American soldiers lacked discipline. There were not enough commissions in the American army for all these aristocratic majors and colonels, few of whom spoke English, and no money with which to pay them. For their part, the Americans found their saviours arrogant. There was a humorous moment when a Bostonian grandee offered to give a banquet for the French. Informed that they lived on frogs and salad, he sent his servants to scour the surrounding swamps, and when the soup plates were handed round, each was found to contain a large green frog.

As Franklin's popularity in Paris grew, and his likeness began to appear on medallions and snuffboxes, and the Comédie-Française staged two little-performed Molière plays because he expressed regret that he had never seen them, so his ceaseless lobbying began to pay off. A French expeditionary force of 8,000 men, under the Comte de Rochambeau, was despatched to fight alongside the rebels, now increasingly hungry, cold and ill-equipped.

In the spring of 1777 Arthur was 29. Quick-tempered and enthusiastic, he embarked on 5 April with the 1,400 men of his regiment, flying their distinctive white and green flag of harp and crown, in a squadron destined initially for the West Indies. Lucie was allowed to go with him as far as Amiens, but Thérèse-Lucy and Mme de la Rothe accompanied him to Brest, where the French fleet was assembling, and where the Archbishop blessed the ships as they sailed out of harbour. Among the officers on the *Diadème* and the *Annibal* were four Dillons; three others sailed with a regiment led by Lauzun.

On their way home from Brest, Mme de Rothe and the Archbishop purchased the entire cargo of a ship that had just put in from the Far East, and returned to Hautefontaine with porcelains from China and Japan, chintzes from Persia, hangings, silks, damasks of every colour. On wet afternoons, Lucie and Marguerite went and watched as the crates and bales were unpacked and sorted out in a warehouse. 'I was often told,' she wrote later, 'how it would indeed one day all be mine . . . But some presentiment of which I said nothing kept me from dwelling too much on future splendours. My young imagination was more inclined to dwell on thoughts of ruin and poverty.'

With her father's departure, Lucie had lost an ally. Her mother, who had taken up her appointment as Marie Antoinette's lady-in-waiting, was often away at court. With Arthur in America and without companions of her own age, she became ever quieter and more reserved. Mme de Rothe's moods remained tempestuous.

* * *

Arthur was a good soldier, brave and resourceful. His superior officers said of him that he was 'intrepid and swashbuckling'. He fought at Grenada, leading a column storming British fortifications above the harbour; though he lost 106 men dead and wounded, he captured or killed 700 English, together with their flags and cannons, which earned him a Croix-de-Saint-Louis. By September 1779, he was at Savannah, part of a French force of 3,500 men, where he led a pre-dawn attack against the British lines through a swamp, fired on from both front and flank. Conditions at Savannah were appalling: there were almost no tents and the men sat out the three-month siege in deep mud. Fighting alongside his close friends, the Comte de Noailles and Théodore de Lameth, Arthur was decorated again, rising to the rank of brigadier, though not without complaints by his superior officers that he was too prone to get into quarrels. By 1781, Arthur was in Tobago, preparing for a surprise raid on Saint Eustatius.

The war was in its closing stages. In June 1781, the French troops marched through Connecticut to join Washington's men and the two armies proceeded south to New Jersey, Philadelphia, Maryland and Yorktown, where the French were waiting with naval support. And it was here, after a three-week siege, that General Cornwallis and the English surrendered. When news of the American victory reached Paris, a confectioner created a model of the blockade in spun sugar.

By now, Arthur and most of the liberal aristocratic French officers had come to like and appreciate their American fellow officers, who took them fox-hunting when the hostilities permitted. When they returned to France, after two years' campaigning, they brought with them praise for a society both virtuous and egalitarian, in which land was owned without restrictions, even if they considered the Americans somewhat insensitive to beauty. The essence of art in all its forms, promoted and nurtured through the court at Versailles and the salons of Paris, was not, they claimed, to be found in Boston or Philadelphia.

For the French, the American war had been expensive. By early 1782, France had lent or given outright to the rebel cause $28 million. Another $6 million would follow. It was money France could ill afford. The previous year, Necker, whose strategy of no new taxes, no state bankruptcy and money to be levied through loans on international money markets had failed, had resigned after his attempts to limit spending had been foiled by the irresponsible nobility. He had also made himself unpopular by his attacks on venality and lack of accountability. For all his professed transparency, his famous *Compte Rendu du Roi*, an overview of

France's financial position, suggested a totally erroneous surplus instead of showing the actual enormous deficit, and had been ridiculed as a *conte bleu* (a fairy story). Crucially, Necker had concealed not only the vast sums eaten by the American war, but the extremely parlous state of current finances. Interest on the loans was already proving almost impossible to meet.

* * *

News of the American victory at Yorktown reached Paris soon after Marie Antoinette, after 11½ years of marriage, at last gave birth to a son. A daughter, Marie-Thérèse, known as Madame Fille du Roi, had been born in December 1778 (after childless years caused, it seemed, by the sexual awkwardness of the King); but girls could not succeed to the throne of France. Louis XVI ordered Paris to be illuminated, so that for several weeks the normally darkened city was visible for miles around. Masses and celebratory concerts were held and delegations of artisans arrived at Versailles, tailors with tiny uniforms, shoemakers with minute boots, musicians with child-sized instruments. The baby had been welcomed into the world, as custom and etiquette demanded, in a room full of people: the King, the royal family, the Princesses of the Blood, and a number of noble women with '*honneurs*', certain rights at court. This time, Marie Antoinette had not fainted, as she had at her daughter's birth, from heat and the press of people. The Princesse de Guéménée, wife of Mme Dillon's friend the Prince and governess to the royal children, had then paraded the newborn baby through clapping crowds. He was given the name Louis-Joseph-Xavier-François.

The Archbishop had excellent connections at Versailles. Madame Adélaïde, Louis XV's favourite daughter, painted in her youth by Nattier as a reclining nymph in a woodland setting, found his company amusing. It was through this friendship that the Archbishop was able to obtain favours at court for some of his impecunious Dillon relations, a matter of great import at a time when preferment ruled the fortunes of much of France's nobility.

Marie Antoinette had never made any effort to conceal her dislike for Archbishop Dillon, and an even greater distate for Mme de Rothe. It was a measure of the Queen's real fondness for Lucie's mother that she had nonetheless taken Thérèse-Lucy into service at the court, where she constantly urged her to stand up to her domineering mother. Lucie would later say that Marie Antoinette, who appreciated high spirits and charm, had been dazzled by her mother's many admirers. And when a ball was held at Versailles, by the *Gardes du Corps*, the royal bodyguards, to celebrate the birth of the Dauphin, Lucie was allowed to accompany her mother to the palace. It was her first visit to Versailles. She was 11 and she was expected to wear a miniature version of her mother's full court dress, including a hooped skirt and powdered wig. In the Grande Salle des Spectacles, she watched as Marie Antoinette opened the dancing in a blue dress dotted with sapphires and diamonds. It was a spectacle Lucie never forgot. The young Queen had been so 'young, beautiful and adored by all'. Many years later, when writing her memoirs, Lucie was haunted by the thought of how short a time Marie Antoinette had left to her.

The court at Versailles was still extremely youthful. Marie Antoinette was 26, the King 28, his brothers 27 and 25. But Versailles had retained much of the formality of earlier reigns. The court itself consisted of some 5,000 people, their separate roles filling 156 pages of the *Almanach de Versailles*, and their routines, duties and prerogatives were minutely observed. Who carried what, sat where, ate how, followed or preceded whom, wore what on which occasion, was all listed and followed. Women were admitted to court only if they could prove titles of nobility dating back to 1400. The ritual of the *lever* and the *coucher*, semi-public events in a royal life that was conducted on a permanently public stage, continued, and the morning *toilette de la reine*, at which Thérèse-Lucy, the other ladies-in-waiting and princesses of the royal family assisted the Queen to dress, remained a fixed point of Marie Antoinette's day.

Clothes, like meals, were elaborate – Louis XVI was famously greedy – costly, and subject to rigidly orchestrated rules and fashions. For the men, this meant special uniforms worn to accompany the King to particular residences – green at Compiègne, green and gold at Choisy – and to the hunt: blue, silver and red for deer, blue and crimson, with gold and silver lace, for boar. For the regular Sunday reception at court, the King put on the Order of the Holy Spirit, in diamonds. Wigs of all kinds, worn even by

A duchess at the court of Louis XVI wearing a dress with paniers.

children at court, in all shades, natural, powdered or dyed, had been court uniform since the days of Louis XIII, who became bald when young. Arthur Young, the English agriculturist, complained during a visit to Paris, that once in silk breeches, stockings and powdered hair, it was impossible to 'botanise in a watered meadow'. Young had a keen eye for what he called 'trifles', saying that 'they mark the temper of a nation, better than objects of importance'.

For Lucie and her mother, as for all the women and girls who attended court, it was considerably more cumbersome. The enormous *paniers*, or hoops, named after a type of wicker frame under which hens were kept, and which were *de rigueur* for formal wear, meant that women had to enter rooms shuffling sideways, and sit like puppets, their feet sticking out. The wigs favoured by Marie Antoinette, elaborate pyramids stuffed with horsehair, sustained by gum arabic, tallow and hog's grease and a forest of pins, sprinkled with flour and held in place at night by swathes of bandages, were often so tall that the women underneath them had to stick their heads out of carriage windows. They also itched unbearably and quickly smelt rancid. On top of this edifice was a pompon, named after Mme de Pompadour, composed of feathers, flowers and diamonds.

At court, guests were meant to glitter in jewels: diamonds, pearls, rubies, sapphires and emeralds fashioned into bouquets of flowers, worn in the hair, around the neck or sprinkled over dresses. There were ribbons, fans, gloves, muffs made of silk, feathers or fur. For walking, shoes were in leather; for an evening at court, in damask, trimmed with gold or silver braid, with narrow heels 3 inches high that made Lucie feel as if she was standing constantly on tiptoe. It made dancing, she wrote, 'a form of torture'. The material used for the dresses changed with the seasons: flowered silk for spring and autumn, satin for summer, damask for winter, all of which, given the muddy state of the streets in Paris and Versailles, needed the constant attention of several maids. Since gold, silver and gauze could not be washed, a primitive form of dry cleaning with the vapour of sulphur was used. The heavy dusting of powder for wigs, delivered as a fine spray by an *houppe de soie*, left the room and everyone in it coated in white flour. Smell, ever a problem of 18th-century life, was countered by scented soaps, pellets, 'odiferous balls' and powders, the mouth washed out with rose water and a paste made of irises, though doctors warned that too strong odours could exhaust the psyche and cause anxiety.

By the time Lucie paid her first visit to Versailles in 1781, the court was enjoying Marie Antoinette's new passion for flowers, which now decorated tables, filled sachets, perfumed gloves, fans and handkerchiefs, and were worn draped over the head or wound around bodices, kept alive in artfully concealed 'bosom bottles'. Musk, once popular, had been abandoned in favour of honeysuckle, ranunculus and hyacinth, lily of the valley and convolvulus. And some of these flowers at least made their way into Versailles's gardens, in which Lucie strolled with Thérèse-Lucy when court duties allowed her mother a few hours off.

It was under Louis XIV's landscape gardener Le Nôtre that Versailles had been transformed from what the Duc de Saint-Simon, the celebrated memoir writer, called 'that most dismal and thankless spot' into the most extravagant and influential garden in European history. As it grew, so it reflected the Sun King himself,

Coiffure au Globe, c.1780.

his power, his concept of monarchy, even his love affairs. Versailles evolved into a vast and ever-spreading geometry of intersected walks, landscaped circles and squares, paths, parterres, copses, lakes and fishponds. Louis XIV's gardens became his court, with a theatre, a concert hall, a conservatory and pleasure grounds, and fabulous water displays, all designed to distract his jaded court and keep them loyal. As at Tivoli near Rome, where water obsessed Hadrian, so Louis XIV had been in thrall to his cascading waterfalls, bubbling artificial springs, jets that spurted far above basins and fountains. Versailles was designed to provide a model of tranquillity and order in the tradition of great Renaissance gardens, reflected in regular avenues, in contrasts between light and shade and in the tensions of sudden vistas. Like Louis XIV, Le Nôtre hated flowers.

Reduced in scope and splendour first by the bankruptcy of Louis XIV's last years, then by the Regent's dislike of formal gardens, Versailles, by the time of Lucie's first visit, was once again in the process of changing. Under the influence of Rousseau's appeal for a return to the natural, sharp angles were being replaced by winding walks, formal parterres abandoned in favour of softer new arrangements of plants, and lakes and rivers created, artificially, to convey the simple, artless life.

This taste for simplicity found favour with Marie Antoinette who, not long after becoming Queen, asked Louis XVI's permission to let her take over Le Petit Trianon, Louis XV's gift to his mistress Madame de Pompadour. There Marie Antoinette set about creating a garden, in what was then known as the English style, of canals, winding paths and curving lakes, with screens of trees and trellises to preserve an impression of intimacy, and the sound of caged singing birds and tinkling water. It was in Le Petit Trianon and its famous hamlet, with its fake Norman farm buildings, that she escaped the stuffiness of court; and there that she created a theatre, with blue and gold papier mâché boxes, for the amateur theatricals in which she took the parts of shepherdess and village maiden.

* * *

Though the Prince de Guéménée had often tried to warn Thérèse-Lucy about the scandals increasingly engulfing the court, urging her to take great care not to become embroiled, Lucie's mother, like Lucie, had a trusting and almost innocent nature. While others of Marie Antoinette's favourites were busy furthering their own fortunes, through graft and financial deals, she played no part in their intrigues. But it was not only a matter of temperament and honesty: Thérèse-Lucy's health was failing. She had not in fact felt well since the birth of her first child, Georges, when she was 18. Now, approaching her 30th birthday, she had little appetite, something the doctors ascribed to a 'lacteal humour' that had settled on her liver, but which Lucie ascribed to Mme de Rothe's nastiness. In spite of being told that her blood was thin and inflamed, Thérèse-Lucy made little effort to take better care of herself, preferring to sing with Piccinni and to ride and hunt with the Prince de Guéménée in the forests around Hautefontaine.

Early in 1782, while Arthur was still in Tobago, where his military exploits had earned him the post of governor, Thérèse-Lucy began to cough blood. Mme de Rothe, maintaining that this was just an excuse to prevent her going to Hautefontaine, refused to believe that there was anything very seriously wrong with her. 'Her invincible hatred and her suspicious nature,' Lucie wrote later, 'led her to see in my poor mother's every action a calculated attempt to free herself from her authority.' It was not until a doctor, diagnosing in Thérèse-Lucy's now repeated haemorrhages a stomach complaint, insisted that she take a cure, that Mme de Rothe reluctantly consented to a visit to Spa, a fashionable health resort in the Ardennes, between Aix-la-Chapelle and Liège. Here frail patients were thought to benefit both from the calm life and the combination of acids and gases in the waters. Even so, it was Marie Antoinette, and not Mme de Rothe, who gave Thérèse-Lucy the money to make the journey.

Despite the efforts of the Enlightenment philosophers to clarify and categorise disease, it remained common in 18th-century France to attribute most illness to bad air, stemming from decomposing organic matter or 'putrid exhalations' rising out of the earth, or even to sorrow. Smallpox, measles and mumps were all known diseases, but 'fevers' could be bilious, autumnal, ephemeral or malignant, and whether 'terminal diarrhoea' meant gastric upset, dysentery, parasites or food poisoning, no one was sure. Equally, how disease was transmitted from one person to another remained a mystery. For the poor, and particularly the city poor, weakened by malnutrition, living in damp, dark rooms with bedding infested with vermin and using copper pots to cook with, illness was a long and baffling sequence of coughs, fevers, rashes, scabs and infected sores, most often ending in death. Water was polluted and alive with germs. Delicate children were doomed. Even the rich could not be cured of tuberculosis. As Thérèse-Lucy grew weaker, two of Lucie's aunts were also coughing blood.

On the way to Spa, Thérèse-Lucy and Lucie stopped in Brussels, where Charles Dillon, Arthur's brother, lived, not daring to return to London on account of vast debts at the gaming tables. Lucie liked his beautiful young wife, who had visited Paris the previous year to attend a ball at Versailles, and she enjoyed the company of her two small cousins. Thérèse-Lucy was very fond of her sister-in-law and in her letters addressed her as '*ma chère soeur*', my dear sister. The Low Countries belonged to Austria, and Lucie and her mother called on the reigning Archduchess, Marie-Christine, Marie Antoinette's sister. At Spa, a place much frequented by European royalty, Thérèse-Lucy took her small daughter dancing and soon the town was talking about the precocious French child who could dance the gavotte and the minuet with such grace. It was there, wrote Lucie that 'I tasted for the first time the heady poison of praise and success'. It was also one of the rare moments in her childhood when she had her mother to herself, and Lucie, who was reading a romantic novel by the Abbé Prévost, extolling the virtues of devotion, longed to bestow her own loving feelings on her mother. But Thérèse-Lucy, evidently alarmed by her sickness and fearful that her daughter might catch it, kept her at a distance. 'I often wept bitter tears,' wrote Lucie, 'because she would not allow me to nurse her, and I had no inkling of the cause of this strange aversion.'

Spa did not improve her mother's health; rather, the haemorrhages increased. Thérèse-Lucy was very reluctant to return to Hautefontaine and to her mother's tantrums and dreamt instead of travelling down to Naples, where the warmth and change of air might do her good. The spring and summer of 1782 were exceptionally cold in northern Europe and it rained ceaselessly. But when Thérèse-Lucy and Lucie reached Paris, it was clear that she was too ill to travel further.

Only now did Mme de Rothe realise that her only daughter was dying. Her manner underwent a profound transformation: from spiteful and tyrannical, she became tender and solicitous, insisting on giving up her own better rooms to the patient and personally seeing that she lacked for nothing. All this was witnessed by the 12-year-old Lucie with disbelief, and it was only many years later, looking back on the events of that terrible summer, that she understood that Mme de Rothe was a woman of passion and extremes and that her generosity now was simply another facet of her domineering character. For her own part, she continued to mind not being allowed to nurse her mother. One day, Marie Antoinette came to the rue du Bac to visit her former lady-in-waiting; and as Thérèse-Lucy grew weaker, so she sent pages from Versailles every day to enquire about her health. Arthur did not return to Paris.

Early on the morning of 8 September 1782, Lucie was woken by Mme Nagle, a friend of her mother's, and told that Mme Dillon had died in the night, in Marguerite's arms. Mme Nagle had come to counsel her. Lucie was to go, immediately, to her grandmother, throw herself at her feet, and beg for her protection and care, without which, with her father away and in any case not in favour, she might well find herself banished to a convent, like her cousin Charlotte, and also very likely disinherited.

Though Lucie was repelled by the need to fawn and deceive, finding it 'utterly repugnant', she did as she was told. Dry-eyed and frozen with grief in the face of Mme de Rothe's hysterical tears, she was accused of being cold and hardhearted. Later, she would wonder why, in this clerical household, there had been no chaplain on hand to give her mother the last rites. At the time she was conscious only of glimpsing, with a sudden adult understanding, a vision of the 'long years of deceit into which I was being forced'.

CHAPTER THREE

A Sparkling Picture

Soon after the death of Lucie's mother, the Archbishop set off south for his annual visit to Narbonne and Montpellier. As one of 158 senior prelates governing the immense French Church, he had presided since 1764 as a kind of viceroy over the states of the Languedoc, one of the largest and most independent provinces, stretching from the Pyrenees to Lyons. Some of the more spiritual bishops chose to remain in their sees, but since Louis XIV had dispensed with obligatory residence, Archbishop Dillon preferred to rule from afar, pressing the interests of the Languedoc at court, and descending to Montepellier only for the two months, November and December, when the Estates convened. It was then that the all-important question of the taxes voted by the clergy, the *don gratuit*, or free gift, were debated, along with public works for the region, for the Church was both extremely rich and very powerful.

There was little theological debate, for the great doctrinal battles between the Jansenists, with their puritanical stress on Augustine austerity and the innate corruptibility of human nature, and the Jesuits, who claimed that a formal observation of church practices was sufficient to achieve God's grace, lay mostly in the past. A skilled and canny administrator, the Archbishop was frequently criticised for a lack of piety, but he was widely admired for what he did for the Languedoc. 'With Monsignor Dillon,' remarked a sharp-tongued local priest, 'it was always the statesman first, the churchman second', something that Lucie, even as a small child, had instinctively understood. But during the quarter-century of his administration, more roads, bridges and canals were built, and more schools to train engineers, miners and hydrologists opened, than in the entire previous century.

Left alone at Hautefontaine with Lucie, Mme de Rothe, who was still only in her early 50s, grew extremely bored. Few of the aristocratic young officers who had once hunted in the forest of Compiègne had returned from America. In any case, the Archbishop's famous hunt, so envied by the King, had been disbanded in the wake of the charming, philandering Prince de Guéménée's spectacular bankruptcy, rumoured to stand at over 25 million *livres* and to have left penniless 3,000 separate small creditors, shopkeepers, bookmakers and servants.* 'Most of our great lords,' noted the Marquis de Bombelles drily in his journal, 'believe that they can get away with anything.' The Duc de Lauzun paid some of the debts, and the King bought up part of the Guéménée estates, but the size and suddenness of the bankruptcy stunned Versailles, where the Prince was not the only courtier with debts. His wife, the somewhat eccentric Princesse, of whom it was said that she believed that her lapdogs were in touch with the spirits, was forced to step down as governess to the royal children. She retired to her father's estates, and the Queen soon ceased to mention her name. This callousness came as no surprise to Lucie, who had learnt that the Queen, apparently so fond of Thérèse-Lucy, had entirely forgotten her within days of her death, to the extent that she had planned to go to the Comédie-Française on the day of her funeral. A reproachful courtier had to remind her that her carriage would have crossed with the cortège and the coffin.

Having nothing to distract her, Mme de Rothe took out her irritation on Lucie. When Thérèse-Lucy had died, she had taken possession of all her daughter's papers and correspondence, and Lucie thought it likely that she had found extremely unflattering references to herself. 'Her despotism,' Lucie wrote, 'ruled my entire life.' Her tutor M. Combes told her over and over again that as a 12-year-old girl, whose mother was dead and whose father was still away in America, she could at any moment find herself disinherited. 'And so,' wrote Lucie, 'I had to resolve to endure the daily trials which were the inevitable consequence of the terrible nature of this woman on whom I was dependent.' For the next

* Just before the revolution, in a good year, a loaf of bread cost 3 sous; 1 *livre* = 140 sous and £1 = 23 *livres* 3 sous. A family in Paris considered itself comfortably off with 6,000 *livres*.

five years, she said, 'not one day passed without my shedding bitter tears'.

Wishing to move closer to Paris, Mme de Rothe now bought a small property at Montfermeil, near Livry, called La Folie Joyeuse – *folie* after *folia*, the Latin for 'leaves', the currently fashionable name for country residences, and Joyeuse after a M. Joyeuse, who had completed only two very pretty wings with pavilions, before running out of money. The estate had a park, enclosed by a wall with gates leading directly out into the forest of Bondy. Carts, piled high with furniture, were despatched from Hautefontaine, but no further work was started on the house for, under French law, the manorial owner of the land, in this case the Comte de Montfermeil, could demand the house for his own at any time in the first year.

Mme de Rothe, distracted by the architects with whom she planned to build, found less time to criticise Lucie; who was in any case fascinated by their work and eager to be involved. Lucie was becoming fonder of her weak but clever great-uncle, the Archbishop who, discovering that she had a talent for numbers, took to discussing with her the plans and drawings for the new house, and asked her to 'calculate and measure with his gardeners the slopes and other surfaces . . . and go through every detail of the estimates, checking the figures'. To pass the hours of her solitary day, Lucie set about learning: to sew and to embroider, to cook and to iron. Along with reading she was acquiring a taste for hard practical work. 'I found time for everything, losing never a moment, strongly aware in my mind of all that I was taught and never forgetting it.' Something about the bleakness of the household made her resolve to equip herself for anything that might come her way. 'This prophetic instinct was always present in my mind and made me want to learn all the handicrafts necessary to a poor girl, and drew me away from the usual occupations of a young lady and an heiress.' There was already something shrewd and resilient in Lucie. When visitors came to Montfermeil, she listened closely to their conversation, storing away bits of knowledge and information for later use.

Life was yet more lonely when they returned to Paris, where the Archbishop had been persuaded to move into a house in the nearby rue de Bourgogne, taking Mme de Rothe and Lucie with him, and letting 91 rue du Bac to Baron de Staël, Sweden's ambassador to Paris. In theory, both the house in the rue du Bac and 4,000 francs in bonds on the Hôtel de Ville were Lucie's inheritance, but no money seemed to come her way, not least because the Archbishop himself, living in a style far above his considerable ecclesiastical revenues, was fast getting into debt. None of this was known to Lucie. In Paris, Mme de Rothe was systematically driving a wedge between Lucie and her childhood friends, 'by a refinement of cruelty' making it seem that the break was instigated by Lucie herself. One of these girls was Rosalie-Sabine de Rochechouart who had recently been married, at the age of 12, to the 17-year-old Comte de Chinon, though she was not expected to live with him until she was older. At the slightest provocation, Lucie was reminded of the fate of one of her Dillon aunts, who had been sent to a convent at the age of 7 and had never left it.

* * *

In the late autumn of 1783, when Lucie was a tall, bookish, independent-minded girl of 13 who spoke good English and had perfected the art of concealing her feelings, Mme de Rothe decided that they would accompany the Archbishop on his annual journey to the south. Lucie had never been further from Paris than Amiens. The party, which included the Archbishop's secretary, four servants, a ladies' maid, Lucie's own English maid Miss Beck, two footmen, a butler and a chef, travelled in two berlines, enormous, cumbersome six-horse coaches. They took with them 18 horses and three couriers to act as outriders for there was a constant threat of highwaymen. M. Combes was sent on ahead to prepare their rooms at the inns, travelling by turgotine, a rapid one-passenger wagon on shafts, named after Turgot, the reforming minister of whom it was said that he was always in a hurry.

France's roads, in the 1780s, were appalling: rutted, uneven and stony, long stretches virtually impassable. There was a constant risk of carriages overturning, and the party's progress was frequently delayed when heavy rains flooded the route and the carriages had to be floated across swollen fords. Inside, Lucie and Mme de Rothe held up their long skirts to prevent their getting wet. Montpellier, their destination, lay over 600 kilometres from Paris. They travelled from before dawn each morning until early evening when they stopped at a post house for dinner, ordered in advance by M. Combes, and to which their own chef put the finishing touches, having brought with him meat jellies and sauces prepared before leaving Paris. Lucie shared a room, and a bed, with Mme de Rothe and the scolding was merciless. 'I was never allowed to go to bed on arrival,' Lucie wrote later, 'despite the fact that each evening I was exhausted with weariness, for she would not allow me to sleep in the carriage or even to lean back.' Mme de Rothe's ruthless insistence on polite behaviour could be sadistic.

The Archbishop preferred to spend no time in Lyons, a natural stopping place, because he did not admire the current incumbent, Archbishop Montazet, considered too godly by the wordly Versailles cleric, and so the two great coaches and their outriders lumbered on, down the valley of the Rhône, to Montélimar. At La Palud, the cortège stopped to allow the Archbishop to change into his purple robes, his shoes with gold buckles, his tricorne hat with gold acorns and the Order of the Holy Ghost. Entering the Languedoc proper at Pont Saint-Esprit, they were greeted by an entire garrison in ceremonial dress, and all the local civil and ecclesiastical dignitaries in full regalia. The Archbishop gave a short speech; he was an elegant figure, if somewhat haughty, and his voice was pleasant. Arrival at Montpellier, after a night at Nîmes, was timed for after sunset, to avoid the necessity of a salute of guns, which would have offended the feelings of the Comte de Périgord, commander of the Languedoc, who was not entitled to one. Even so far from Versailles, the minutiae of etiquette and rank were jealously observed.

Medieval Montpellier was a rich city, a centre for wine, wool and verdigris, and it had traded in eau-de-vie and leather with the Levant for many years. It had a renowned medical school, a Royal Society of Science, a flourishing theatre with monthly comic operas, and a powerful masonic lodge. Its Société des Beaux Arts ran classes in drawing and engraving. In the dark, narrow streets, little altered since the Middle Ages, tailors sat cross-legged in their windows, and in the counting houses merchants weighed coins. Though Narbonne was neither the grandest nor the most opulent of the bishoprics, the Dillons lived and entertained in considerable grandeur.

Archbishop Dillon's palace in Narbonne.

Wherever they were, the household followed the same, impeccably organised ritual, the servants in their places, liveried and powdered, Lucie properly turned out, her hair dressed, within hours of arrival. Every year, the Archbishop rented the same palatial house, furnished in crimson damask, with Turkish carpets and vast stone fireplaces; the dining room sat 50 guests. He set aside his mornings for work, for which he wore a clerical robe in crimson velvet; as president of the Estates, he was left free to carry out his duties and enjoy his revenues and exalted status without interference. Lucie was allowed to study physics with the Abbé Bertholon, helping in his laboratory with his experiments, while her English maid Miss Beck cleaned and dried the apparatus. Though harsh towards her granddaughter when it came to behaviour, Mme de Rothe's desire for her to receive all the benefits of Enlightenment education was genuine.

Dinner was at three o'clock, in full dress and jewellery, the Archbishop by now having changed into black velvet, with diamond buttons. Whenever there were Englishmen present, Lucie was placed next to them; it taught her, she wrote later, the art of conversation, deciding which subject would most interest her neighbours. With the cool and appraising eye that would later mark her memoirs, she noted that they were often people of 'importance', and that just occasionally they were 'of learning too'. Each of the male guests brought his own servant, who waited on him, fetching wine from the sideboard – it was seldom kept on the table – and finding dishes his master liked from among the dozens that covered the long table. The food, on such occasions, would all be put on to the table at once, for maximum effect, the grandest dishes in the middle, artfully arranged so that they complemented one another. Jellies, moulded and dyed in blue or violet, were created with expensive indigo. With the help of their own servants, the diners worked their way towards the centre. As Voltaire had remarked, there was *bon ton* in food as in all else, and a man of taste would recognise in an instant the good from the bad.

Lucie's servant wore her own livery, which should have been blue but was in fact red, since her blue English livery was considered too close to the Bourbon blue reserved for the royal family. He also dressed and powdered her hair. Lucie and her grandmother were the only women present, something much remarked on in the town, where there was talk of a 'harem', the gossip of Paris having preceded the party south. 'The Archbishop,' recalled a visiting prelate later, 'divided his time in two: in the mornings, he chased a Cardinal's hat; in the afternoons, he looked for amusement.' Despite an unfortunate rumour that he had had a hand in the murder of an unruly Jansenist, who refused to be evicted from his monastery, the Archbishop was well liked for the prosperity he brought to the area. He was also feared, for he could be intimidating. 'Monsignor,' he told one troublesome bishop, who had foolishly voiced some contrary opinion, 'this is not a parliament: our assemblies permit of no discussion.' The political economist Adam Smith, visiting France to study the administration of the Estates and stopping in Montpellier, came away impressed; the Archbishop, he declared, was a most effective administrator.

For the Languedoc, the annual meeting of the Estates was the peak of the social year. In the mornings, Mme de Rothe and Lucie set out along the narrow streets, carried in sedan chairs, to pay visits; in the evenings, there were balls and receptions; on Sundays, after Mass in the cathedral, walks along the Promenade du Peyrou. Processions, an important part of 18th-century French life, were colourful and imposing affairs, with trumpeters, mace bearers, halberdiers with spears, orphans in coarse uniforms of the poorhouse, consuls in ceremonial robes of scarlet with purple hoods, magistrates in black silk soutanes and ermine hoods, and crosses borne high above the crowds, a long river of wealth, privilege, silk, uniforms and corporate order, weaving its way down the narrow streets. As Archbishop of Narbonne, Dillon was allowed to wear pink. Never had the splendour of the immensely rich and powerful French Church seemed so entrenched.

* * *

The signing of the peace treaty between England and America finally brought Arthur home from St Kitts, where he had ruled briefly as governor before handing the island over to the English. When Lucie returned to Paris at the beginning of 1784, just before her 14th birthday, she found him waiting for her. Arthur was now 33; he had been away almost five years. Her great pleasure in his return was, however, very brief. Pausing in Martinique on his way home, Arthur had met and become attached to the 31-year-old widow of a naval officer, the Comtesse de la Touche. She was a rich Creole, with two small children, Elizabeth and Alexandre, and a large plantation. She was also, Lucie noted crisply, amiable and good, though weak, 'with the careless good nature of all Creoles', a dangerous trait which she had unfortunately passed on to her children.

When Arthur arrived back in Paris he brought with him not only news of his engagement, but his betrothed herself, and he was eager for the marriage to take place as soon as possible, as he wanted a son. Mme de Rothe had always assumed that were he to marry again, he would choose a bride from among his English Catholic relations. She found the idea of a Creole abhorrent. Despite the fact that the King and Marie Antoinette had consented to sign the marriage certificate and the new Countess Dillon was presented at court, Lucie was forbidden to attend the wedding. 'This was,' Lucie wrote later, resorting to the terseness to which she was prone when faced with painful events, 'a great grief to me.'

Arthur, like everyone else, was afraid of Mme de Rothe, and in any case he was conscious that Lucie's future lay with her inheritance in France, and not with her stepbrother and stepsister in the West Indies. Writing to his brother, he reported that he had attempted to see Lucie several times, and that he was trying to obtain from Mme de Rothe a clear understanding of his daughter's financial position, but that she was never at home when he went to call. For himself, he wrote, he was determined to be 'happy and independent', and his second marriage was exactly what he needed. In England, where Arthur took his bride, there were several references in *The Times* to his honourable and generous conduct towards the vanquished English after the battle for Saint Eustatius, and the couple were fêted by his relations and received at court. As his sister Lady Jerningham observed, Arthur was a very good-hearted man.

Arthur had long hoped to be made governor of Martinique or San Domingue, large, prosperous islands rich in plantations and slaves. After his excellent record in the American war, he felt that one or other was his due. But Mme de Rothe was not without influence, and after the Archbishop refused to plead his case with the King, Arthur was obliged to accept the considerably less prestigious Tobago.

Lucie was only permitted to meet her stepmother once. Then her father departed for the West Indies, taking with him his wife and their new baby, Fanny. They left behind in Paris the two older children, Elizabeth, always known as Betsy, in the convent of the Assumption, Alexandre in a school with a tutor. But the cruellest blow, for Lucie, was the decision that her own much-loved tutor, M. Combes, who had protected and cared for her for the past seven years, would accompany Arthur as his secretary.

Before leaving, Arthur, who had finally managed to see Mme de Rothe, spoke to her about a young man whom he believed might make a good husband for Lucie. His name was Frédéric de Gouvernet and they had become friends during the American war, where Frédéric served as a young aide-de-camp. Frédéric, said Arthur, was not only much liked and esteemed by his fellow officers, but he was the only son of a prominent noble family. Mme de Rothe refused even to consider such a match. She knew Frédéric by reputation, she said, and he was ugly, short and a 'bad lot'. The amiable but weak Arthur did not insist. As he left, he gave the Archbishop full powers of proxy to arrange whatever marriage for Lucie he thought best. But Lucie herself, who had talked at length to her father about Frédéric, had been intrigued by the idea of the young officer. She discussed him with her cousin Dominic Sheldon, who had come to live with them, and who had met Frédéric several times in Paris. But when she learnt that Frédéric had indeed been very wild in his youth, she resolved to think no more about him. Her curious and unhappy childhood was producing a girl of strong character and determined ways.

In the spring of 1794 Lucie turned 14. She was no prettier in a conventional way than she had been as a small child, but she was robust and energetic, and her liveliness and intelligence made her very attractive. With every month that passed, she minded her grandmother's scoldings less, 'either because I had become used to ill-treatment, or because . . . the calm with which I met the calumnies she spread in every direction . . . forced her to hold me in a certain degree of respect'. She was now also taller than her grandmother. Later, she wondered whether there was not an element of fear in Mme de Rothe's newfound restraint, lest her granddaughter, who would soon be entering society, tell the world about what she had endured. Mme de Rothe had foolishly never tried to conceal her dislike for Marie Antoinette, nor her resentment at the Queen's evident fondness for Thérèse-Lucy; the thought that Lucie might soon be repeating her disparaging remarks at Versailles must have been disquieting.

* * *

And Lucie, at last, was allowed greater freedom to move around Paris. The city had never been so beguiling. Though the winters of 1783 and 1784 were bitterly cold, the Seine icing over and the temperature never rising above freezing for weeks on end, and though wolves were spotted prowling around the outskirts, the mood in the capital was buoyant. Efforts to control public spending had been defeated and a new policy, spending more and borrowing more, had been adopted. Around the perimeter of Paris, an immense wall, 3 metres tall, was rising to encircle the city, as a way of preventing the flow of untaxed goods and in the hopes that the increased excise revenue would boost the dwindling reserves of money in the government coffers. Money was being poured into sugar refineries, coal mines, armaments and dock-yards.

The new rich, or those who now felt rich, looking for ways to spend their money, were commissioning pictures, and particularly portraits. Elisabeth Vigée-Lebrun, one of the very few women in a field dominated by men, was at work on what would become 30 sentimental portraits of the Queen and her children, in which she took care to play down Marie Antoinette's narrow Habsburg face, pale protruding eyes and thick lower lip. In the hierarchy of desirable art, flowers were at the bottom, followed by landscapes, animals and human beings, the high ground taken by allegorical compositions. Jacques-Louis David was the rising star of the neo-classical movement. In the newly fashionable Chaussée d'Antin, the rich widow of a banker from Geneva opened her *hôtel particulier*, built by Leroux, to visitors. Those prepared to buy tickets could wander among its frescoed rooms, where Bacchus, Hercules and Cupid sported with Pleasure and Liberty, or stroll on the terrace among grottoes, statues and Corinthian peristyles.

No quarter of Paris was in a greater state of upheaval than the streets surrounding the Palais-Royal, opposite the Tuileries. Originally built for Cardinal Richelieu in 1629, the Palais-Royal itself was occupied by the Duc d'Orléans, first Prince of the Blood, and his son, the Duc de Chartres. Louis XVI, pious, serious, awkward with people, and his cousin, the indolent, good-looking Duc de Chartres, had never liked one another.* With houses in Brighton and London and with horses that he raced at Newmarket under his pink and black colours, the Duc de Chartres was rapidly running out of money. To remain solvent, he had decided to develop the Palais-Royal by surrounding the gardens with buildings and arcades and renting them out. By the time Lucie returned to Paris in the spring of 1784, these gleaming white stone arcades housed cafés, shops and clubs, where members dined, played chess and discussed politics. And, since the Palais-Royal was the private preserve of the Orléans family, the city police were virtually banned, with the result that the arcades were becoming a haven for private publishers of forbidden books and pamphlets.

Since America, England and France had signed the peace treaty

* The Duc de Chartres, whose libertine ways were widely remarked on, was reported to have drawn up a list of all the women he knew under seven headings: beautiful, pretty, passable, ugly, frightful, hideous and abominable.

of Versailles in September 1793, relations between the French and the Americans had become extremely close. Lafayette, returning to Paris, was greeted as a hero. The English, estranged from France for the four years of the American war, now hastened back across the Channel. Anglomania soon reached heights of absurdity. The Duc de Chartres had his gardens at Monceau filled with a 'profusion of English delights', though few of them could actually be said to be English: a Dutch windmill, a fort with a drawbridge, ruins of a Temple of Mars, a minaret, an obelisk, a Tartar tent and a merry-go-round with a dragon. People who had never spoken English before took to reading Shakespeare – sometimes written 'Sakespear' – drinking punch, and wearing coats with triple lapels. 'Anglomanes' carried walking sticks, wore small neat wigs and ate 'rostbif'; they played whist and dice, and drove a high-slung, two-wheel carriage called a wiski. When they began heating their bedrooms and installing the first flush lavatories, they called them '*à l'anglaise*' as a tribute to what they perceived as the English attention to comfort. Their wives read translations of English novels, with plots that usually revolved around French lovers fleeing wrathful parents to England, the land of freedom and serious thought, or bloodthirsty milords, cold beauties and phlegmatic heroes. 'We seem to want at all cost to be English,' complained the gloomy Baronne d'Oberkirch, 'we are trying to forget our past in order to forge a new future, getting rid of our fashions and our customs in order to become like our neighbours, whom we loathe.'

With these English came Dillon friends and relations, among them the young Arthur Wellesley and his sister Anne, to visit Paris and congregate round the Archbishop. Lucie was able to practise her English. She was particularly attached to her aunt, Lady Jerningham, who had been made godmother to her new half-sister, Fanny.

Not everyone admired what they found. The English agrarian reformer Arthur Young noted that, compared to London, Paris was exceedingly dirty, inconvenient, full of terrifyingly fast cabriolets which made the pavementless streets a constant hazard, while the thick clinging mud meant that 'all persons of small or moderate fortune and who could not afford carriages' were forced to wear black, with black stockings. 'Such a mixture of Pomp and Beggary, filth and magnificence,' wrote 19-year-old Francis

'The Flying Coquette' costume during the craze for ballooning.

Burdett to his aunt in London. The British, accustomed to substantial meals, with a great deal of meat, were highly dubious about fricasseed frogs and unidentifiable sauces. Mercier, ever critical of his countrymen, remarked that while the French indulged in 'silly luxury, which kills true happiness and wastes energy and money', the English offered a model of 'peacefulness and decent conduct of domestic life'. Parisian women, he observed, made particularly poor wives and mothers, though they could be good friends.*

Paris, meanwhile, was consumed with balloon fever. Two brothers, the Montgolfiers, having experimented with bags made of linen, lined with paper and filled with heated air, built a 'vast machine', 60 feet high and 43 feet in diameter, and decorated it with painted paper. On its first outing, the balloon rose into perfect blue skies above Versailles and travelled slowly over Paris for 25 minutes at a height of 100 metres. There were other

*Mercier's great work, which would ultimately run to 12 volumes and 2,000 chapters, had begun to appear in 1782.

launchings, with a duck and a dog. After the Montgolfiers' balloons, others took to the skies, prompting fears among young women that marauding Turks or '*Barbares*' would descend and pluck them from their gardens. A Mr Blanchard and a Dr Jeffries crossed from Dover to the French coast in two and a half hours, bringing with them letters from 'people of distinction' addressed to the French nobility. Entrepreneurial dressmakers turned out hats shaped like balloons, writers composed plays and verses with balloons in them, artists used balloons with which to decorate fans and snuffboxes. When the craze for balloons palled, it was replaced by one for a fashionable Viennese quack, Franz Anton Mesmer, who held that all illness resulted from an imbalance of magnetic fluids. Mesmer, who received his patients in a lilac silk dressing gown and gold slippers, claimed to be able to rebalance them, using a special iron rod, and to cure gout, asthma and even epilepsy. And after Mesmer, the curious and the gullible moved on to the occult, to sorcerers, somnambulists, alchemists and hypnotists. For a while, Cagliostro, the son of a Sicilian Jew, charmed the Faubourg Saint-Germain and Versailles with his deep, metallic voice and his black embroidered coat, as he gazed into his crystal ball and promised them gold and diamonds.

And then there were the exotic animals, brought back by the merchant ships dealing in sugar, coffee, indigo and slaves with the Caribbean, Africa and the Far East. After Bougainville's circumnavigation of the globe between 1768 and 1771, artists and naturalists were eager passengers on ships to the undiscovered parts of the world. The specimens they carried back with them were displayed at fairs and in travelling menageries. Posters advertised fabulous beasts, such as the great Tarlata of Tartary, which turned out to be a hairless bear, wearing clothes and performing 'disgusting caresses'. An elephant able to uncork and drink bottles of beer inspired a book, *Mémoires de l'Éléphant*, narrated in the first person.

When the Ménagerie at Versailles had been built, under Louis XIV, it was constructed in the shape of a large octagon, with doors opening on to balconies from which visitors could gaze down into eight enclosures, each with an exotic species of animal or bird. Under Louis XVI, given the vogue for natural history, these animals had acquired scientific interest. For the naturalist Buffon, the elephant, gentle and obedient, stood at the pinnacle of the animal hierarchy, second only to man in intelligence, 'at least as much as matter can approach spirit'. By the time Lucie began to visit Versailles, the King's Ménagerie housed a panther, a hyena, badgers, a mandrill and an elephant which, soon afterwards, broke out and drowned itself in the nearby lake. There were other large beasts and the cost and the difficulty of transporting them were prodigious, the tiger having needed fresh meat for 10 months at sea and 24 days overland. The rhinoceros, being vegetarian, was cheaper, but even so it had grown '*fort et méchant*', strong and nasty, on the long sea voyage, and had to be conveyed to Versailles in a cart, restrained by a leather collar and iron shackles, looked after by two keepers – who happened to be butchers – and an overseer, who kept its skin supple by rubbing it with fish oil. It was, wrote Thomas Blaikie, a Scottish gardener working for the French royal family, 'an huge Animal the skin of which is of a looss scaly nature without hairs', and its horn had been cut off because 'it tore up the walls'.

It was all of it, the animals and the spectacles and the luxury, as the Comte de Ségur, companion to Arthur in the American war, would later write, like a sparkling picture, composed of thousands of colours, made up of magnificent castles, laughing landscapes and rich harvests – except that even then it was clear that the very lightest of breezes would be enough to blow it all away. 'We were proud to be French,' he wrote in the 1820s, looking back on the years leading up to the revolution, 'and even prouder to be French in the 18th century, which we regarded as a golden age handed to us by the Enlightenment philosophers . . . Each of us believed that we were advancing towards perfection, without worrying about obstacles and without fearing them.'

* * *

In the autumn of 1785, Lucie once again accompanied her grandmother and Archbishop Dillon to Montpellier. On the way, they spent a month in the Cevennes, where Lucie was taken to see coal and sulphate mines and spent her evenings discussing with the engineers who came to dine all that she had learnt from M. Combes about physics and chemistry. The young men of the Cevennes had formed themselves into a guard of honour for the Archbishop and in the evenings, wearing the red and yellow uniform of the Dillon regiment, they came to dine and dance in the château where Lucie was staying. Among the young men were several whose names were proposed to the Archbishop as possible suitors. Neither he nor Mme de Rothe considered them of sufficient grandeur.

At the end of the winter, the Archbishop accepted an invitation to visit Bordeaux, where the neo-classical architect Victor Louis's fine theatre had recently opened. A new ship, the *Henriette-Lucy*, was to be launched, a 600-ton vessel belonging to an Irishman, Mr MacHarty, and he had asked if Lucie would name her. Before the ship set sail for India, there was a splendid lunch party on board. A few days later, as they were preparing to return to Paris, Lucie's servant asked whether he might have leave to visit friends at the nearby château of Le Bouilh, belonging to a Comte de la Tour du Pin Gouvernet, and to rejoin the party as they passed by on their journey northwards. Lucie agreed. She realised that the estate must belong to Frédéric de Gouvernet's father, and hoped that when they paused to collect her servant before crossing the Dordogne at Saint-André-en-Cubzac, she might catch sight of the château from the road. To her annoyance, it was shielded from sight by tall poplars. As her carriage skirted the estate she told herself 'over and over again that I might have been châtelaine of all that beautiful countryside'. Frédéric – once proposed by her father as a possible husband for her – was now back in her thoughts, but why she felt so intrigued by the idea of this unknown young man, she could not say.

* * *

Many years later, looking back on the events that befell France in 1789, Napoleon would say that the French Revolution dated, not from the fall of the Bastille, but from the opening night of *The Marriage of Figaro*.

In 1775, the playwright Beaumarchais had won great acclaim with *The Barber of Seville*. Its sequel, *The Marriage of Figaro*, was finished three years later, but the plot, revolving around a licentious count and a cast of duplicitous servants who triumph over their masters, was judged too subversive and shocking, and it was not until April 1784 that it finally opened in Paris, with the action shifted to Spain and its anti-clerical passages removed. It delighted the Parisians. And though Beaumarchais was briefly detained, *Figaro* earned him 60,000 *livres* in the next three years, a phenomenal sum at the time, when even Laclos's daring and successful *Les Liaisons Dangereuses* earned just 1,600. At heart, Beaumarchais was not opposed to the monarchy: he was a royal watchmaker, a member of the nobility, and he had bought himself a number of the many posts for sale at court, which entailed no duties but brought status to the holder and revenue to the government. But *Figaro*, making its way from Paris to provincial theatres, was soon perceived as a call for greater freedom of speech, and a warning to Louis XVI – who very publicly condemned the play – that he could not rule by fiat over theatre, court and kingdom.

Then, in the summer of 1784, just a few months later, came the scandal of Marie Antoinette, the Cardinal de Rohan, and the diamonds.

Louis, Prince de Rohan, Cardinal Bishop of Strasbourg and Grand Almoner to the King, was a greedy, arrogant man who craved acceptance at court and knew that to win it he needed Marie Antoinette's approval. Aware, as was all Versailles, of the Queen's appetite for jewellery, and made gullible by ambition, he fell for a swindle that could have come straight out of Beaumarchais's play. It involved a young woman called Jeanne de la Motte, a member of the minor nobility who had fallen on hard times and who, like the Cardinal, was desperate to make her way at court.

Some years earlier, Bohmer and Bassenge, Paris's leading jewellers, had made a *rivière*, a many-looped necklace with 647 diamonds. It was designed for Mme du Barry, Louis XV's mistress, but the King had died before they could deliver it. The only likely customer for this extremely expensive item – said to cost 1.6 million *livres* – was Marie Antoinette, who had bought several pieces of jewellery from them in the past. But Marie Antoinette was not interested.

On 10 August 1784, Jeanne de la Motte, who may have been the Cardinal's mistress and certainly knew of his ambitions, dressed up a young, blonde woman in the kind of white muslin dress that Marie Antoinette favoured, and placed her late at night in the Grove of Venus at Versailles. The girl was told that she was to press a single rose into the hand of the man who approached her, with the words 'You know what this means'.

The Cardinal, lurking in the bushes, saw what he thought to be the Queen, approached and took the rose; he was overcome with pleasure, all the greater because Marie Antoinette had previously made no effort to conceal her dislike for him. And when, not long afterwards, Jeanne de la Motte told him that the Queen, eager to own the famous necklace, wished for the Cardinal to act as her intermediary with the jewellers in order to negotiate a series of deferred payments, he readily agreed. The jewellers were informed that the Queen had changed her mind and now wanted the necklace, and they gave it to Jeanne de la Motte to pass on to her. The necklace disappeared into the hands of La Motte's lover, who quickly broke it up into many different pieces, sending some to London.

The days passed, and still the Cardinal did not see Marie Antoinette in the magnificent necklace. Weeks, then months, went by. Finally, Bohmer and Bassenge began to press for payment, and the whole farcical affair unravelled. Summoned to see the King, the Cardinal was forced to admit that he had been taken in by the hoax; he begged Louis XVI to cover up the scandal. The King, furious at the light it shed on Marie Antoinette, refused and had the Cardinal thrown into the Bastille.

A hat, the *cardinal sur paille* – in the gutter – was soon being worn all round Paris, and ivory snuffboxes were sold, with a single black spot in the middle to indicate that the Cardinal would be unlikely to emerge from the affair entirely white. When the trial against him, La Motte and their various accomplices came to a close, Cardinal de Rohan, wearing the deepest purple for mourning, was divested of all his offices; Jeanne de la Motte was ordered to be stripped naked, beaten by the public executioner, branded with a V for *voleuse*, thief, and sent to serve a life sentence in the prison of the Salpêtrière. While branding her shoulder, during which Jeanne struggled and screamed, the executioner's hand slipped and the V ended up on her breast.

But the real casualty was the Queen's reputation. For she emerged as spiteful, wishing to destroy the Cardinal, and a spendthrift. Worse was the implication that the Cardinal could actually have believed that Marie Antoinette, Queen of France, was so without moral scruples that she was prepared to have a secret tryst with him in a garden late at night. In the *libelles*, the scandalous, clandestine pamphlets that fed the French public daily with tales of corruption and licentiousness at court, Marie Antoinette was portrayed as stopping at nothing to satisfy her appetites. Some hinted at lesbian relations with Jeanne de la Motte. There had been attacks on the Queen's morals in satirical cartoons from the moment she came to the throne, but they now rose to new heights of vitriol; Marie Antoinette and her coterie were portrayed as debauched, riddled with venereal disease and as sexually degenerate as they were politically corrupt. None of this would have been quite so damaging had it not been for the excesses for which the Queen was known.

When Louis XVI and Marie Antoinette were crowned in 1774, there had been genuine approval of their simple, unfussy manners after the solemnity of the court of Louis XV. For a while, the people enjoyed their pretty young Queen, and Lucie's mother's time at court had coincided with a moment of popularity for the monarchy. But by the mid-1780s stories began to circulate about excessive gambling and frivolity, and about a 'queen's party' of pleasure-seeking and horse-racing aristocrats such as Thérèse-Lucy's supposed admirer, the Duc de Lauzun. The names of Arthur Dillon and a good-looking young Swede, the Comte de Fersen, were also mentioned. After the Prince de Guéménée's bankruptcy and his wife's retirement from court, Marie Antoinette had taken into service as royal governess a beautiful but chilly young woman, the Countess Yolande de Polignac. The Polignacs, like the de Rohans, were perceived as ambitious and venal, as the Prince de Guéménée had once tried to warn Thérèse-Lucy.

The Queen was now surrounded by a group of courtiers very different from those who had filled Versailles under Louis XV. The starchy princesses and countesses, elderly and jealous guardians of ritual and prerogative, had departed, offended, to Paris or to their estates. Their places had been taken by light-hearted young women, with a taste for amateur theatricals in which they dressed as dreamy Watteau shepherdesses, for parties with magicians and tightrope walkers, and for the gaming tables. They were eager to spend, prodigiously, on clothes, jewellery, trinkets and anything else that might provide them with pleasure. 'Virtue in men and good conduct in women,' Lucie would write, 'became the objects of ridicule and were considered provincial.' At Le Petit Trianon, where the servants wore the Queen's own livery of red and silver, rather than the blue, silver and red of the King, etiquette was largely forgotten. On coming to the throne, Marie Antoinette had replaced the staid, ageing footmen with flamboyantly dressed tall young men.

It was this world that Lucie was about to join, for Marie Antoinette had indicated that, when she came of age, Lucie was to take her mother's place at court as one of the 12 Ladies of the Queen's Household. Lucie was both intrigued and nervous. Her thoughts were concentrated on how to escape her grandmother, and both Versailles and marriage promised early release. Reflective and knowing beyond her years, and with a sternness that verged on prudishness, she was also extremely wary. She mistrusted the way that, in the Queen's circle, 'gaming, debauchery, immorality' were now all 'flaunted openly'.

The King, meanwhile, hunted. The vast forests of the Ile de France were his hunting grounds, full of stag, boar, partridge, pheasants and hares. In the saddle, this serious-minded, short-sighted, ungainly man lost his gaucheness. He hunted at Fontainebleau, at Compiègne, at Marly and Choisy, and later at Rambouillet and Saint-Cloud, which he bought from the Duc d'Orléans, soon after the death of his father, Louis-Philippe le Gros, whose appetite was so voracious that he was said to eat 27 partridge wings at a single sitting. And when he was not hunting, which was seldom – the phrase 'the King is doing nothing today' had come to mean that he had not gone out hunting – Louis XVI shut himself away in his special workshops, to pore over his scientific instruments, clocks and locks.

Versailles itself, still extraordinarily splendid on the nights of Marie Antoinette's Wednesday balls, was often deserted, the nobility more and more preferring to spend their time in salons in Paris or on their estates when not actually summoned to a court many found exaggerated and frivolous. The public, on the other hand, flocked: the palace of Versailles and its park were open for visitors to wander and stare at the royal family. Thomas Blaikie, the Scottish gardener, was taken to see the King at breakfast and marvelled to find him so available and little guarded, 'dressed almost like a country farmer, a good Rough man'.

Marie Antoinette had a weakness not just for jewels, but for clothes. Three rooms in her private apartments were set aside for her wardrobe, where several times a week the royal milliners and dressmakers came to show their wares. Every season, every event, every ritual in the royal day had its change of dress, of material and of trimming. Gold was the colour for frosty days. After the King remarked that brown, the shade of the moment in the summer of 1775, looked 'flea-coloured', everything was known by a different hue of flea – flea's back, flea's belly. Trimmings, too, incorporated every '*caprice du moment*'. As the Evangelical philanthropist Hannah More wrote to a friend, during a visit to Paris, the fashionable material of the day was silk with 'a *soupçon de vert*, lined with a *soupir étoffe et brodée de l'espérance*; now you must not consult your old-fashioned dictionary for the word *espérance*, for you will find that it means nothing but hope, whereas *espérance* in the new language of the times means rose-buds.'

When Marie Antoinette had first come to the throne, her dress allowance of 96,000 *livres* was raised to 200,000; but it was never enough. In the hands of Rose Bertin, her increasingly influential milliner, who irritated the women at court with her mixture of fawning and haughtiness, Marie Antoinette sped from one fashion

Coiffure celebrating American Independence and the triumph of liberty.

to another; and the rest of the court followed. Rose Bertin became very rich. There were Tyrolean balls, Indian balls, Norwegian balls, each necessitating costumes for the several thousand guests; and when the snow was deep enough to take out the horse-drawn sledges, the grooms wore special Russian hussar outfits. After the first zebra arrived at the Ménagerie, coats, waistcoats and even men's stockings were striped.

Nowhere, however, did fashion reach such ludicrous heights as over hats. They were enormous, grotesque and fanciful, gigantic scaffoldings of gauze, papier mâché, silk, flowers, paint, ribbon and feathers, fashioned to mirror the topic of the day. The craze for the hat as riddle, allegory, social comment or celebration, started in 1775, when the high price of flour led to bread riots in Paris and someone had the idea of creating a '*bonnet à la revolte*'. After that, no major event passed unmarked: there were hats *à la Montgolfier*, *à la Suzanne* (*The Marriage of Figaro*), *aux insurgents* (the American War of Independence), *à l'inoculation* (vaccination of the King against smallpox). Léonard Autie, the Queen's hairdresser, described what he called the '*ménagerie*' at court. 'Frivolous women,' he wrote, 'covered their heads with butterflies, sentimental women nestled swarms of cupids in their hair; the wives of general officers wore squadrons perched on their forehead; melancholic women put a sarcophagus and cinerary urns.' But none, it was universally agreed, was more spectacular than the landscape sported by the Duchesse de Lauzun one night in Mme du Deffand's salon. The Duchess arrived with an entire tableau consisting of a stormy sea, ducks swimming near the shore and a man with a gun sprouting from her head. Above, on the crown, stood a mill with the miller's wife being seduced by a priest, while over one ear the miller could be seen leading his donkey. It made Madame de Matignon's display of fresh vegetables, broccoli topped with artichokes, look very tame.

It was not surprising, perhaps, that the *libellistes* found much to satirise, nor that when Marie Antoinette was discovered to have ordered 172 new gowns in a single year, she was quickly nicknamed *Madame Déficit*. Marie Antoinette, noted the Marquis de Bombelles, surrounded herself with people too weak and irresolute to act as sensible guides, leaving her to flit from idea to idea, from 'one branch to another'.

When Marie Antoinette started spending more time at Le Petit Trianon, she asked Rose Bertin to design dresses and hats such as shepherdesses might wear. White, said to have been introduced originally by the Creoles of Saint Domingue arriving in France, became the Queen's favourite colour, whether in linen, lawn or calico. The mood for absurdity and exaggeration was at last passing. Marie Antoinette was reported to have decided to 'renounce plumes, flowers and pink'. A spirit of economy was sweeping Versailles, where the Queen's candles had until now been blown out and replaced whenever she left a room, and new ones lit, even if she was only gone for a few minutes. The King declared that he would reduce the expenditure of his immense stables by cutting the number of horses from 2,400 to 1,125; but there was not much he could do about the vast costs of the separate households of his sisters, aunts, brothers and nephews, all borne by him.

The cutbacks in spending, the spirit of sobriety and sensibleness, all came, however, too late. Versailles, for long a centre for cliques and a source of gossip and scandals, was widely perceived as too outrageous, too extravagant. Marie Antoinette herself, a little stouter and her health not very good, had never recovered from the affair of the diamond necklace; she now inspired both dislike and lack of respect among people who thought her foolish, indiscreet and selfish. *Le parti de la reine*, the Queen's coterie, was seen as not only pro-Habsburg but meddlesome in the affairs of state. At Le Petit Trianon, the mistrusted Polignac family were thought to be too powerful. Paradoxically, even the Queen's newfound taste for simplicity in dress was not what the public wanted of a queen.

* * *

Louis XVI had come to the throne in 1774 professing his intention to be responsive to the needs and desires of his people. He might, perhaps, have steered a successful middle course, somewhere between the autocracy of his grandfather and a more egalitarian form of governance. As it was, he vacillated between despotic rule and attempts at modernisation; and, as the 1780s wore on, so a financial crisis began to overwhelm France, where, across all fields, from the army to the Church, the sciences to the civil service, men were speaking and acting together, exploring the new ideas handed down by the *Encyclopédistes* in ways that they never had before. By attacking intolerance, fanaticism and superstition, by advocating the removal of education from the clergy and putting it in the hands of intellectuals, by placing new emphasis on individual values and freedoms – all ideas supported by Frédéric and his friends – Diderot and his fellow authors had effectively challenged the most enduring received wisdoms of the century.

Though the heady days that followed the Treaty of Versailles, when the country had seemed on the brink of a glorious future, lay only four years in the past, France itself had never been in greater disarray than it was now, its diplomacy stalled over ill-advised deals, its economy in desperate need of rescue and unable to float further loans. The witty, articulate Charles-Alexandre de Calonne, Controller General since 1783, had gambled on the future by borrowing and some 45 million *livres* had been added to the already immense interest payments left by Necker. There had been a number of spectacular bankruptcies, furthering fears of chaos. Under the twists and turns of economic policy – there had been 10 finance ministers in 15 years – the country was floundering.

Calonne and Louis XVI now took a radical step. They announced that they would invoke the Assembly of Notables, an archaic body chosen by the King from France's three 'Estates' – the clergy, the nobility and the 'third' estate, the people – in order to lay before them a package of reforms based around a new and permanent land tax on all landowners, regardless of class or rank. This new land tax, it was argued, would be both simpler and fairer, in that it would draw on the wealthy, who notoriously escaped taxation, and at the same time bypass the venal magistracy. The Assembly had last met, in response to the wishes of Cardinal Richelieu, in 1626. Archbishop Dillon, regarded as one of France's three leading 'administrative prelates', was named as one of its 144 members and the Assembly was to meet at Versailles, where Lucie and Mme de Rothe would join him. Already there were protests from senior clerics and courtiers, who saw in these reforms a dangerous challenge to the idea of an absolute monarchy and an all-powerful Church, hitherto set in an immutable social order.

The Assembly opened on 22 February 1787, with the wholehearted backing of the King, and each of its seven committees was chaired by one of the royal princes. Away from the official sessions, the notables used their time to meet informally and to make deals; Mme de Rothe presided over long evenings while the Archbishop and his supporters talked and argued. 'And so,' wrote the anonymous author of the *Correspondance Secrète*, 'the time passes in cabals, in intestinal fermentations, when the affairs of state should demand total concentration.' Archbishop Dillon, he added, was a 'prelate ever ready to talk, having an abundance of words and a penury of wits'. Though opposition to the land tax was almost unanimous among the nobles and the immensely rich clerics, not even they could fail to be shocked by the terrifying size of the deficit in the country's finances: it stood, Calonne informed them, at 112 million *livres*, of which well over a third was going to service loans. The economic picture was worse, far worse, than anyone had suspected.

Looking back, almost half a century later, Lucie would say that, for her, the opening of the Assembly of Notables marked the first day of the French Revolution.

* * *

In the spring of 1786, Lucie celebrated her 16th birthday. Mme de Rothe informed her that she was making arrangements for her betrothal to the Marquis Adrien de Laval. 'The name,' noted Lucie, 'sounded well in my aristocratic ears.' Adrien's older brother had recently died and he was heir to a considerable fortune; he had therefore left the seminary where he had been studying and gone into the army. Better, a quick marriage would release her from her grandmother's clutches. 'I was no longer a child,' Lucie wrote. 'My education had begun so early that by the time I was 16, I was as mature as most girls are at 25, and my grandmother was making my life miserably unhappy.' Maturity, for Lucie, had brought with it increasing determination and a definite confidence in her own worth; and she had no intention of making a rash or unworthy match. There was already something formidable in her refusal to be buffeted by the wishes of others.

Lucie was widely regarded as an excellent catch. Sole heiress to Hautefontaine, Montfermeil, the house in the rue du Bac and the rents on the Hôtel de Ville, she was known to have been promised a position at court. And 'to be at court', as she noted, 'was a magic phrase'. Many years later, describing herself at the time, she coolly recorded that she was tall, with a 'dazzlingly clear and transparent complexion', thick ash-blonde hair and a good figure, which between them made her overshadow far better-looking girls, particularly by day when her unblemished skin could best be appreciated. 'My best feature,' she wrote, 'was my mouth, for my lips were well shaped and had a fresh look. I also had very good teeth.' Good skin and teeth, in an age of smallpox and primitive dentistry, were among the features of a girl most often remarked on. Her forehead was high and her nose she described as 'Grecian, but long and too heavy at the tip'. She suspected, however, that people nevertheless found her ugly, deducing this from the way they compared her to certain other girls, whose looks she considered 'hideous'. This realistic appraisal of her own looks enabled her, she wrote, to avoid all the pangs of 'demeaning jealousy' that affected prettier girls. 'I set myself a code,' Lucie noted somewhat smugly in old age, 'and I have never departed from it.'

Already, her tone could be almost unnervingly detached. The years of unhappiness under Mme de Rothe's tyrannical tutelage had crushed none of her curiosity, but they had made her level-headed and highly critical, both of herself and of others. What made her most angry with herself was that she remained very shy on social occasions, so shy that her legs wobbled. This was a weakness and 'cowardly'. However, she derived a great deal of pleasure from a friend made at around this time. This was the 85-year-old Maréchal de Biron, father of Arthur's friend the Duc de Lauzun, one of the last surviving members of the court of Louis XIV. The Maréchal had taken a liking to Lucie and at the grand dinners in his magnificent house in the Faubourg Saint-Germain, to which she sometimes accompanied her grandmother, she would find herself seated next to him. When you grow old, he said to her one day over dinner, never either bore young people or try to dress in ways unbecoming to your age. She remembered and heeded his words. One day, she heard the Maréchal tell her great-uncle that were he not still married – he had been estranged from his wife for many years – he would have asked for Lucie's hand, despite the very great difference in their ages. Lucie watched and admired the ease and courtesy with which he handled his vast acquaintance; and later she would recall him with admiration and affection.

The betrothal to Adrien did not take place, though they were to remain friends all their lives: Adrien's grandfather, the Maréchal de Laval, chose for him another bride, with a grander title and a larger dowry. Mme de Rothe's next choice was the young Vicomte de Fleury; Lucie turned him down on the grounds that he lacked intelligence, distinction and a sufficiently 'illustrious' family. Then came Espérance de l'Aigle, whose father's estate, Tracy, lay close to Hautefontaine, and with whom she had often played as a child. Lucie refused him too. Mme de Rothe was not pleased, though she made clear that she was determined to do everything in her power to bring about an excellent marriage for her only granddaughter. Only years later did Lucie come to realise that Mme de Rothe had quite other reasons for this sudden desire to make her happy: like the Prince de Guéménée, Archbishop Dillon was now severely in debt, and he had convinced her grandmother that more money needed urgently to be brought into the family. The fact that he had effectively squandered Lucie's inheritance was never mentioned. 'Violence and duplicity' was how Lucie would later describe the situation.

Lucie herself, however, had other ideas, unusual at a time when young girls were expected to look favourably on the marriages their parents arranged for them. She had still never actually set eyes on Frédéric-Séraphim de Gouvernet, whose father's château, Le Bouilh, she had passed by without seeing. But something in her increasingly strong and stubborn nature, combined perhaps with a desire to thwart Mme de Rothe, who thought him a very poor match, now resolved her to accept no one else. She told her cousin, Dominic Sheldon, that she had made up her mind and he tried hard to argue her out of what he called an 'obsession'. She refused to listen.

Frédéric, meanwhile, having heard much about Lucie from her father, decided that the moment had come for him to take a wife. Through his cousin, the Archbishop of Auch, he approached Archbishop Dillon and offered for Lucie's hand. Mme de Rothe protested, but Marie Antoinette herself let it be known that she was in favour; and other members of Frédéric's large and prominent family came forward to put pressure for the match to take place. Mme de Rothe expected – and wished – Lucie to refuse. Without hesitating, Lucie accepted. 'It was an instinct,' she wrote later, 'a guidance from above . . . I felt that I belonged to him, that my whole life was his.' Negotiations now opened between the two families. Frédéric, like Lucie, was heir to a considerable fortune in feudal dues, several estates, invested money, leases from mills and a toll river crossing, in all bringing in some 30,000 francs a year.

Frédéric was an interesting young man. He was 11 years older than Lucie, and had served with distinction in America with Lafayette, which was when he had impressed Arthur with his intelligence and courage. What was more, he came from a remarkable family, with a long line of illustrious ancestors – Lucie noted and appreciated illustrious forebears – descending from the dukes of Aquitaine, and bearing as their coat of arms two winged griffins and several dolphins. His maternal grandfather, M. de Monconseil, page to Louis XIV, had once set the King's wig on fire while conducting him back from visiting his mistress, Mme de Maintenon. M. de Monconseil had risen to become commander of the army in Upper Alsace, then used a fortune made gambling to buy a vast estate at Saintonge. Though he had recently died, his widow, Mme de Monconseil, was known to be anxious for the marriage to Lucie to take place. She was a friend of Voltaire's, and she was generally referred to in her circle as '*une grande intrigante*', a great schemer. Former lady-in-waiting to the Queen of Poland, and remembered for the spectacular parties she gave at her country house, Bagatelle, Mme de Monconseil was 85, a witty, forceful woman, cynical in the manner of the early 18th-century salon hostesses.

Frédéric's mother, Cécile, was Mme de Monconseil's eldest daughter, but after a series of scandals many years earlier she had been shut away in a convent, as was still the custom in many families. There she had been enclosed for the past 20 years. Her sister Adélaïde, Frédéric's aunt, had been married, at 15, to the Prince d'Hénin. Mme de Monconseil and the young Princesse d'Hénin held a perpetual salon in the fashionable Chaussée d'Antin, with beds and sofas in the antechambers, so that guests simply came and stayed for several days; mother and daughter were regarded as powerful figures in Paris society. The Princesse d'Hénin, observed the Marquis de Bombelles, who was to publish

71 volumes of diaries before the end of the century, was pretty, fashionable and no less scheming than her mother.

The Princesse d'Hénin would play a central part in Lucie's life. She was not an easy woman, being both impetuous and irascible; but she was a generous and devoted friend. Her own story was not unlike that of many of the women of the nobility at the end of the 18th century. She had been born in 1750, a very late baby, after Mme de Monconseil had been married for 24 years, and as a small child she was always known as Bijou or '*la seconde mademoiselle*', the first, Frédéric's mother Cécile, being already 15. Having made her first appearance in society at the age of 7, dressed as a little peasant girl at a reception for King Stanislav of Poland, Adélaïde had spent the next seven years enclosed in a convent in the rue du Chasse Midi, before being married to a captain of the Guards. The Prince d'Hénin was a sickly, ugly, dissolute young man with the reputation of being profoundly boring. Bijou, who was extremely pretty and argumentative, and who never sat still, was known as '*la merveilleuse Princesse d'Hénin*'. But not for long. Shortly before her 17th birthday, she caught smallpox; she recovered, but was left with scars all over her face, which wept (though some said that it was shock at a scandal that erupted one night at a fancy dress ball that had given her herpes and ruined her skin).

The Princesse had not, however, lost her good teeth, her pleasing figure, her thick hair and shining eyes. The Prince, after a few weak attempts at fidelity, returned to his mistress, an actress, and the Princesse, once she had got over the loss of her beauty, chose from her many admirers the Chevalier de Coigny, principal equerry to the King, as her *cavaliere servante*. What she lacked in looks, she made up for in elegance: she ordered her gowns from the renowned Mlle Couvert, her hats from Rose Bertin, her pomades and scents from a special perfumer at Versailles. The silk for her dresses came from China. Every summer, she moved from Paris to her mother's château, Bagatelle, and later to a rented house in Passy, in which the Chevalier was given his own set of rooms. The Princesse was one of the 12 Ladies of the Queen's Household and, at 32, renowned for her wit and her sharp tongue.

* * *

Before the marriage contract had been agreed, Lucie heard of the sudden death of Mme de Monconseil. To her great relief, Frédéric informed the Archbishop and Mme de Rothe that he still wished to proceed with the marriage. As soon as he was able to get away, he hastened south to Le Bouilh for his father's formal permission, returning with it in record time, which Lucie considered to be in the 'very best taste'. They had still not actually met, the custom being to delay any encounter until after the signing of the marriage contract. It was from behind a heavy curtain that she first set eyes on her future husband, getting out of a cabriolet drawn by 'a fine and very spirited grey', soberly dressed in deep mourning for his grandmother; she found him not ugly at all, as Mme de Rothe maintained, but resolute and energetic. His military hat, denoting his high rank, combined with his youthful appearance, made him look 'very dashing'. 'His assurance and air of decision pleased me immediately,' she wrote.

The next step was a visit to the Dillon house in the rue du Bac by the Princesse d'Hénin. Terrified at the impression she might make, Lucie, summoned to the drawing room, was unable to speak. Her legs trembled. Her English aunt, Lady Jerningham, come to Paris to remove her daughter Charlotte from the convent in which she had spent three enclosed years, tried to smooth over the moment of presentation. But it was not until the Princesse d'Hénin, having scrutinised Lucie carefully all over, pronounced, 'Oh what a pretty figure. She is charming. My nephew is so lucky', that Lucie found her voice. For this first encounter, Lucie had defied Mme de Rothe and insisted on choosing her own dress: a plain white muslin gown, with a waistband of dark blue ribbon and a fringe of brightly coloured silk along the hem. 'They said,' she wrote rather smugly, 'I was as pretty as a picture. They looked at my hair which was really very beautiful.' With her engagement, Lucie was enjoying her new powers.

From this moment on, Frédéric came to Lucie's house every day. Once he brought with him his brother-in-law, the tall and solemn Marquis Augustin de Lameth, married since 1777 to his only sister, Cécile-Suzanne, and two of de Lameth's brothers, Charles and Alexandre. The Marquis de Lameth owned the château of Hénéncourt, near Amiens in Picardy, and was an officer in the King's Regiment; like Frédéric he was a man with a passion for fairness and discipline. The more time she spent with Frédéric, the more Lucie was convinced that her choice had been right: she found his tastes and his ideas to be much like her own and was sure that the dissoluteness of his youth was a thing of the distant past. On the contrary, he seemed to her serious, conscientious and very clear in his views. 'We became,' she wrote later, 'increasingly certain that we were made for one another . . . trying to understand and know each other, each one of us studying the opinions and tastes of the other . . . what pleasant plans we made for our future.' She also enjoyed listening to him as he argued over dinner about what was being discussed in Paris, though she found the political content boring. She was still just 17.

It was now that she was at last introduced to her future father-in-law, Jean-Frédéric de la Tour du Pin Gouvernet, head of one branch of his large and powerful family. Like his son, he was short, with an upright, military bearing and fine eyes and teeth, something Lucie immediately remarked on. Like Frédéric, too, he had been a soldier since early boyhood. Since his appointment as commander of the provinces of Saintonge and Poitou, he came only rarely to Paris, preferring to spend what time he could get away from his duties at Le Bouilh, where he was adding to the château and making a garden. Lucie found him easy and very charming, and felt that everything about him belonged to an earlier, more formal age. She admired him all the more when she was told that he owed his many promotions in the *Grenadiers de France*, a regiment made up of the cream of all other regiments, not to intrigues and favours at court, but to his own merits, though this had earned him the contempt of Mme de Monconseil, his mother-in-law, who had also resented his severity towards her errant older daughter.

Both Frédéric and his father regarded themselves as liberals; they were monarchists, but they shared a dream of a reformed liberal monarchy, with far greater taxation of the wealthy Church and aristocracy and far more spending on France's impoverished people. Together with other liberal aristocrats, men such as the de Lameth brothers and the Duc de Lauzun, they were ready to question the very foundations on which 18th-century France was built, and to attack privilege as part of the 'debris of an irrelevant, pernicious gothic world'. But they did not see a need to get rid of the King.

Marie Antoinette, learning that the marriage had been settled, asked to see Lucie. The honour was pleasing to all except to Lucie herself, who could think only of the Queen's heartlessness at the time of her mother's funeral. What she feared now was a display of sentimental hypocrisy. It was as she expected. After Lucie was taken to Versailles, and had kissed her hand, Marie Antoinette was full of affection and kindness. Lucie found herself frozen with shyness and dislike and unable to respond. Many years later, she regretted the bad impression that she had made that day, and wished she had made more efforts to please the Queen, and later had been more able to offer her useful help.

At Montfermeil, where the marriage was to take place in the Folie Joyeuse, Lucie was given an apartment, newly furnished with hangings of Indian calico, decorated with a pattern of flowers and birds and lined in bright green silk. In the cupboards was Mme de Rothe's trousseau of household linen, and lace and muslin gowns. The Princesse d'Hénin sent a tea table and a Sèvres porcelain tea service in silver gilt, and Lucie's English grandfather, Viscount Dillon, some valuable earrings. Frédéric's presents were generous and imaginative: jewellery, ribbons, flowers, feathers, gloves, lace, hats, bonnets and capes, as well as 70 volumes of English poetry and some framed English prints, along with a jardinière filled with exotic and rare plants.

The wedding took place at midday on 22 May. Lucie wore white crêpe and Brussels lace, with two streamers flowing from her bonnet and orange blossom in her hair. The guests included Frédéric's mother, allowed out of her convent for the day and whom Lucie met for the first time. Lady Jerningham and her husband Sir William Jerningham, Charlotte, Dominic Sheldon and various other English relations were there, as were the de Lameth brothers, the Princesse d'Hénin, a number of senior clerics and several government ministers. Lucie crossed the courtyard from the house to the chapel on the arm of 16-year-old Edward, her youngest and favourite Jerningham cousin, to whom she later presented a sword. A 7-year-old nephew of Frédéric's, Alfred de Lameth, helped to hold the wedding canopy. In her old age, remembering her wedding day, Lucie dwelt on the elegance and the splendour of the occasion, simply because of its contrast with everything that followed.

Her great-uncle did not officiate at the Mass, but he gave a nuptial blessing and preached a short sermon, and Lucie was then embraced by all the ladies, one by one, in order of kinship and age. After the ceremony came the handing out of knots, favours, cords and fans to the wedding guests, expensive trimmings in gold and embroidery carefully tailored to rank and status, essential for any such grand occasion. While waiting for the wedding dinner, which took place at four in the afternoon, Lucie went to visit the long tables set nearby, one for the liveried servants in their many-coloured uniforms, the other for local peasants and farmers, in whose company, escaping Mme de Rothe, she had spent many happy hours. The evening finished with a concert. Next morning, the Princesse d'Hénin informed her that, five days later, as the new Comtesse de Gouvernet, she was to be formally presented at court.

CHAPTER FOUR

The Colour of Hope

There was very little time for Lucie to prepare for the ceremony of presentation at court that had continued more or less unchanged for over a century. Dress, deportment, jewellery, timing, all had to be impeccable; any slip or breach would provide Versailles with malicious gossip for days. To Mme de Rothe's annoyance, Lucie was entrusted to the Princesse d'Hénin to be schooled for the coming ordeal, which would include not merely her reception by the Queen but the regular Sundays at court, when manners and ceremonial were still rigidly orchestrated.

In Paris, the Princesse d'Hénin took her to M. Huart, dancing master to the nobility, a large man who looked absurd in his powdered wig and, worn around his generous middle, the hoop he put on to act the part of Marie Antoinette. M. Huart taught her how to curtsey, how to walk backwards in her own hoop and very long train, and when and how to remove her glove and bend to kiss the hem of the Queen's gown, and then to recognise the sign which indicated that she should rise.

Five days of rehearsals later, Lucie and the Princesse d'Hénin arrived at Versailles. Full court dress required a special bodice, laced at the back, a lawn chemise which left the shoulders and neck bare, and much cream-coloured lace. Because she was still in half-mourning for Mme de Monconseil, Lucie's dress was white, embroidered with pearls and silver. Round her neck, she wore eight rows of large diamonds, lent to her by Marie Antoinette, with more diamonds in clusters in her hair. The three required curtseys went well; after this, came presentations to the King and the Princes of the Blood. That night, Lucie was required to attend the '*jeu*', the playing of chess, cards and backgammon. The whole day, Lucie would write, 'was very embarrassing and exceedingly tiring. It meant being stared at by the whole court and torn to shreds by every critical tongue.'

Marie Antoinette had decided that Lucie would not take her mother's place in her household until she reached 19, but that whenever she was at court for the formal Sunday ceremonial, she could be included in the ritual of the Queen's *lever*. As with the presentation, every movement at the *lever* was regulated according to status and etiquette, the various *princesses*, *marquises*, *duchesses* and *comtesses* advancing and retreating as they proffered first one article of clothing and then another. Sometimes, so many women were present, puffed out in their immense hoops, their dressed and powdered hair towering above them, that they were obliged to bunch closely together to allow the Queen space in which to move. All but the very elderly and the very pregnant had to remain standing. Marie Antoinette treated Lucie kindly, though not without flashes of somewhat malicious teasing, her barbed compliments painful to a shy 17-year-old. When Lucie's glowing adolescent complexion was perceived to be uncomfortably more attractive than the Queen's, an old friend of her mother's suggested to her that she should avoid, when in the Queen's presence, standing in the full light of day.

On Sundays, the King appeared towards the end of the morning. He was so short-sighted that he seldom recognised his courtiers. When he did, he was friendly. Lucie, with her irreverent eye, noted that he looked 'like some peasant shambling along behind his plough', constantly fiddling with his hat and sword, but that in full court dress and jewellery he was capable of being 'magnificent'. Processing to the chapel through the long Salon of Hercules under Lemoyne's frescoes, fading from the glow of innumerable candles, required another terrifying ordeal of rank and procedure. Walking became a sort of shuffle, so as not to step on the train of the woman in front. Once through the salon, it turned into a frenzied, but subdued, scramble, the trains whipped smartly over the hoops as the women rushed for the best places in the chapel, places where they would be seen, and where they found their servants waiting with missals in red velvet bags. A number of older aristocratic ladies came to Versailles from Paris for the Sunday receptions, dressed in court gowns with skirts longer than those worn by Lucie and the younger girls; they were nicknamed the 'stragglers'.

At the royal dinner that followed the Sunday Mass, Louis XVI and Marie Antoinette ate alone, on two green thrones set side by side. On the other side of the table, on low stools, perched the grandest of the courtiers; the rest stood. The King, noted Lucie, 'ate heartily'; the Queen 'neither removed her gloves nor unfolded her napkin'. Everyone watched. As soon as they could escape, the courtiers bowed and curtseyed, then hurried away to pay their visits to the royal princes, racing from one end of the palace of Versailles to the other, the women clutching their trains, easing their hoops through narrow doorways, perched on heels that were 3 inches high, scattering powder and pomade from their artfully dressed hair. Later came the obligatory attendance at the '*jeu*'. Those ladies fortunate enough to have their hair done by Léonard, sometimes hours before dawn such was his popularity, would have to spend the day concentrating on keeping their heads still, for fear that the carefully concocted edifice would crumble.

Of all the royal visits Lucie was obliged to pay, those she most enjoyed were the ones to the charming and good-looking 30-year-old Comte d'Artois, the King's youngest brother, whom the Princesse d'Hénin knew well. Lucie had already noted that it had become fashionable to pronounce oneself bored at Versailles, bored with the court, the uniforms, the rules, the company, and that whenever they could, the courtiers escaped to Paris. 'All the ties,' she observed, 'were being loosened, and it was, alas, the nobility who led the way. Unnoticed, the spirit of revolt was rampant.'

As the Minister for War, the Maréchal de Ségur, who had been a guest at Lucie's wedding, had given Frédéric a month's leave from his regiment, they remained at Montfermeil. 'We were aware,' Lucie would write, 'of a deep conviction that however great the reverses we might have to endure, we would find in our mutual love the strength to withstand them unfalteringly.' They spent their days hunting in the surrounding forests, Lucie riding a 'fine grey' given to her by her great-uncle, wearing the most fashionable English riding dress and hat, sent specially from England; sometimes they paid visits, in an elegant little carriage also given to her by the Archbishop. She was happy, in a way she had never been before, even if to Mme de Rothe's habitual dislike of her granddaughter was now added jealousy, for Frédéric seemed to get on irritatingly well with the Archbishop. Frédéric also made it plain that he knew, from friends at court, about Mme de Rothe's mean-spiritedness, and he did all he could to shield Lucie from her attacks. Mme de Rothe, in return, never missed a chance to speak ill of him behind his back. 'She delighted,' wrote Lucie later, 'in all kinds of painful, hurtful remarks.'

When Frédéric was obliged to return to his regiment, he arranged for Lucie to move out of Mme de Rothe's house and join the Princesse d'Hénin in her official rooms in the Courtyard of the Princes at Versailles. There, standing at the windows which looked across the great central courtyard, she observed the bustle of the court. Sometimes she took refuge with her new sister-in-law Cécile at Hénencourt, a pale pink-red brick château, with a colonnaded front, curved courtyards, a moat and a dovecot. The gardens had been laid out by Le Nôtre, and beyond a pond in the shape of a four-leaf clover lay fields and forests. When they were on leave from their regiments, Frédéric and Augustin de Lameth joined them. 'To live with my husband and his amiable sister,' wrote Lucie, 'seemed to change my whole existence . . . Never had I enjoyed myself so much.'

* * *

In the late summer, Lucie discovered that she was pregnant. To her great pleasure, this meant that she would be unable to accompany the Archbishop and Mme de Rothe on their annual journey to Montpellier. Since Frédéric was again with his regiment, she went to stay with the Princesse d'Hénin who, after her mother's death, had taken a house in the Faubourg Saint-Germain where she spent much of the winter. The Queen gave them permission to use her boxes at the Opéra, the Comédie-Française and the Comédie-Italienne, where the new popular light operas were sung in French. In each theatre, the Queen's box, just above the stage, was furnished like a drawing room, with dressing and writing tables, and two warm, well-lit anterooms behind, guarded by footmen in the King's livery. From here, Lucie had fine views of the Indian ambassadors sent by Tippoo Sahib, the francophile ruler of Mysore, to ask the French to support him against the English. The three richly dressed Indians with long white beards sat in a special box, propping their yellow leather slippers over the side, to the delight of the audience below.

The exodus of the nobility from Versailles was becoming more marked. '*La ville*', Paris itself, was turning into the real centre of the social world, much of it grouped, as in the past, around a number of salons, presided over by powerful women, each with her own political bent. Now, however, the talk was less of '*politesse*' and '*esprit*' and more of politics and the economy, how to cope with France's national debt which continued to mount month by month.

Both the Princesse d'Hénin and her companion the Chevalier de Coigny had become fond of Lucie and since it was the custom for a new bride during her first year in society never to go out unaccompanied, Frédéric's aunt became her willing chaperone. According to the starchier ladies at Versailles, Lucie should also have had a liveried footman constantly at her side, but Frédéric had little time for such conventions. Like the Princesse, the Chevalier was witty and charming, with a wry, mocking tone. Lucie, who called him a 'stout, gay and amiable knight', enjoyed his fund of anecdotes about court life, to which she listened closely, believing that they would prove helpful to her later with Marie Antoinette. 'A knowledge of the past,' she had decided, 'would be very useful to me.' But when, one day, she and the Princesse drove out to Longchamps during the annual fashionable parade, she was shocked to be overtaken by an identical matching carriage, bearing the identical Hénin livery. It turned out to be the Prince's brazen mistress. It was absurd to have been so surprised, Lucie would later remark, with a touch of the snobbery that occasionally coloured her remarks, since the 'common people', who had 'no shades of feeling', were all the time being 'set such a bad example' by the arrogant and ill-mannered nobility. Between them, the Prince and the Princesse d'Hénin spent and borrowed prodigiously, paying off their vast debts in small, reluctant amounts.

The *libellistes* and the pamphleteers, who delighted in society scandal, often reported on the activities of the Hénins, and particularly on those of the Princesse, who, together with her friends the Princesse de Poix, the Duchesse de Lauzun and the Comtesse de Simiane, were known as the '*princesses combinées*'. These four friends, close since childhood, followers of Rousseau, Voltaire and the *Encyclopédistes*, saw themselves as arbiters of '*bon ton*'. The Duchesse de Lauzun, unhappy wife of Arthur's faithless friend, was a gentle, sweet-natured woman, left for months on end by her husband, who, like the Prince d'Hénin, preferred an actress at the Comédie-Française. She owned a library of rare books, including a manuscript, in Rousseau's own hand, of *La Nouvelle Héloïse*; it was in her library that Lucie came across a letter that Rousseau had written, explaining the reasons for putting his children into a foundling home, reasons that struck her as ignoble. The Princesse de Poix, who had been a close friend of Lucie's mother, was a clever and studious woman who had been left partially paralysed after the birth of her second child.

In the wider circle of these '*princesses combinées*', to whom Lucie was now introduced, there was also the Princesse de Bouillon, married while still a child to an extremely rich and well-born man, who happened to be mentally deficient, and so never appeared in public. She was, thought Lucie, one of the most distinguished and charming women she had ever encountered, but also extremely ugly, thin as a skeleton, with 'a flat Germanic face, a turned-up nose, ugly teeth, and yellow hair'. But when she drew up her stick-like legs, crossed her emaciated arms and began to speak, she would sparkle, from behind that 'collection of fleshless bones', as Lucie would write, 'with so much wit, originality and amusing conversation that one was carried away on the wings of enchantment'. Even the skeletal Princesse de Bouillon had a lover, the tall, somewhat insipid Prince Emmanuel de Salm, and the little girl seen around her house bore a great resemblance to them both. The Princesse was evidently much taken with Lucie, despite the 22-year difference in their ages, and to Lucie's great pride and pleasure, 'allowed me to visit her as if I had been her contemporary'.

There was another princess, not one of the '*combinées*' but a family friend, whom Lucie now met again. This was the Princesse de Beauvau, for whom the Chevalier de Boufflers, returning to Paris from his post as governor of Senegal, had recently brought back a small black child as a present, along with a parakeet that spoke both Senegalese and French for the Queen, and an ostrich for the Duc de Nivernais. The Princesse de Beauvau was not the only 18th-century Parisian to have a black child, bought, and kept perfumed, as a pet. Often compared to monkeys, and dressed as blackamoors, or rajahs, or in tiny military uniforms, these children were said to rank in their mistresses' affection along with 'parrots, greyhounds, spaniels and cats'. In the Palais-Royal, little Scipio, who could walk on his hands, carried his mistress's parasol. In portraits of the time, small black turbaned children were portrayed with baskets of fruit, playing with jewels, or handing their mistress a letter or a fan.

Lucie was discovering that it was wise to be accepted by these witty *princesses combinées*, who 'upheld one another, defended one another, adopted one another's ideas, friends, opinions and tastes', for they protected any young woman they liked against the malice of the court and Paris society. What she really liked and admired about them was not only their loyalty to one another and their erudition, but the fact that they were for the most part very discreet about their affairs, and usually faithful to those they loved. The ladies of the court, by contrast, 'flaunted' their liaisons 'shamelessly'. 'I knew,' wrote Lucie, 'how important it was for me to make friends with older women, who in those days were all-powerful.' Though content with her new life, Lucie was already looking to a successful future.

Around these witty *princesses combinées*, in whose circle Lucie spent an increasing amount of time, gathered a coterie of amusing men, such as Frédéric's brothers-in-law, the de Lameth brothers, and the Duc de Guines, later renowned for wearing rose-coloured jackets and short tight trousers. They welcomed Lucie and seemed to take pleasure in her evident enjoyment of their company. It was the Duc de Guines who told his daughter, as she was about to be presented at court: 'In this country vices do not matter, but ridicule kills.'

But it was not all about frivolity and wit. Most of the habitués of these salons were friends and followers of the Swiss Protestant banker, Necker, whose wife Suzanne, a tall, very pale woman who was said to rest only in her bath, also held a salon. Mme Necker had been a governess and was reputed to be extremely snobbish. Lucie referred to her as 'priggish' and 'pedantic'. 'When God created Mme Necker,' wrote Baronne d'Oberkirch in her memoirs, 'he dipped her first, inside and out, in a bucket of starch.' But since Mme Necker was also witty and had an excellent cook, Parisian society flocked to her drawing room.

In 1786, the Neckers' only child, Germaine, had married the Baron de Staël, the Swedish Ambassador to Paris, bringing with her a dowry of 650,000 *livres*, magnificent diamonds, a splendid trousseau and a carriage and four horses. Germaine, a childhood friend of Frédéric's, now lived in Lucie's old house in the rue du Bac, which she had filled with magnificent chandeliers, obelisks and marble columns. It was not a happy marriage. Germaine said of her husband that he was a 'totally honest man, incapable of saying, or doing, anything stupid' but that he lacked resourcefulness and energy. Germaine was not pretty, but she had a large bosom and strong features, and shining, very expressive eyes, and she had grown up listening not only to Voltaire, but to Gibbon, Hume and Walpole, sitting on a footstool by her mother's chair. In 1787, when Lucie met her, she was 21, a forceful, self-assured young woman who spoke good English and revelled in the art of conversation. Not naturally given to subservience, she liked to argue and to analyse.

It was an extraordinary moment to be young and to be French. Paris was alive with ideas and arguments, rumours and opinions. Never had the salons been so lively nor their guests more outspoken and opinionated. 'All heads were turned,' wrote the Duc de Levis. 'Soldiers talked about government, magistrates gave up the law and dreamt of politics. Men of letters wanted to make laws, clerics talked finance and women spoke of everything . . . It was enough to sparkle among the women, since they directed opinion.' Love, added Thomas Jefferson, writing to a friend in London, had 'lost its part in conversation'. Though not for Lucie: even if he was frequently away with his regiment, her marriage to Frédéric was turning out to be extremely happy.

* * *

In 1784, Jefferson, one of the authors of the American Declaration of Independence, had agreed to join Benjamin Franklin and John Adams in Paris as Commissioners with the task of convincing the French government that the United States remained a good financial risk, with great promise of growth and development. The war and subsequent break with the British had left America saddled with debts for goods ordered from merchants in London and Glasgow, while the traditional markets for its tobacco, whale oil, rice and wheat had been cut off. In Paris, Jefferson's view that happiness resided ultimately in the elevated principles of a well-ordered state found ready listeners among Frédéric and the young liberal nobles who had returned from the American campaigns excited by its vision of equality. As Mercier remarked, America was proof of what 'Man can do when he adds knowledge to a generous heart'. When these liberal aristocrats met, the conversation turned naturally to how the American experience might somehow be applied to France.

Listening to Frédéric and his friends, Lucie was acquiring a political education, even if she often found the incessant talk about the precariousness of French society tedious. What astonished her, when writing her memoirs almost half a century later, was that even when discussing the growing sense of unrest, the word 'revolution' was never uttered. 'Had anyone dared to use it,' she would write, 'he would have been thought mad.'

In the exalted and high-minded salon of Mme Helvétius, widow of the philosopher and close companion to Franklin, in whose drawing room in Auteuil prowled 18 angora cats dressed in satin, the Commissioners met and made friends with the liberal Frenchmen of the day. There was the mathematician and philosopher Marie-Jean Condorcet, a tall, ungainly, dishevelled figure, with very white skin. Condorcet was a visionary optimist, standing at the liberal end of the philosophers and believing in the utopian reform of French society. There was also the Duc de La Rochefoucault, a reflective, urbane gentleman-farmer, who shared Jefferson's interests in scientific agrarian experiments and had a model farm in Normandy. During the American war, with Franklin's help, La Rochefoucault had translated various key documents and published them in clandestine periodicals in France. By 1787, 25 Frenchmen had been elected members of the prestigious American Philosophical Society. Others, like Arthur Dillon, had been made members of the Cincinnatus Society,* a group of young officers, French and American, who had fought together for George Washington.

The three Commissioners, with the help of George Washington's agent in Paris, an articulate, patrician New York businessman called Gouverneur Morris, presided over gatherings for the growing American community and their French liberal friends.

---

* Cincinnatus was the Roman consul who, after the triumph of the republic, had given back his dictator's sword and returned to the fields.

There they discussed, long into night, the rights of man, the abolition of slavery and equality for women. Jefferson himself, a tall, sinewy, rather stiff man, with greying strawberry-blond hair and red cheeks, kept a lavish table and served excellent wines, saying that good wine stimulated the intellect and conversation.

By 1787 the Americans in Paris had been joined by Thomas Paine, already known for the French translation of his book, *Common Sense*. But they now lost Franklin, who was 79 and ailing, and in such physical discomfort that Marie Antoinette lent him a special mule-drawn litter in which he could recline on cushions to carry him from Paris to the ship that was to bear him to America. Franklin was greatly missed, by the many women he had charmed and by the French liberals. His parting gift from the King was a portrait encrusted in diamonds, and his luggage, as he lumbered across Brittany behind his mules, included 23 cases of books, a dismantled printing press and various sets of printer's type, a cabriolet, a crate of fruit trees and two of Mme Helvétius's angora cats. For the Americans left behind debating ideas with Frédéric and his friends about prosperity and enlightenment, the question was how to transform a country of some 26 million people, at least 25 million of whom had never known either prosperity or liberty, into a functioning republic.

Lucie was drawn into this American world not only by Frédéric and the de Lameth brothers, but by her excellent spoken English. Teased by the young men she met for her unfeigned devotion to her husband, she was conscious of the need to win her way back into the favour of her mother's friends, both in Paris and at court, alienated since Thérèse-Lucy's death by Mme de Rothe's sniping and abrasive manner. One of these was the Rochechouart family, where Lucie had practised the piano as a child, and where she returned to sing at musical gatherings still held every week in the rue de Grenelle. Her friend Rosalie-Sabine had contracted a wasting illness and had grown into a misshapen hunchback, and neither her good nature, nor her intelligence, nor her remarkable singing voice, could reconcile her young husband to living with her. He had fled to Russia, to join the Imperial army.

Soon after settling in Paris, Lucie was taken by Frédéric to call on Mme de Montesson, one of the older women he most admired and in whose house he had lived as a young man. She had not long before put off her mourning for her husband, and returned from a convent to her house in the Chaussée d'Antin. Mme de Montesson – for reasons of protocol – was not received at court, but in Paris her invitations were much sought after. She had always been devoted to Frédéric and now treated Lucie as a daughter, introducing her to her formidable niece, Félicité de Genlis. Mme de Genlis was tutor to the young Princesse d'Orléans, known as 'Mademoiselle', and to her three brothers. Soon after being appointed – her curious title of tutor explained by the fact that the post was usually held by a man – Mme de Genlis had started a school in a specially built pavilion at the Convent of Belle Chasse, to which the princes came from the Palais-Royal every day. In appearance more handsome than pretty, she was 41, with a pronounced nose and a large mouth. She was bringing up the princes frugally, to follow Rousseau's precepts about nature, character-training and physical exercise. Buffon declared that she was 'training souls'.

In the pavilion in the Chasse Midi, where Lucie was taken to visit her, history was taught with magic lanterns, supper was conducted in Italian, and time was found for literary criticism, mechanics, botany, anatomy, mineralogy, fencing and book-binding. The walls of the school rooms were decorated with busts of the Roman emperors. Pupils rose at 6.30 and breakfasted on bread and grapes, Mme de Genlis no more believing in greed than she did in holidays. Setting herself up as a beacon of morality, in a society buffeted by scandals, she had founded an Order of Perseverance, whose members wore enamel rings engraved with the words 'candour, loyalty, courage, chastity, virtue, goodness and perseverance'. She also played the harp. Among her pupils was a little girl, *la belle Pamela*, an English child brought at the age of 5 from her penniless parents to be a companion to Mademoiselle, though it was rumoured in Paris that she was in fact Mme de Genlis's own daughter by the Duc d'Orléans.

Lucie soon became friendly with Pulchérie de Valance, Mme de Genlis's grown-up daughter. Three years older than Lucie, and also pregnant, she was married to a man of whom some said that he was her great-aunt's lover. When there were evening dances at the pavilion at Belle Chasse, there was never anything to eat, and only water to drink, Mme de Genlis extending her frugality to her guests. With Mme de Montesson, or with Frédéric, Lucie was also an occasional guest at the Palais-Royal, where the Duc d'Orléans, who was not frugal at all, gave glittering supper parties. Neither she nor Frédéric liked or admired the world of the Palais-Royal, where there was much criticism of the King and Marie Antoinette and where, Lucie wrote later, the Duke 'corrupted everything within reach'.

* * *

In February 1788, while at court one Sunday evening, Lucie heard the clocks strike nine while she was still some way from the Queen's apartments. She was now almost five months pregnant. Fearing she would be late, she began to run. Her hoop caught in a doorway, and she stumbled. Two days later, she miscarried.

When Mme de Rothe returned to Paris soon afterwards, Lucie and Frédéric reluctantly moved back to live with her – as was expected of a young couple with no establishment yet of their own – but relations between her jealous, angry grandmother and Frédéric soon grew so strained that the young couple were asked to leave. Paris and the court were quick to take sides, some blaming Frédéric for being overly hasty and unaccommodating, others pointing to Mme de Rothe's dictatorial ways. Lucie, who had strong feelings about family loyalty, felt torn.

They were offered an apartment in the Princesse d'Hénin's house in the rue de Vermeuil, overlooking a small, dark garden, and gratefully moved in, taking with them Marguerite, Lucie's much-loved childhood maid. Lucie was soon pregnant again. Whenever they could, they escaped to Passy, where the *princesses combinées* shared a house; but Lucie was often required at court, though Marie Antoinette now excused her from the long walk to Chapel, for fear that she might again slip and fall. Instead, she helped in the elaborate ritual of making the Queen's bed, stripping the sheets and handing them in baskets lined with green silk to footmen. Marie Antoinette herself had lost her daughter, Sophie, just before her 1st birthday. Sophie had always been sickly, but the Queen was heartbroken. Asked by a courtier how she could grieve for a baby so young, of whom she as yet knew so little, the Queen replied: 'She might have become my friend.' Madame Vigée-Lebrun was asked to paint out the baby from the most recent group portrait of the Queen and her children, leaving the figure of the Dauphin pointing to an empty cradle.

In December, Lucie gave birth to a stillborn baby, strangled by the umbilical cord. Labour lasted 24 hours 'of unbearable pain', and she almost died. It was her second lost child, but as with all the most unhappy events in her life, she chose not to dwell on it. Nor were miscarriages or stillborn children unusual; few 18th-century women managed to keep all of their babies.

For the rest of that winter she was confined to her bed, recovering slowly from puerperal fever. When she at last felt better, she returned to see her friend Rosalie-Sabine at the Hôtel de Rochechouart where she sang contralto in musical evenings with singers from the Opéra. Viotti, the celebrated violinist, accompanied them. Rehearsals lasted all day, and for the final performance friends from the Faubourg Saint-Germain provided an audience. Later, talking of this time, Lucie would remember and miss the ease and the good manners, the way the generations mixed and enjoyed each other's company, and regret that the old ladies, who had once set the tone and the etiquette of the salons, were no longer to be found in society.

Lucie was now 18; with her fair hair and light complexion, and her excellent spoken English, she was much in fashion in a city obsessed by all things English. Even Marie Antoinette commented on her English manners, and the way that she remembered to shake hands when English visitors came to court. She was becoming, not only more certain of her opinions and feelings, but bolder; and she was not without vanity. At a ball given by the outgoing English Ambassador, the Duke of Dorset, she decided to ignore the instructions to wear only white and ordered a dress in blue crêpe, with matching blue flowers. To this she added gloves and a fan trimmed in blue and when Léonard came to dress her hair, she asked him to weave blue flowers into the meshes. To her satisfaction, she was much remarked on; the Duke of Dorset observed that the Irish had always been unruly.

Many years later, the Vicomtesse de Noailles, granddaughter to one of the *princesses combinées*, looking back fondly at a moment when Paris seemed so lively, so full of warmth and splendour, would write: 'While we waited for the catastrophe, society, social life, was delicious.' What she did not add was that it was also often very spiteful. Behind the perfect courtesy and the rigid etiquette there was both self-regard and malice; the constant quarrels between the stiff-necked Mme Necker and the sharp-tongued Mme de Genlis kept all Paris amused.

* * *

Away from the elegant talk, the nuances of *bon ton* and *esprit*, the glitter of costume and livery, France was edging towards bankruptcy. Arthur Young, travelling on his horse through the countryside, remarked on the squalor and backwardness, and on the cripplingly hard labour, particularly of the women. Seeing so many without shoes or stockings reminded him of the misery of Ireland. Everywhere, people were poor, and often hungry; for many, meals consisted of little but soup, made from bread, water and vegetables. The peasants, whole families living in a single room, locked into a feudal system, paying much of their harvest in tithes and taxes to absent noble landowners, were also battling a vicious circle. There was never enough grain to feed the animals, and never enough animals to produce enough manure to feed the fields and thus increase their yields. Working from before dawn until after dusk, there was very little to cushion them against illness, drought or sudden calamities.

Children who could not be fed were sometimes smothered. 'The whole parish is poor,' wrote one country priest. 'There are at most twenty households or families who are living decently; all the rest struggle to get by on their wits.' So endemic was poverty that it had spawned a whole new vocabulary: there were the shameful poor, the indigent, the wretched, the professional beggars and the beggars by necessity. As the price of grain kept rising – it had risen by 65 per cent since 1770, the year of Lucie's birth – so France saw ever more abandoned babies, and ever greater number of vagrants of all ages, poaching, stealing firewood, picking pockets. Highway robbery had become so menacing that it was made punishable by death on the wheel.

Paris itself, like a metaphor for the ills consuming the country, had never smelt so rank or so sinister, a putrefying 'miasma' seeping out from graves, basements and underground prisons, a silent fermentation that threatened to engulf the entire city. The houses, warned Mercier, were now 'leaning over abysses'. In the gardens of the Palais-Royal, the stench of urine and faeces was overwhelming; along the Avenue de Saint-Cloud at Versailles dead cats lay in stagnant water, while livestock, wandering into the great gallery, left behind trails of dung. Paris, observed a visitor, the world centre of the arts, fashion and good taste, had become the 'centre of stench'. No amount of 'odiferous balls', of lemons studded with cloves, of sachets of rue, mint and rosemary, could drive it away.

The Assembly of Notables, with its 144 nobles and clerics, had not proved biddable. After many false starts it had convened, only to refuse to endorse Calonne's proposed land tax, prompting Lafayette to remark that the Notables might perhaps better be called Not-Ables. So futile was the whole exercise perceived to be, with its grandiose flights of rhetoric, that in one satirical cartoon, a monkey was shown addressing a group of chickens: 'My dear creatures,' it was saying, 'I have assembled you here to deliberate on the sauce in which you will be served.' Calonne was dismissed, and Loménie de Brienne, the Archbishop of Toulouse – whom the King had turned down as Archbishop of Paris on the grounds, so it was said, that he thought that the primate of the French Church should at least believe in God – was appointed Principal Minister. Brienne was an excellent administrator, keen to implement a number of sensible and liberal policies, intended to stabilise the state; but again the Notables rejected them. This time, however, it was they who fell, and the Assembly was disbanded.

Brienne now tried to get at least some of the measures through the Paris parliament, promising economies in government as well as at court (the cravat holders and wolf hunters of the royal household were to go), but the parliamentarians, sensing loss of power, fanned the flames of dissent, gaining support from the people and from the provincial parliaments. Discontent with the King, spearheaded by the nobility, who argued that they were protecting France from ministerial despotism, was spreading throughout the country. On 7 August 1788, with serious crop failures caused by hailstorms, a run on banks and the collapse of government stocks, the King agreed to convoke the Estates General. It was Brienne's turn to be forced from office, despite support from the powerful clerics on whom he had showered patronage. Necker was recalled as Director General of Finances, and he announced that all reforms were to be postponed until the Estates General met.

There had been no Estates General for more than 170 years.

This archaic body, first convened by Philip III in 1302 to give counsel at times of crisis, had not actually met since 1614. The crucial question was how it was going to be composed. The First Estate, the clergy, representing the 10,000 members of the Catholic Church, combined with the Second, the nobility, speaking for 400,000 nobles, both of which enjoyed numerous privileges and rights, hoped to vote together. This way they could block any proposed reforms put forward by the Third Estate, who represented France's 25 million people. However, the First Estate included large numbers of parish priests, far closer to their parishioners than to the aristocratic bishops and cardinals; and the Second, many liberal nobles, reared on the Enlightenment and genuinely hoping for reform. With these liberals were Frédéric and his father, the de Lameth brothers, the Duc de Lauzun, Lafayette and the wild, unruly Comte de Mirabeau, the writer and polemicist, men who had recently founded a Society of Thirty to promote reform.

Among these aristocratic liberals, meeting in the salons of Mme Necker, in the Palais-Royal or in Mme de Genlis's pavilion to talk about the rights of man, the abolition of privileges or the sale of Church property, was Charles-Maurice de Talleyrand-Périgord, recently made Bishop of Autun, a man better known for his scheming and amorality than for his piety, but clever, witty, full of irony and charm. Like Archbishop Dillon, Talleyrand saw no difficulty in having mistresses and was already the father of a son; he was a fine-looking man, with a genial if slightly sly expression, and a full head of hair, curled back in the fashion of the day. He was also lame, with a severely deformed right foot. True philosophers, declared Talleyrand, should 'adore God, serve Kings, love mankind'. He would play an important part in Lucie and Frédéric's lives.

These liberals, among them not only Frédéric but also his father and his brothers-in-law, wished to double the representation of the Third Estate from 300 to 600, thus making them equal to the first two Estates combined and weakening the entrenched autocracy of Church and nobility. When, two days after Christmas 1788, the Royal Council met to consider their proposal, Necker voted in favour, and the King reluctantly agreed.

* * *

Much of the population of France, by 1789, could not read, or only barely. This did not prevent an outpouring of pamphlets, allegorical compositions and political caricatures from flooding the capital. *Libelle* literature, purporting to use precise descriptions, dialogue and phrases from letters to create the illusion of an accurate picture of the scandalous and licentious life at court, as seen by an invisible fly on the wall, was being circulated all round Paris. To this were now added tracts, journals, almanacs, pamphlets and posters, all reflecting the growing state of turmoil and uncertainty. Every day, observed the Parisian journalist and author Nicolas Ruault, it was 'raining pamphlets and brochures'. Along with the caricatures came scurrilous songs and plays, drawing on an ancient vein of satire against the clergy. In *Charles IX*, a popular play by Marie-Joseph Chénier, the King was depicted as a devious and amoral halfwit, surrounded by cardinals and bishops plotting to exterminate good citizens.

In January 1789, a radical priest called the Abbé Sieyès, whose essays would mirror the coming revolutionary ferment, produced a pamphlet attacking what he described as the parasitic nobility and clergy. What, asked Sieyès, is the Third Estate? 'Everything. What has it represented in the political order until now? Nothing. What is it asking for? To become something.' '*Qu'est-ce-que le Tiers Etat?*' ('What is the Third Estate?') became the most famous essay of the day. Everyone knew what the First and Second Estates were. The question, now, was just what, exactly, the Third Estate might become.

Early in 1789 began one of the most ambitious exercises in political consultation ever undertaken. A hundred thousand copies of a leaflet, explaining in detail the composition and the purpose of the Estates General, were printed and distributed. It was addressed neither to women, nor to servants, nor to actors, but even so its scope was vast, and people were encouraged to speak out through *cahiers de doléances*, books of grievances. All through the late winter and early spring they did so, loudly, coherently and very firmly, demanding a radical review of all forms of taxation and an end to seigneurial and ecclesiastical dues.

The winter was bitterly cold. The Seine froze almost all the way to Le Havre. In the Channel, fish died and oysters were brought up from the depths in Brittany frozen solid. In Paris's fashionable streets glided sleighs in the shape of dragons and sirens, their passengers enveloped in fur-lined velvet cloaks trimmed with gold braid, their coachmen decked out as moujiks with long false beards. The spring brought floods. The price of both bread and firewood almost doubled in six months. Twenty-five thousand silk workers were out of work in Lyon, and 10,000 textile workers in Rouen. Necker used what very little money he could still raise to buy grain abroad in order to prevent famine, but there were attacks on bakeries in Brittany and riots in the Midi, the Dauphiné, Provence and the Languedoc. On the streets of Paris, thousands of beggars clamoured for food and some were put to work, for 20 sous a day, digging the hills of Montmartre. Everyone was waiting to see what the harvest of 1789 would bring.

Candidates for nominations to the Estates General put themselves forward. To Lucie's immense relief – for she had a presentiment, she wrote later, that disaster was looming – Frédéric, who stood for the Second Estate for both Nemours and Grenoble, failed to get himself elected. Archbishop Dillon, perceived as too venal and too partisan, also failed to win a place among the First Estate. But Lucie's father-in-law, M. de la Tour du Pin, was elected to represent Saintonge, and the de Lameth brothers, Talleyrand, Lafayette, Mirabeau and Condorcet all secured places.

At the court, and among the Parisian nobility, the spring arrived with its customary round of social diversions. Lucie, who remarked later that never had Paris seemed less anxious or more intent on pleasure, was constantly either at the theatre or at balls. 'Many upright and honourable people', she wrote, continued to believe that France, albeit reformed but with its monarchy, landed aristocracy and powerful Church basically unchanged, was about to enter on a 'Golden Age'. She herself was no longer as certain. On 27 April, she accompanied Pulchérie de Valance to the races at Vincennes to watch the Duc d'Orléans's horses, with their English jockeys, run against those of the Comte d'Artois. They travelled under the Duc d'Orléans's livery, since M. de Valance was First Equerry to the Duke. On their way back into Paris in the afternoon, they were stopped by a menacing crowd but, once the people recognised the Duke's livery, there were cheers of 'Long live d'Orléans', for d'Orléans, unlike most of the rest of the royal family, remained popular with Parisians for his outspoken, reforming opinions.

Only later did Lucie hear that there had been an attack on the wallpaper factory of a man called Réveillon, sparked off by a false rumour that he had been planning to reduce his workers' wages. Riots spread across the Faubourg Saint-Antoine, and Réveillon's factory was burnt to the ground, along with his entire stock of printing blocks, machines and warehouses. The *Gardes* were called out to confront the rioters and by nightfall had killed some 50 people. The incident, Lucie noted later, was all the more unfortunate because Réveillon was in fact a generous employer, who, earlier in his life, had saved one of his workmen from debtors' prison and his family from starvation.

Once the elections to the Estates General were over, and 1,196 men selected to represent constituencies from all over France, Lucie, Frédéric and the Princesse d'Hénin moved to Versailles for the opening ceremony. To absorb the delegates, their families and servants, the town had become a vast hotel, people crammed into every room and attic. The Princesse d'Hénin still had an apartment on an upper floor, above the Galerie des Princes, which overlooked the Orangery and the gardens to the south. The governor of Versailles, the Prince de Poix, offered them his house at the Ménagerie at the far end of one arm of the canal, opposite the Trianon, almost half a mile from the palace itself. Here, next to the elephant and the rhinoceros, looking out across the park, the water and the rides, Lucie and the Princesse installed themselves in great comfort, with footmen, a cook, horses and Lucie's English groom. Versailles was extremely lively. There was still, remembered Lucie later, no real concern about events, only a sense that the Estates General would give birth to a new, better, more egalitarian and more prosperous France. 'Amid all these pleasures,' she wrote, 'we were laughing and dancing our way to the precipice.' And, she added, while such blindness was pardonable among the young, it was 'inexplicable in men of the world, in Ministers and above all, in the King'.

Frédéric was so annoyed at having failed to be elected to the Estates General that he decided to rejoin his regiment and refused to wait for the opening session on 5 May. But Lucie watched it all. The day before, a magnificent day of early summer, with bright sunshine and sparkling light following a night of heavy rain, the royal family, the court and the 800 delegates who had already reached Versailles, had attended a solemn Mass in the Church of Notre Dame. The procession of the Holy Sacrament to the church had been a brilliant display of scarlet, purple and gold for the two upper Estates, of black and white for the Third. In front were mounted falconers in medieval dress holding hooded birds, and heralds on white horses, in purple velvet embroidered with fleurs de lys, blowing trumpets. Of the royal princes, only the Duc d'Orléans walked apart, preferring to march with the Third Estate. The King was loudly cheered; but the Queen was not. Lucie, standing with the Princesse de Poix at the windows of the royal stables, thought she looked sad and cross.

On the morning of 5 May, 3,000 people crammed themselves into the Salle des Menus Plaisirs, the former storeroom for scenery and props, redecorated with white and gold Doric columns and hung with magnificent Gobelin tapestries. Behind the King's throne on its golden dais hung purple velvet hangings, picked out in gold fleurs de lys. The three Estates were seated separately. The King and the Princes of Royal Blood wore robes of the Order of the Holy Ghost, the King's, noted Lucie, more 'thickly encrusted with diamonds and more richly embroidered'. She decided that he looked very undignified, standing awkwardly and walking 'with a waddle'. All Louis's movements, she wrote later, 'were abrupt and lacking in grace'; and, since he refused to wear glasses, he screwed up his eyes in order to see. The Queen, sitting on a platform just below his purple and gold throne, wore white satin and a cloak of purple velvet. From the way she used her fan, she seemed to Lucie very agitated. She appeared to be scouring the faces of the seated ranks of the Third Estate, among whom she knew she had many enemies, as if she were searching for something. Gouverneur Morris, also studying the Queen's face during the King's speech, observed that she looked as if she were crying. Her eldest son, the 7-year-old Dauphin, was very ill, in the last stages of tuberculosis.

There was an awkward moment when Mirabeau, the renegade philosopher count, already emerging as the most inventive and brilliant orator of the revolution, and one of a handful of nobles who had secured a place to represent not the Second, but the Third Estate, that of the people, arrived to take his place. When he moved towards the middle of one of a row of benches, those close to him moved away; throughout the hall could be heard a low, hostile hiss from the nobles who regarded him as a traitor. As he took his place among the lawyers and shopkeepers, Lucie thought his smile looked 'contemptuous'. She found Necker's two-hour opening speech about tax and administrative reforms extremely boring. Seated on a wide, backless bench reserved for those ladies of the court not actually in attendance on the Queen that day, she felt obliged to 'maintain an impeccable attitude' throughout interminable sentences that to her 19-year-old ears had little of meaning or interest.

There was also a moment of comedy, when the Third Estate, unaware of the age-old ritual of the doffing of hats by the King and the nobility, began doffing their own, unleashing a round of further doffing and more replacing of hats, until the King put an end to the confusion by firmly replacing his 'Henri IV', a great white plumed beaver hat with an enormous diamond. After many speeches, some of them inaudible to much of the vast audience, the King left the hall, urging the delegates to avoid 'dangerous innovations'. 'Here drops,' noted Gouverneur Morris, 'the curtain on the first great Act of this drama.'

From the first, the meetings of the Estates General were extremely noisy, with the delegates all eager to speak at once, and rancourous. Germaine de Staël, sitting near Lucie, remarked that the tone was also a curious mixture of frivolity and pedantry. Most of the Third Estate were men who had in some way been touched by the Enlightenment, and they had come to Versailles envisaging not rebellion but reform and believing that a king, even if not this king, was central to the idea of a state. Among the Second Estate were 4 princes, 16 dukes and 83 marquises; many were also soldiers, and most were very rich. No one had liked Necker's speech, neither the people, who felt that he regarded them as little other than provincial administrators, nor the nobles, who were against the proposed new taxes. Weeks passed in deadlock, all progress halted by disagreement over how they should even meet or vote.

As the proceedings dragged on, early on the morning of 4 June, the Dauphin, Louis-Joseph, died. According to custom, he lay in state at Meudon, where deputies from all three of the Estates came to sprinkle holy water on the small corpse. Later his heart, in an urn, was taken to the Benedictine convent of Val-de-Grâce, while his coffin, covered in a silver cloth, with the crown, sword and orders of the Dauphin of France laid on top, made its way to the crypt of Saint-Denis. When members of the Third Estate insisted on a meeting with the King three days later, Louis, sombre and uncommunicative, commented: 'So there are no fathers among the Third Estate?' Marie Antoinette had lost two of her children in two years. Two were left, Marie-Thérèse, who was 11, and 4-year-old Louis-Charles, who now became Dauphin.

On 10 June, still locked in disagreement, the Third Estate invited the nobility and the clergy to meet them in a general session. This was refused. On the 17th, they issued a unilateral declaration that they were henceforth to be known as the National Assembly. Necker, hoping to isolate the extremists, urged the King to consider a package of reforms, to be debated at a plenary session. A misunderstanding developed. When, on 20 June, the Third Estate arrived at the Menus Plaisirs, they found the doors locked and barred: the King, assuming that nothing more would happen before the plenary session, had simply suspended the proceedings. Suspecting a royal coup, the Third Estate repaired to the nearby *Jeu de Paume*, Louis XIV's tennis court, from where they issued the famous Tennis Court Oath: they would not leave, they declared, until a new constitution was agreed.

Three days later the King agreed to endorse most of Necker's reforms. He promised to abolish the hated *lettres de cachet* and to give greater freedom both to the press and to individuals; and he accepted that he would govern, in the future, with an elected assembly. However, he refused to break up the separate Estates, which effectively meant that the clergy and the nobility, voting together, could continue to block motions put forward by the people. When the moment came to disperse, the Third Estate refused to leave the building. 'We are here by the power of the people,' declared Mirabeau, 'and will leave only at the point of a bayonet.'

Old procedures and old assumptions were unravelling, one after the other. Nobility, clergy, bourgeoisie and the people were all indignant and angry; together, they were discovering a vocabulary of protest in the Enlightenment's ideas about social contracts and the rights of man. While a new constitution was being drafted, the King, influenced by the Queen and some of the court, summoned troops under the Duc de Broglie to gather round Paris and Versailles. Frédéric, garrisoned at Valenciennes 200 kilometres away, was not among them. Mirabeau and Necker, warning that this would send a very threatening message to the Third Estate, urged the King to show more restraint. Instead, he took the dangerous step of dismissing Necker, with orders that he should leave France in secret.

When the news of his disgrace was confirmed, the theatres in Paris, as if after a national disaster, closed. Two hundred thousand people took to the streets, shouting his name. Lucie, who was staying at a country house just outside Paris with the *princesses combinées*, had lived through Frédéric's many discussions about the perils facing France, through the bread riots and the attack on the wallpaper factory, all without great fear of the future. Now she was suddenly struck with a sense of foreboding. It was not simply Necker's resignation, but the speed with which he had fled the country and the volatile mood of the crowd who called for his return, that seemed to her so ominous.

Paris was full of rumours. In the coffee houses, in the streets, orators were making speeches about liberty. With every twist and turn, every rumour and conjecture, there was more talk, more ill temper. In the Palais-Royal, the shops selling political pamphlets were mobbed. Home to the defiantly hostile Duc d'Orléans, who seemed to have distanced himself from the rest of the royal family, the Palais-Royal had become the centre for much of the growing disorder. The Duke had provided food and shelter to those driven on to the streets by hunger, but both Lucie and Frédéric suspected that his real aim was to increase his own popularity with the people by ostentatious shows of generosity. The King, by contrast, Lucie observed, was 'hidden away at Versailles, or busy out hunting in the nearby forests', suspecting nothing, foreseeing nothing, and believing nothing that he was told. 'The court,' noted Lucie, 'stricken sublimely blind, could not see disaster approaching.'

On 12 July, in the celebrated Café Foy in the Palais-Royal, with the crowds clamouring for Necker's return, Camille Desmoulins, a fiery 26-year-old journalist from Picardy, jumped on to a table and shouted: 'To arms, to arms', and, seizing up some leaves from one of the Palais-Royal's chestnut trees, called out, 'Let us all take a green cockade, the colour of hope.'

The following day, Lucie, who had been invited by Mme de Montesson to her country house at Berny, two hours from

Versailles, decided to send her saddle horses with her English groom on ahead, by way of Paris. Next day, leaving the Princesse d'Hénin at the Ménagerie, and taking with her just a maid and a manservant, she set out by carriage for Berny through the forest of Verrières, a long, lonely drive. There had been a violent storm, and there were fallen trees along the way. So calm was Versailles, so sure that the King's troops were in control in the plain of Paris and its surroundings, that there was no anxiety about her journey. 'That fact alone,' Lucie wrote, 'shows the degree of our extraordinary unawareness.' Frédéric was still at Valenciennes with his regiment.

Lucie was surprised when she arrived at Berny to find the château deserted and the doors locked. Eventually a doorkeeper appeared, in a state of great agitation. Mme de Montesson, he told her, had sent word to say that she had been unable to leave Paris, that the city was in a state of turmoil, and that the gates had been barricaded. The Guardsmen, far from standing up to the rioters, had gone over to the mob. Lucie, overriding the fears and objections of the grooms and servants, ordered her carriage to put back to Versailles 'at a fast gallop'. Along the way, she heard more stories about an uprising on the streets of Paris, with the rebellious *Gardes* apparently now all sporting Desmoulins's

Anonymous French cartoon. *Adieu Bastille (Goodbye Bastille)*, 1789.

green cockades. At the Ménagerie, she found the Princesse d'Hénin still in bed, unaware that anything had happened. What worried Lucie most, at this stage, was the fate of the English groom she had despatched to Paris, of whom there was no news.

It was only the next morning, 15 July, that Lucie and the Princesse went to the palace and found Frédéric's father, M. de la Tour du Pin. He told them that the Bastille, the fortress that had become the symbol of despotism and the hated *secret du roi,* had been stormed by local people, disaffected soldiers and members of the *Gardes*, with weapons and two cannons seized from the royal storehouse. Its seven prisoners had been freed, but 83 people had died in the fighting. The prison governor, Bernard-René de Launey, had been spat on, knocked down, then murdered with knives and bayonets. His head, hacked off with a pocket-knife, was being paraded through the streets on a pike by a crowd of chanting, drunken men and women. Calonne's 10-feet walls and customs houses, encircling the city, had also been attacked. Paris was now in the hands of the insurgents. The *Gardes* had been disbanded and the disaffected soldiers sent to garrisons in the provinces where, Lucie would write, 'they spread that same fatal spirit of insubordination that they had learned in Paris and which thereafter could not be stamped out'. Next day, Lafayette, hero of the American wars, was put at the head of a newly formed *Garde Nationale,* in order to try to restore order to the city.

Two days earlier, the Duc de Liancourt, one of the more liberal members of the nobility, a friend of Frédéric's and a founder of the Society of Thirty, had come to Versailles to report to the King on the disturbances in Paris. 'Is this a revolt?' the King had asked. 'No, sire,' the Duke had replied, 'this is a revolution.'

CHAPTER FIVE

The Dismantling of Paris

Even as the revolution was unfolding, most Frenchmen still believed that nothing drastic or fundamental would befall France. In the eerie but brief calm that immediately followed the fall of the Bastille, Frédéric obtained leave from his regiment to take Lucie to the waters at Forges-les-Eaux, in Normandy. She had never fully recovered from the birth of her stillborn second baby and the doctors, fearing that damage to her kidneys might mean that she would not be able to have more children – 'a possibility that reduced me to despair' – had recommended a month in the countryside. Later, Lucie would remember these days as some of the happiest of her life. Forges was surrounded by forests. They spent their days riding through sunlit clearings and along grassy tracks; in the evenings, while Lucie sewed, Frédéric read aloud, setting a pattern they would follow all their lives. She was always eager to learn; Frédéric, who was an 'indefatigable reader', was a good teacher.

Their tranquillity did not last long. In the countryside, a sense of panic was catching fire. It was prompted by anxiety about the forthcoming harvest, by the expectations raised by the Estates General, and by a spirit of simmering hostility against the nobility and the clergy that had found expression and legitimacy in the *cahiers de doléances*. Towards the end of July, these grievances burst out into what became known as the Great Fear. It was grounded not in facts but in rumours and fantasy – that the Austrians were attacking from the Netherlands, that Spaniards were marching on Bordeaux, that brigands had been recruited by the nobility to destroy the livelihoods of those calling for reform. But while it lasted it was unpredictable and terrifying. A sudden dust storm, an unfamiliar rider, a deranged beggar, all were enough to cause women and children to hide in cellars and attics, and men to arm themselves with pitchforks and sickles. At one village in Champagne, men mobilised to confront a reported gang of bandits: it turned out to be a herd of cows.

In Forges, Lucy and Frédéric had taken rooms on the first floor of a house on the main road between Dieppe and Neufchâtel. At seven o'clock on the morning of 28 July, precisely two weeks after the attack on the Bastille, Lucie was standing at the window waiting for Frédéric, who had gone to the waters, when a crowd burst into the square below, the women wailing, the men shouting and gesticulating. In their midst was a dishevelled rider, on a lathered, dappled horse. The man began to harangue the crowd, saying that the Austrians were advancing on the nearby town of Gaillefontaine, and would soon reach Forges; then he galloped away to spread the news.

By temperament not given to panic, Lucie hurried outside and tried to reason with the crowd, pointing out that France was not at war with Austria. At the church door, she encountered the local priest, about to ring the tocsin, and she was still there clutching his cassock, trying to prevent his getting to the bells, when Frédéric, alerted by Lucie's English groom, arrived and took charge. Telling the assembled people that they would ride to Gaillefontaine themselves to find out what was happening, Frédéric, Lucie and the groom, who complained that the French had clearly gone mad, set off at a canter.

It took them an hour to reach Gaillefontaine. As they entered the town, they were challenged by a man with a rusty pistol who demanded to know whether the Austrians were at Forges. On being assured that they were not, he led them to the main square, where another agitated crowd awaited. At this moment, a prosperous-looking villager, pointing at Lucie, set up a cry: 'It's the Queen!' Immediately, Lucie's horse was surrounded by angry, menacing women. Fortunately, a young locksmith's apprentice had recently been in Versailles and explained, as Lucie wrote later, that Marie Antoinette was at least 'twice as old and twice as large'. Lucie and Frédéric were released and hastened back to Forges, where the worried inhabitants were still waiting for news of the Austrian advance.

They had escaped very lightly. Others were not as fortunate, though how true the stories of violence were no one could be sure.

'Vive la République'.

All over France, the politics of paranoia were feeding into the settling of old scores. Châteaux were set on fire and their contents looted; the symbols of the *ancien régime* had become targets, and none more so than the nobility and their possessions. It was rumoured that a 94-year-old marchioness was thrown on to a smoking stack and died watching her servants distributing her linen, furniture and porcelain. One countess was said to have been strangled; another to have had her teeth broken. A princess and her two young daughters were reported to have been tied naked to trees. When Mme de Montesu, held and tormented all day by men and women she had known all her life, and whom she believed liked her, begged for a drink of water, she was – so it was said – dragged across the courtyard and drowned in the pond. A collective frenzy gripped France. Even in normal times, the 4,000 men of the provincial constabulary, the *maréchaussée*, would have been totally inadequate. As it was, they did almost nothing.

* * *

At Versailles, M. de la Tour du Pin, Frédéric's father, had been appointed Minister for War. Regarded by the liberal aristocrats as loyal and level-headed, he was pleased to be offered the position because he had long been a critic of corruption in government and genuinely believed that aspects of the *ancien régime*, if reformed, might yet prove the basis of a legimate, functioning monarchy. A portrait of him, painted around this time by Greuze, showed a genial-looking man, with an oval face, heavy dark eyebrows and a small, neat wig. Her father-in-law's appointment was the beginning of Lucie's own public life. Still only 19, but accustomed to the Archbishop's large and sumptuous receptions, she was put in charge, together with her sister-in-law Cécile de Lameth, of the household in the ministry on the first floor of the south wing of the Court of Ministers. Every week, the two young women, seated at either end of a long table, played hostess to members of the rechristened Constituent Assembly, taking care to place the most important guests on either side of them. Wives were not invited.

Versailles was emptying. The first to go, encouraged by the King and Queen, who felt it prudent for the objects of public hatred to distance themselves from France, were the much-loathed Polignac family, who left for Switzerland. Soon after, to Koblenz, went the King's brother, the Comte d'Artois, and his wife, and after them the Princes of the Blood, Conti and Condé, and their families. 'Emigration,' noted Lucie drily, 'became all the vogue.' Carriages were seen trundling out of the palace, full of retainers and servants, laden high with baggage. 'The Queen's entourage have dispersed and become fugitives,' wrote a diplomat in a despatch. 'Several of her women have abandoned her, in the cruellest manner.' There would certainly be misery in the capital before long, he went on, with so many rich customers departed. 'I do not believe that the winter will pass without bitter scenes.' Paris, he added, looked deserted, with an air of having been 'dismantled'. Terror 'is painted on every face'.

Among the nobles who had, for the moment, no intention of emigrating, were many who, like Frédéric and his father, were genuinely liberal in their aspirations for France, and believed in the possibility of a constitutional monarchy, along English lines, in which the king would govern, rather than rule, under a new constitution respectful of all people's rights. These '*gentilhommes démocrates*' included not only the de Lameth brothers, but Lafayette and Mirabeau, and many of the young officers who had served in America with Arthur Dillon and Frédéric. 'Never had the aristocracy,' the historian Hippolyte Taine would write in his history of the revolution, 'been more liberal, more humane, more in sympathy with useful reform.'

Lafayette, still only 32, a youthful figure on his white charger and with his large military epaulettes, had been the first, even before the fall of the Bastille, to propose a Declaration of Rights, on the American model. All through July and August 1789 the new Constituent Assembly haggled over the question of how you could invest a nation, rather than its ruler, with political sovereignty, and how dismantle the ancient privileged fabric of Bourbon polity and put in its place a set of laws based on liberty and equality.

But the nobles, too, were now carried away by the flood of their own rhetoric. On 4 August, in what the Comte de Ferrièrres would later describe as a 'moment of patriotic drunkenness', dukes, marquises, counts and bishops voted to give up tithes, dues, benefices and proprietary regiments. In the Assembly, Trophime-Gérard de Lally-Tollendal, the Princesse d'Hénin's new lover, sent a note across to the President. 'They are not in their right minds, adjourn the session.' But there was no halting the drive to collective abnegation. The deputies would never forget the night on which they gave away their patrimony. By mid-August, France's feudal *ancien régime* was in pieces. It had been manifestly unfair; but the speed with which it was dismantled was terrifying.

The frenzy of dispossession hit Lucie hard. 'It was,' she said later, 'a veritable orgy of iniquity.' Much of the considerable de la Tour du Pin fortune lay in its seigneurial dues. 'Everything,' Lucie would write, 'was swept away', though for the time being neither she nor Frédéric realised the extent of their loss, not least because they were both convinced that a fairer France would emerge. As Minister for War, M. de la Tour du Pin still received a handsome salary; Lucie's receptions at Versailles continued, to which she added smaller dinner parties on two other nights each week. Despite her young age, Lucie was treated kindly by the wives and daughters of the other ministers.

Since the recall of Necker in response to public pressure, Mme

Necker was again presiding over a political salon, at which her daughter Germaine de Staël was the rising star. Lucie found her a curious mixture of 'virtue and vice' and noted shrewdly that Mme de Staël, though genuinely intellectual in her interests, was far more pleased by attention to the 'beauty of her embrace' and would abandon herself, instantly and without a struggle, to passion. At their frequent meetings, Mme de Staël repeatedly asked Lucie why she did not take more pleasure in her excellent figure and her unblemished complexion, saying that had she possessed them she would 'have wanted to rouse the world'. Lucie replied with characteristic directness that she could see no point in dwelling on them, since they would so soon disappear with age. What Mme de Staël found hardest to understand, she told Lucie, was the younger woman's excessive love for Frédéric and her willingness to 'act in accordance with the ideals of devotion, self-sacrifice, abnegation and courage', in short, to sacrifice herself to his every wish. 'It seems to me,' she said to the younger girl, 'that you love him as a lover.' Even now, with the revolution turning upside down all ideas about society, the nuances of 18th-century fidelity hung on. To show such evident love for one's husband was unusual, even a little absurd; but among her contemporaries Lucie was unusual, sometimes disconcertingly so.

Lucie was, to her immense pleasure, pregnant for the third time. Feeling well and healthy, refusing to dwell on her two lost babies, she pronounced herself confident that this one would survive. M. de la Tour du Pin had been given a stable of 12 horses, which he seldom took out himself, and she and Cécile spent the fine afternoons of late summer driving out in the forests of Versailles. It was sometimes as if the Bastille had never fallen. One day she was asked to take the collection at a ceremony for the blessing of the standards belonging to the newly formed *Garde Nationale* of Versailles. It was, she noted, a 'very magnificent and solemn ceremony' and it was attended by the entire military corps of Versailles, but it left her uneasy. Though she put on a 'pretty toilette' and was much complimented at the dinner she gave afterwards, she was deeply mistrustful of the *Gardes*, drawn from among the messengers and staff of the various ministries, now armed to the teeth and clearly averse to any kind of discipline. Even musicians were in uniform. One night a well-known singer came to perform a motet at court in the uniform of a captain. There was also a children's battalion, nominally commanded by the Dauphin; they wore tiny grenadier bearskin caps and manoeuvred light 1-pound cannons around the park.

From Paris, while the Assembly in Versailles was struggling to fashion a constitution of rights 'for all times and all nations', there came daily news of small riots, sparked by hunger and the growing shortage of flour, and fanned by an outpouring of pamphlets, papers and journals. With the proclamation on 24 August of the Declaration of the Rights of Man and Citizen, and its assertion that men were born and remained 'free and equal in rights', the old system of censorship was breaking apart. The subculture of Enlightenment publishing, so long underground, was surfacing into the light of day from back alleys, prison cells and hidden attics, and from across foreign borders. Pamphlets and news-sheets, so ephemeral that many lasted no longer than a couple of issues, were produced by printers on hand-presses working all day and long into the night by candlelight. They were read, handed around and discussed, noisily and with passion, in the hundreds of cafés opening all over the city, visited, for the first time, by women as well as men. As Jacques-Pierre Brissot, editor of the *Patriote Français*, put it, 'it was necessary to enlighten ceaselessly the minds of the people, not through voluminous and well-reasoned works, because people do not read them, but through little works . . .'. In 1789 alone, 184 new journals appeared. They would serve, said Brissot, to 'teach the truth at the same moment to millions of men', something that had never happened in such a way before, and they would lead men to discuss them 'without tumult' and to reach calm decisions. The lack of tumult would soon prove little more than an illusion.

Paris and Versailles remained calm, if on edge. Lafayette, on his white charger, and the National Guardsmen, in their new uniforms of blue coats, white facings, lapels, vests and leggings, with their red trim, patrolled and kept the peace. For a moment, it began to seem as if the threat of the collapse of organised authority could be contained. Two executive committees, designed to centralise policy on appointments and security, were set up, no one perceiving that this would effectively pave the way for a revolutionary police state and a network of spies and informers. The tricolour cockade, made of dimity cotton, was everywhere and the staunchly nationalistic Mercier greeted it as a fitting emblem for the new 'citizen-warrior'. Hunger and suspicion were, so the thinking went, to be allayed by patriotism and a show of armed discipline.

Frédéric, recently appointed by his father as second-in-command to the *Garde Nationale* of Versailles, was soon called upon to act. Like his father and father-in-law, Frédéric was anxious for reform, but like them he feared it was coming too fast, too soon. When, one day in late August, two men, convicted of plotting to create a shortage of food, were due to be executed, the Commander-in-Chief of the Versailles *Gardes* refused to return from Paris to confront the mob gathered to free the plotters. It was Frédéric who had to rally the *Gardes*, threaten them with dismissal if they disobeyed his orders, and insist that the hangings proceed. Though the actual executions distressed him, he was convinced that firmness would help the fragile sense of order. Like Lucie, he believed in discipline and clarity. Very little trace of his wild youth remained.

* * *

There was a long-established custom at Versailles that at the end of the summer, on the feast day of St Louis, a deputation of market women came from Paris to pay homage to the Queen. Wearing neat white gowns, they brought bunches of flowers, and curtseyed. But the *poissardes*, the market women and fishwives of Les Halles, who descended on Versailles in October 1789, were in quite another mood.

On 2 October, a banquet was held in the great theatre of the château to welcome the Flanders regiment, summoned to Versailles as a precautionary measure to protect the royal family and the Assembly. Towards the end of dinner, Lucie and Cécile went to watch the scene. Unwisely, given the general sense of precariousness, the King and Queen decided to make an appearance, bringing with them the boisterous 5-year-old Dauphin; they were received with cries of '*Vive le Roi*' and the little boy was paraded around the hall by one of the Swiss Guards. A young courtier, the 19-year-old Duchesse de Maillé, foolishly decided to distribute white ribbons, the colour of the Bourbon kings, to some of the soldiers. Frédéric, in a whisper, said to Lucie that he feared that inflammatory remarks were being made.

Next day, rumours duly spread round Paris that an 'orgy' of treason and gluttony had been held. Stories of the sumptuous feast seemed outrageous to those queuing outside empty bakeries. On Monday the 5th, however, Versailles woke to a calm, rainy day. The Assembly continued with its deliberations; the King went hunting in the forest of Verrières; the Queen visited Le Petit Trianon; and Lucie went driving with Pulchérie de Valance, who was about to give birth to her second child. As they crossed the main avenue of the park, a horseman galloped past them, shouting: 'Paris is marching here with guns.' They hastened to the château to find Frédéric frantically despatching riders to search for the King and bring back the Queen, while disposing his *Garde Nationale* in battle order before the iron gates leading into the Cour Royale. The gates were closed and locked, the doors and entrances to the château barricaded. The Swiss Guards and the Flanders regiment took up positions at various strategic points, all facing towards the Grande Avenue, up which the attackers were expected to come. There was a lull. It continued to rain.

Just before three o'clock, the King and his suite arrived at a gallop up the avenue. Saying nothing encouraging to the soldiers standing in the pouring rain, and whose mood was already uncertain, he shut himself away in his apartments. Lucie, standing at her windows above the courtyard, watched. 'The *Gardes*,' she noted, 'were getting their first taste of war.'

Around four o'clock, the leading column of fishwives, exhausted, some of them drunk, could be seen advancing through the misty autumn early evening up the avenue. The neat white gowns they usually wore to Versailles were filthy and the women carried not flowers but muskets and pikes, ransacked earlier from the Hôtel de Ville, as well as broom handles and kitchen knives. Prevented from entering the château by Frédéric and his men, they pushed their way into the Assembly, where they harangued the deputies and demanded an audience with the King. At this stage, though somewhat inebriated, most of the women were still in a good humour. They had come to demand food from the King, and they were confident that they would get it. But from her window, where Lucie remained all day, she saw the *Gardes* stationed in the courtyard begin to grow restless; soon, in ones and twos, they drifted away to join the women, in spite of Frédéric's attempts to maintain order. One Guardsman, suddenly losing his temper, aimed his musket at Frédéric: the shot missed him but hit another officer, breaking his elbow.

A small delegation of women, with a rather pretty and fairly clean 17-year-old at their head, was nonetheless admitted into the King's apartments; he listened to their complaints and promised to release grain stocks and have them delivered to Paris. By six o'clock, Louis had also promised to sign the decrees voted by the Assembly, as well as the Declaration of the Rights of Man and Citizen. It looked as if the women would now withdraw. But there was a sudden commotion. A group of fishwives, steered that way, Lucie would later say, by an unknown treacherous insider, discovered a small door leading on to a staircase in the Cour Royale, rushed through it and up the stairs into the ministers' apartments. Lucie, who was in her rooms above, found herself surrounded by angry, gesticulating women. She was rescued by Frédéric, who led her to the Great Gallery of the château, now thronged with agitated, anxious courtiers and their families. No one knew what would happen next.

The King had been hesitating about whether to take Frédéric's advice and depart for the safety of the château at Rambouillet. When he finally decided to go, and the waiting carriages were brought round, it was too late: the *poissardes*, with furious cries of 'The King is leaving', unhitched the horses and led them away. 'This good Prince,' Lucie later wrote, 'repeating over and over again "I do not want to compromise anyone" lost precious time.' M. de la Tour du Pin now offered his own carriages; but the King refused. In the Great Gallery, the courtiers continued to pace up and down. Lucie, excited and restless, hoping to find Frédéric or her father-in-law, wandered from darkened room to room, past ladies sitting whispering on stools and perched on tables. 'The waiting,' she wrote, 'seemed unbearable. I was so agitated that I could not remain still a moment.'

At about midnight, Frédéric came to the Gallery to say that the *Garde Nationale* of Paris, with Lafayette a 'prisoner of his own troops', had just arrived. They, too, were soaking. Lafayette had tried but failed to prevent them setting out; he was reluctant to accompany them, and only did so in the hopes that his presence might act as a brake to any violence. Admitted to the King's presence, he brought with him a plea to allow the Paris *Gardes* to protect the royal family, rather than the Swiss Guards and the Flanders regiment. The King agreed. All now seemed calm.

The fishwives, having eaten everything they could lay their hands on in the château, stretched out to sleep in the stables, in the coach house, on the floors of the kitchens and on the benches of the Assembly. The King and Queen retired to their bedrooms, the candles were blown out in the Great Gallery, and Frédéric accompanied Lucie back to the Princesse d'Hénin's apartment, since their own was full of wet, sleeping women. The sight of them, noted Lucie, was 'most revolting'. Versailles was quiet. Frédéric begged his father to get some rest. Making a last tour of inspection of the courtyards and the passages, he found the *Gardes* at their posts.

The attack came as dawn was breaking. Frédéric, keeping a vigil at the windows of the Ministry of War, heard the sound of tramping feet. Through the dim, uncertain light he could just make out people advancing with axes and sabres, pushing their way through a gate which should have been locked. By the time he reached the courtyard, the guard on duty was dead and the mob was racing across the Cour Royale; a group of some 200 people broke away and stormed up the marble staircase towards the royal apartments. A troop of bodyguards, hearing the din, took refuge in the guardroom, leaving one of their men locked outside to be torn to pieces.

There was only one man on duty outside the Queen's rooms. He just had time to call through the locked door to her bedroom that people were coming to kill her, when the women fell on him, shouting that they had come to tear out the heart of the 'Austrian whore' and to '*fricasser*' her liver. His unconscious body, blocking the doorway, held them up long enough for the Queen to escape along a secret passageway to the King's apartments. The *poissardes*, bursting into her bedroom, plunged their pikes into her mattress.

Lucie, deeply asleep in Princesse d'Hénin's apartment, was woken by Cécile, who said that she could hear shouting. When the two girls leaned out of the window, then climbed on to the ledge for a better view, they could still see nothing, but cries of 'Kill them! Kill them! Kill the bodyguards!' were clearly audible. At this point Marguerite appeared, trembling and terrified, saying that she had just seen a man with a long beard hack off the head of one of the guards. She also said that she had seen the Duc d'Orléans, a man none of the liberals trusted and whom Lucie was convinced was behind the disturbances, among the rioters.

As it grew light, the King, at the urging of Lafayette, agreed to appear on the balcony overlooking the courtyard, where some 10,000 women, disaffected Guardsmen and men with pikes, were milling around and shouting. The Queen, who came to join him, made as if to bring out the Dauphin and his sister, but was prevented from doing so by cries from the crowd of 'Not the children'. What the people wanted was for the royal family to go to Paris.

At 12.30 on 6 October a ragged, mournful cortège duly set off, through blustery winds and heavy rain. The carriages of the King, the Queen, the royal children, the King's sister Mme Elisabeth, and his brother and sister-in-law the Comte and Comtesse de Provence were followed by several cannon, decorated by the fishwives with laurel leaves, by wagons containing flour from the King's stores, and by a large crowd of women, singing and shouting. The heads of two bodyguards danced before the royal carriage, impaled on the top of pikes

Lucie missed the closing hours of the attack on Versailles. Frédéric had insisted that she leave for the safety of a house at the Orangerie, and there she waited, in mounting fear and anxiety, for news. She was alone, Cécile having gone in search of her children; Marguerite was at the Ministry of War packing up their belongings. 'I do not think,' Lucie wrote 30 years later, remembering the events of that day, 'that I have ever in my life passed such cruelly anxious hours. The cries of people being murdered rang in my head. The slightest noise made me tremble.'

When Frédéric finally arrived, he told her that the King's parting words to him had been: 'You are completely in charge here. Try to save my poor Versailles for me.' Lucie begged to be allowed to stay with him, but Frédéric was adamant that she should accompany the Princesse d'Hénin to the safety of the Château of Saint-Germain, where her lover Lally-Tollendal kept rooms. Lucie's presence, Frédéric said, would only 'paralyse the effort which it was his duty to make to justify the King's trust in him'.

Before leaving, Lucie returned to the silent and deserted Ministry of War. The only sound to be heard in the vast palace of Versailles, that she had known only as a bustling, crowded village of colour and movement, was the banging of doors and shutters, many of which had not been closed in decades, and which were now being barricaded against looters. Chairs and tables lay on their sides, knocked over by the rushing crowd. Discarded clothing was scattered around the floor. Versailles, for the first time in almost a hundred years, was empty. The Comte d'Hézecques, who as a boy had been a royal page dressed in crimson velvet embroidered with gold, walked, much as Lucie had, through the empty and echoing rooms, discovering corridors and whole apartments he knew nothing of.

* * *

The 3-hour drive to Saint-Germain, in a poorly sprung carriage, was bumpy and painful. When Lucie reached Lally-Tollendal's apartment she collapsed. For a while, it looked as if she would miscarry. But she rallied, and both she and the unborn baby survived a ferocious cupping.

By the time that Lucie, still very pale and alarming Frédéric by her lack of appetite, arrived back in Paris towards the end of October, M. de la Tour du Pin was preparing to move into the new Ministry for War in the Hôtel de Choiseul. A pleasant apartment was being done up for Lucie and Frédéric, with its own separate entrance and with windows that opened on to a garden. Frédéric was working closely with his father, and Lucie, with the help of the Princesse d'Hénin, again became hostess at their official dinners, though they were smaller and less grand than at Versailles. They dined at four in the afternoon, after which Lucie returned to her own apartment or went out visiting. On moving to Paris, Marie Antoinette had given up her boxes at the various theatres and Lucie, in the middle months of her pregnancy, was too fearful of the mood of the Parisian crowds to go without the safety of a guarded box. The Queen's decision, she thought, had been a mistake, for it served only to isolate her still further from the already hostile people. Marie Antoinette, she wrote, 'was gifted with very great courage, but little intelligence, absolutely no tact and, worst of all, a mistrust – always misplaced – of those most willing to serve her'.

Arthur had recently returned from the West Indies to represent Martinique at the Estates General, and was living not far away, at 9 Porte Saint-Honoré. He had left his new family behind, and was constantly short of money. When not at the Assembly, speaking on colonial and naval matters, he spent his days trying to collect dues and a settlement on the Dillon regiment, which, together with all other proprietory regiments, had been integrated into the French army. For almost the first time in her life, Lucie was able to spend time with her father. She had not seen him for nearly five years. They grew close.

Having been hauled back to Paris by the *poissardes*, the royal family had been installed in the Tuileries, the collection of buildings started by Catherine de' Medici in the 16th century. Overlooking the Seine, and made up of several wings and pavilions and 368 separate rooms, the Tuileries were dark, decrepit and unwelcoming. None of the doors shut properly. With the precipitous arrival of the royal family, the buildings had been cleared hastily of their occupants, most of them actors, artists and the families of court servants. After a first night, with courtiers sleeping on floors and tables, apartments had been arranged for the King, the Queen, the Dauphin and his sister. The children had a new governess, Mme de Tourzel, the 41-year-old widow of a Grand Prévôt of France, a woman whose impeccable probity was intended to counter the reputation of the compromising and frivolous Duchesse de Polignac, now safely in Switzerland. Mme Elisabeth, the King's sister, was allocated rooms on the ground floor of the south wing; she complained that the *poissardes* came to scowl at her through the windows.

Gradually, as the mirrors and Gobelin tapestries were brought from Versailles to decorate the royal apartments, the court re-established itself. Rose Bertin appeared with swatches of material; Léonard came to dress the royal hair. Greater efforts were made to economise and live more simply, but otherwise the royal day passed curiously unchanged: the *lever*, attendance at Mass, lessons for the children, tapestry for the Queen, billiards with the King, the *coucher*. The Dauphin played in the Tuileries gardens. The King went hunting. Lucie, returning to court, found that full court dress with hoops was still required. When guests came, they often wore lilies and white ribbons, symbols of the Bourbons; but when they left, they exchanged them at the gates for a tricolour cockade, without which they were liable to be arrested. The King, no longer 'above the law', but subject to it, was now simply *Roi des Français*,

King of the French, and not *Roi de France et de Navarre*, a linguistic distinction meant to highlight his dwindling powers.

The Assembly had sat for its last session among the Doric columns and royal portraits of the Menus Plaisirs in Versailles on 14 October. Three days later the deputies resumed their deliberations in Paris in a hall of the Archbishop's Palace, considerably shaken by the violence they had witnessed, though no one had actually been injured in the tumult. A few took the opportunity to slip away. Their confidence was not increased when a baker was lynched, and his head impaled on the now customary pike. Paris, as Gouverneur Morris gloomily observed, 'is perhaps as wicked a spot as exists. Incest, Murder, Bestiality, Fraud, Rapine, Oppression, Baseness, Cruelty . . .'

To the pleasure of many of the deputies, their formal dress had been abolished and they were arrayed in a bewildering variety of outfits. They looked, Morris complained, extremely shabby. In the first issue of the new *Journal de la Mode et du Goût*, the truly democratic man, dressed *à la Révolution*, was decreed to be one who wore, with his plain black cloth coat, a red waistcoat and yellow breeches. If he was very patriotic, he wore buckles *à la Bastille,* for fashion continued to mirror life. Clothes everywhere but at court were becoming simpler and more comfortable: for women, no hoop, no heels to their shoes and hair worn loose and long, falling in curls to the shoulder.

For a majority of the deputies, especially those from the Third Estate or the lower orders of the clergy, this was their first visit to Paris. They had come to Versailles from all over France expecting a stay of just a few months; they had already been away from home for six. As the winter grew colder, they sent for warm clothes, for supplies of wine and cheese, and even for their families. The Assembly met every day, from eight in the morning until after ten at night, with a couple of hours off to eat. With a thousand people packed into a space so small that not all of them could sit down at once, they were soon complaining of the bad air; many caught colds and wrote home to their wives that Paris was filthy, its streets deep in mud, rubbish and sewage.*

---

* Before they finally disbanded 20 months later, 14 nobles, 13 men from the Third Estate and14 members of the clergy would die of natural causes, and 30 more emigrate or resign.

On 9 November, they moved into the Salle du Manège, the indoor riding school of the Tuileries, where they complained of the heat.

The deputies were confronted with a daunting task: that of drafting a Constitution in which it was clearly laid down where power resided. Nothing was made easier by the extraordinary process of change going on around them. 'Everything is new to us,' said M. Clermont-Tonnerre, a moderate royalist who would for a while side with the Third Estate. 'We are seeking to regenerate ourselves; we are having to invent words to express new ideas.' The old and arbitrary privileges, the powers of ancient lineage, the patchwork of illogical and overlapping jurisdictions, all had to be swept away and replaced by rational, equalising institutions. The very word 'revolution' had acquired new meaning, that of the total transformation of all areas of life, opening on to a limitless future. Even death would be more equal. On 28 November, a deputy called M. Guillotin proposed that all executions, regardless of class or crime, be henceforth carried out by a single plunging blade.

When the Assembly settled in Paris, several of the deputies, fearful of the confusions of this enormous, volatile city, decided to look for a place where they could meet and talk, not too far from the Tuileries. For 400 francs a year, they rented one wing of the Convent of the Jacobins, in the rue Saint-Honoré. At first, they called themselves the Club Breton, after the 40 deputies from Brittany, and their early meetings were attended by some of the monks, who sat together at the back of the room in their white habits and black hoods. As their numbers grew, to include journalists and lawyers, the meetings became noisier and more passionate, and were soon overflowing into the library. To welcome their new friends, they renamed themselves the Société des Amis de la Constitution; but soon they were simply known as the Jacobins.

The meetings in the Salle du Manège were also passionate and so noisy that the delegates at the back had trouble in making themselves heard. They were dominated by a group of the most forceful members, most of them lawyers from Paris or other large cities and all but a few Jacobins. One of these was a rigid, fastidious lawyer from Arras called Maximilien Robespierre, who was repelled by all forms of vulgarity and disorder and who, with the mind of a grand inquisitor, searched for hidden meanings behind every word. Lucie, at her father-in-law's dinner parties where Robespierre was an occasional guest, was impressed by his apple-green coat and his thick white hair, elegantly dressed. 'Before starting out,' Robespierre declared, 'you must know where you want to end up.' There was Georges-Jacques Danton, another lawyer, a previously rather unassuming man turned street agitator; Danton would later be called the 'Mirabeau of the gutter'. There was Camille Desmoulins, the young journalist from Picardy who had called on the citizens of Paris to rise. The court and the clergy, Desmoulins announced, were 'ne'er-do-wells . . . who, despite their great wealth . . . are merely vegetables'. And there was Jean-Paul Marat, at 46 one of the eldest, a broad-shouldered, muscular journalist with an inflammatory skin disease, barely 5 feet tall; in his paper *L'Ami du Peuple*, Marat was busy inventing the language of the Terror. For his persistent calls for blood, Marat would be branded the 'street-corner Caligula'.

But it was rapidly becoming clear that political harmony in the Assembly was fracturing, even among men whose similar backgrounds might have suggested unity. The 'anti-monarchicals', described as of the 'left' – the first such use of the term – split away from a new grouping, the Club de 1789. Two of the de Lameth brothers, Alexandre and Théodore, joined Robespierre, Marat and Danton in agreeing that the real threat to the revolution would come from a royalist conspiracy; they wanted the state subjugated to the citizen. Members of the Club de 1789, on the other hand, which included Talleyrand, Lafayette, Mirabeau and the Abbé Sieyès, were more frightened of anarchy, and hoped to retain a strong, powerful France with a reformed monarchy. Among their supporters were Frédéric, his father and Arthur.

What gave their debates an edge was the fact that Necker's measures were not paying off. France was running out of wheat. The nobility were being forced to reduce their households and dismiss many of their servants; merchants complained that they had no customers. Eighty thousand families had already left Paris. Writing to George Washington, Gouverneur Morris, who had stayed on in Paris after Jefferson's departure in October, said that though Necker 'understands Man as a Covetous Creature, he does not understand Mankind, a Defect that is remedyless'. What was more, he was a poor financier, with 'feeble and ineptious' plans.

'This new Order of Things,' he added, 'cannot endure.' That winter, Morris dined with Arthur Dillon, 'whose wine is very good'. Even when insolvent, Lucie's father drank well.

* * *

On 10 October, even before the Assembly moved to Paris, Talleyrand, who had defrocked himself in dress if not in title, had proposed that since financial disaster threatened, the Church's vast wealth in lands, monasteries and foundations should be used as collateral for a new loan. Great dangers, he said, demanded 'equally drastic remedies'. Since the Declaration of the Rights of Man and the Citizen had expressly stated that property was inviolable, this prompted fury among the higher reaches of the clergy. Horace Walpole, hearing the news in London, observed that Talleyrand was a 'viper that has cast its skin'. But Talleyrand pressed on, and the Assembly, by 564 votes to 346, decreed on 2 November that all property owned by the Church would be placed at the disposal of the nation – the palaces, the thousands of acres of forest and farming land, the riches in gold and silver and art. As Barnave, the young deputy from the Dauphiné put it, since the clergy only 'existed by the virtue of the Nation, so the Nation . . . can destroy it'.

In February came a further assault on the Church: there were to be no more perpetual monastic vows or vocations, though a few of the teaching and nursing orders would for the time being be reprieved. The new citizen of France, a man of liberty, patriotism and happiness, would not be allowed to surrender his freedom, except for the public good. Even as commissioners arrived to begin the task of drawing up inventories and preparing for sales, monks from the great Cistercian abbeys of Clairvaux, Cluny and Cîteaux took off their cowls and prepared to rejoin the world. France's 51,000 nuns proved rather less willing to abandon their cloisters but they, too, were soon driven out, to settle, forcibly deprived of their habits, with their families or in small groups. Lady Jerningham, Lucie's aunt, whose house near Norfolk included a hidden Catholic chapel, offered shelter to a group of Blue Nuns from Paris. Over the next two years, the secularisation of the Church would go further than anyone had imagined, with churches demolished or turned into warehouses, church bells and plate melted down, religious orders made destitute, and priests turned into public servants.

On 19 May, Lucie gave birth to a healthy, but rather thin, son. He appeared to consist only of 'skin and bones', but a good wet-nurse was found and the baby soon put on weight. She called him Humbert-Frédéric, and had him christened in the Church of St Eustache by Les Halles, appointing as godparents the Princesse d'Hénin and her father-in-law. Humbert filled her with delight. There was talk of Lucie taking the Princesse d'Hénin's place at the court in the Tuileries, but Marie Antoinette decided against it, as it looked as if Frédéric might be appointed Minister for Holland. 'And who knows,' she added, 'if I might not expose her to further dangers?'

* * *

Paris itself, day by day, was changing. The imagination of the men shaping the revolution had been seized by a cult of antiquity, which owed little to Aeschylus or Herodotus, but much to Horace, Virgil and Cicero, writers living at a time when the greatest days of Rome were effectively past, and attributing to that earlier age all the virtues of the simple life. The revolution of 1789 was, as they saw it, a similar moment, a society in which self-made men, Cicero's *homines novi*, new men, might, solely by virtue of their eloquence, rise to hold the very highest positions. Among the painters similarly enchanted by antiquity was David, whose martial and patriotic *Oath of the Horatii* had caused such an uproar at the Salon of 1784, and who now, in an equally famous but unfinished work on the Tennis Court Oath, used the same outstretched hand of the Horatii as the fitting gesture for revolutionary oaths. The Phrygian cap, the red woollen bonnet worn by the freed slaves of Greece and Rome, was beginning to be seen around the streets of Paris, as were haircuts *à la Brutus*. Men abandoned powder and wigs for short, severe cuts.

This obsession with the ancient republicans extended to the theatre, where Talma, the rising star of the Comédie-Française, turned to David for help in designing costumes. Talma, who looked remarkably like a Roman senator himself, took to the boards as Brutus in a toga. From the theatre, antiquity moved to the streets where, by early 1790, classical festivals of public games and displays of gymnastics, as recommended by Rousseau and described by Plutarch, were being staged. With them came catchy new tunes and a new revolutionary musical form, a '*genre hymnique*', easy to sing, mixing antiquity with allegory, and sung by immense choirs, often in outdoor settings, all designed to unite the crowd in communal emotion. Paris resounded to '*Ça ira*', with its ominous refrain '*Les aristocrates, on les pendra! Le despotisme expirera! La Liberté triomphera!*'*

Very early on, Robespierre had understood that revolutions need celebrations. But it was Talleyrand, by June 1790 the only bishop still sitting in the Assembly, of which he was now President, who would star in one of the first and most magnificent of the revolutionary festivals, the celebration, on 14 July 1790, of the fall of the Bastille. It was to be held on the Champs de Mars, the open field used by cadets of the École Militaire to drill, but because no one could quite decide where the King should sit, work did not begin until the end of June. There were just two weeks to transform a stony, uneven patch of open ground into an amphitheatre able to seat an expected 400,000 people. The *Gardes Nationales* from all over France were invited to send deputations. Frédéric had been given the task of organising their lodgings, food and entertainment, and Lucie often went to the Champs de Mars to watch the progress.

It was a formidable task. Two hundred thousand Parisians, drawn from every class and every occupation, were drafted in to carry earth, which was to form a semicircle round a central 'altar of the fatherland' and a triple, arched Arc de Triomphe. Lucie spotted Capuchin friars harnessed to little carts, Knights of Malta with wheelbarrows, nuns with baskets. The nobility lent their horses. Workshops across the capital stood idle. All round the rising amphitheatre, taverns set up tables, laden with free food and barrels of wine and ale. It rained. Sand and gravel were brought to stiffen the sifting mud. The indefatigable and sentimental Mercier lyrically described seeing citizens 'making the most superb picture of concord, labour, movement and joy that has ever been witnessed . . .'. Even Lucie, normally cool on such matters, admitted that it was the most 'extraordinary spectacle' and one that would not be seen again.

---

* 'We'll hang the aristocrats, Despotism will expire, Liberty will triumph!'

On the night of the 13th, Lucie and her sister-in-law Cécile went to sleep in an apartment lent to them in the École Militaire, overlooking the Champs de Mars, from whose windows they could watch the crowds arriving. M. de la Tour sent food with them, so that they could offer lunch to the military officers during the festivities.

The 14th dawned with heavy rain. The Guardsmen in their red, white and blue uniforms, marched, squelching, to the sounds of military bands. The King and Queen, standing on a special platform, held up the Dauphin to the crowds. The child wore the uniform of a little Guardsman. Lucie reported later that Marie Antoinette had said to an officer, pointing to the child's bare head: 'He has not got the cap yet.' No, the man replied, 'but he has many at his service'. Lucie had become 'accustomed to the Queen's various expressions', and thought that she looked displeased, and should have made more effort to conceal her ill-temper. The carefully chosen red, white and blue feathers and plumes in Marie Antoinette's hat dripped. In driving rain and gusting winds, Talleyrand – 'the least estimable of all French priests', as Lucie noted tartly – celebrated Mass in full episcopalian regalia at the high altar and blessed the banners of the assembled troops, before Lafayette, on his white charger, led the chorus of oaths to the constitution and to the new nation. All around the amphitheatre, spectators sheltered under brightly coloured umbrellas, a novelty introduced to Paris not long before.

The ceremony was interminably drawn out. Humbert was not yet 2 months old and Lucie did not feel strong enough to join the crowds, so she remained inside. No provisions had been brought from the Tuileries to feed the Dauphin and his sister, who became very hungry, and Lucie offered to let them share the food prepared for the officers.

Despite the reassuring presence of Talleyrand at the altar, and the King on his special platform, plans were proceeding at considerable pace to dechristianise and make equal France's 26 million citizens. On 19 June 1790, late at night when few of the remaining nobles were present in the Assembly, all titles of hereditary nobility had been abolished, and with them went liveries, coats of arms, which were painted over, and any name that suggested a place rather than a family. The Duc d'Orléans opted for the name Philippe-Egalité. The Assembly pressed on with its destruction of the Church, despite the protests of the King. A Civil Constitution of the Clergy made priests subject to appointment by the state, like all other public officials. Soon, priests would be required to sign oaths of allegiance to the state: those who refused would be branded 'refractory' or 'non-juror' and lose their jobs.

Archbishop Dillon, who had refused to return to Narbonne since the previous winter, on the grounds that he was greatly in debt as a result of the loss of his revenues, retired quietly to Hautefontaine with Mme de Rothe, far from his creditors. The size of his debts – nearly 2 million francs – only became known later, when they were paid out of what remained of Lucie's inheritance. For many people, the new law against the clergy marked a turning point. The primacy of the Pope on matters of faith and morality was a verity recognised since the beginning of time: to deny it was to cease to be Catholic. Out of 160 bishops, only 7 would agree to become 'jurors'. Talleyrand was one of them. Archbishop Dillon was not. It was impossible, he wrote to the King, for him to 'acquiesce in the degradation of the Church'. Retribution followed swiftly. 'It is time,' wrote the new administrator for Church affairs in Montpellier, 'to rid ourselves of this overweening priest who so impudently dares slander the nation's representatives.' Dillon, the last Archbishop of Narbonne, had no further role to play. In the following months, Hautefontaine would become a refuge for refractory priests, men like him who refused to give up their allegiance to the Pope and the old Church. With his administrative skills no longer wanted, the Archbishop retreated to a quiet life of prayer and Mme de Rothe.

The Great Fear had driven many nobles to seek safety outside France, their numbers swelling after each new decree or violent incident. After the attacks on the Tuileries in October 1789, the Princesse d'Hénin, taking her lover, the obedient Lally-Tollendal, with her, departed for Switzerland. Towards the end of July, Lucie, leaving Humbert with his wet-nurse and Marguerite in the Hôtel de Choiseul, decided to join them. Paris was calm.

It was still just possible to believe, as Lucie and Frédéric did, that the great and bold transformation of France into a constitutional monarchy might be accomplished without further bloodshed.

CHAPTER SIX

Deep and Dark Shades

The journey to Switzerland almost proved disastrous. France was in so volatile a state that the smallest event was enough to trigger sudden, irrational violence. Lucie, the Princesse d'Hénin, who had come to Paris for the birth of Humbert, and a young cousin, Pauline de Pully – three aristocratic women accompanied by a cook, a maid and three menservants, travelling in two carriages with uniformed coachmen – were precisely the kind of sight likely to inflame republican tempers.

They had taken care to equip themselves with every form of passport, and believed that these would see them safely to Geneva, where Lally-Tollendal awaited them. But at Dôle, after Mme d'Hénin insisted on changing the route, they found themselves suddenly in a busy market square. They had passed through other towns without incident but Dôle was packed with people and the two carriages were forced to slow down to walking pace. Suddenly, as at Forges, cries went up: 'It's the Queen.' A crowd converged on the carriages, unharnessed and led away the horses, and dragged the three women to the house of the local commander of the *Garde Nationale*. There they found the remains of a delicious meal, but no sign of anyone. Being hungry, the three young women sat down to a stew, a meat pâté and some fruit.

Three hours later a rather solemn man, who introduced himself as the President of the Dôle commune, appeared. After much discussion, Lucie's cook was despatched to Paris to fetch letters to authenticate their story, and the three women were sent to sleep in a nearby house. Next morning, they were subjected to hostile questioning, one man asking Lucie why she needed quite so many shoes for a 6-week stay in Switzerland. The rumour in Dôle was that Mme d'Hénin was the Queen and Lucie her sister-in-law. It was now that Lucie realised that some of the officers who had come to lunch on the day of the celebration at the Champs de Mars were probably stationed in the garrison near Dôle. They were fetched and quickly recognised her and an embarrassed President bowed them on their way, but not before the young officers had terrified her with their professions of loyalty towards the *ancien régime*, and much flowery language. The party crossed the border and reached Nyon late that night; Lucie woke with delight next morning to see the lake of Geneva sparkling in the dawn light. It was her first visit to a foreign country.

* * *

The first French émigrés, dusty, exhausted and weighed down by luggage, had arrived in Switzerland soon after the fall of the Bastille. By the summer of 1790, many of the hotels along the lake of Geneva were full to overflowing with unhappy marquises, counts and senior prelates. Switzerland, being neutral, was attractive, even if each of the 13 cantons had somewhat different reactions to the revolution in France; the Catholic cantons of Fribourg and Soleure were the most welcoming to the nobility and to the nuns and monks forced out of their monasteries. Mme de Staël's father, M. Necker, departing the French court for the third and last time, had bought a property at Coppet, a two-storey manor house overlooking the lake, with vineyards and orchards. Mme de Staël herself loved the property but complained that life in Switzerland was narrow and puritanical and that there was something hypocritical about the Swiss, saying that their love of equality was 'no more than a wish to bring everyone down'.

Lucie spent the first fortnight in Lausanne. She was introduced to Gibbon, a frequent resident of the large English community, and found his appearance 'so grotesque that it was difficult not to laugh'. Gibbon, whose last volume of *The Decline and Fall of the Roman Empire* had appeared not long before, was suffering from a chronic and disfiguring inflammation of the testicles, all too visible in an age of tight clothes. Lally-Tollendal was also in poor shape: he had not only caught smallpox – from which he recovered – but had been forced by the imperious Mme d'Hénin to marry his former mistress, a Miss Halkett, niece to Lord Loughborough, by whom he had a daughter. It was a marriage in name only, for Mme d'Hénin had no intention of parting with him.

In the 18th century, the French aristocracy travelled very little, preferring to stay in surroundings and company they found congenial and considered superior to anything they might find abroad. Growing up in a sheltered world, where position and manners were so carefully prescribed, many were finding exile trying. Some rose to the challenge, and as their money started to run out, turned their hand to giving lessons or making hats. In Lausanne, the Vicomtesse de Montmorency-Laval was embroidering waistcoats, while her son Mathieu, a fervent royalist and friend of Frédéric's, farmed. Forced to give up her carriages and footmen, the Princesse de Conti ordered a pair of stout walking shoes and could be seen striding in the countryside, evidently enjoying a lack of formality denied her by the pretentiousness of the *ancien régime*. Some of these exiles ventured as far as Chamonix, which by the 1780s had become a fashionable resort for climbers.

But among the émigrés, as Lucie soon noted with disgust, there were many who struggled to hold on to the world they knew, complained of their small lodgings and frugal food, and behaved with disdain towards their Swiss hosts, mocking the simplicity of their ways. 'They all brought,' she wrote, 'the airs and insolence of Paris society . . . and were everlastingly amazed that there should exist in the world anything besides themselves and their ways.' In the summer of 1790, the inn at Sécheron became a haven for these ungrateful exiles, and their arrogance and ignorance drove Lucie away to share a house with Pulchérie de Valance and her children at Paquis, near Geneva. As the celebrated novelist Isabelle de Charrière, who had a salon to which she invited the émigrés, remarked: 'These French are unbelievable . . . They are spoiling their cause wherever they go, they destroy all sense of pity for them . . . You can see that the French nobility is nothing but wind, that it isn't worth a thing, that its day is past and it is already being forgotten.' Lucie, observing the insularity and self-absorption of her compatriots, worried that she herself might come across as equally ridiculous. 'I cannot be certain,' she wrote, 'that I did not sometimes fall into these same errors.' She was after all, as she noted many years later, of their class and their world, and these were the manners she had grown up with. But Lucie, though never as clever as Mme de Staël, was shrewd, practical, generous-spirited and far better educated than most of them. More important, perhaps, she was full of curiosity and eager to learn; and she never took herself too seriously.

Soon after reaching Switzerland, Lucie received a letter from Frédéric saying that he was being sent by his father to Nancy, with orders for the Marquis de Bouillé, commander-in-chief in Lorraine and Alsace and a veteran of the American wars, to quell a local mutiny of three regiments. By 1790, sporadic mutinies were flaring up throughout France's army, which like other institutions was riddled with anomalies and ancient privileges. Though the troops were frequently well trained and well led, their officers were still all of noble birth, and most of the appointments were bought. Morale was low and army discipline savage.

At Nancy a group of soldiers, vowing to throw off the tyranny of their officers, had barricaded themselves into their garrison, having first seized the regimental funds, and taken hostage the commandant of the town, M. de Malseigne. A cavalry regiment had then gone over to the rebels. M. de la Tour du Pin's orders to M. de Bouillé, to be transmitted by Frédéric, were to put down the mutiny with utmost severity, but only if certain of victory, because the men were 'in a mood for insurrection'. An example, he and Lafayette agreed, had to be made.

On reaching M. de Bouillé's headquarters, Frédéric was sent into Nancy to negotiate with the rebels. He found M. de Malseigne safe, but the mutineers adamant that they would not capitulate without guarantees of reform. Frédéric refused to make any such promise. He returned to M. de Bouillé and a decision was taken to march on the town, despite fears that M. de Malseigne would pay for the attack with his life. Before they left, however, M. de Malseigne appeared, exhausted and wet, having managed to slip away from his guards and reach a river, pursued by the rebels, across which he forced his horse to swim. Next morning, M. de Bouillé ordered the attack.

As Frédéric and his men neared Nancy, they saw a young officer on the rebel side ordering his men not to fire, and indicating that he wished to talk. Frédéric ordered his own men to hold their fire and then rode ahead to meet him. But as he neared the city gates, soldiers in the town lit the fuse of a cannon loaded with grapeshot. Frédéric's horse was hit. Behind him, many of his men were killed. While Frédéric's soldiers forced their way forwards, entered the town and overpowered the mutineers, his servant discovered him lying bruised and shaken among the dead bodies and carried him away to safety. M. Désilles, the courageous young officer who had tried to prevent the fighting, was badly wounded by shots fired by his own men.

Among the rebels were men from a Swiss regiment, which had retained the privilege of trying its own soldiers. There was a court martial, after which 27 men were executed. The two mutinying French regiments were disbanded and their troops sent to other units; a considerable number were either shot or sent to penal servitude. Frédéric, who soon recovered, was despatched to Paris with news of the assault on Nancy. Arriving at the Ministry of War in his dirty uniform, he was taken straight to the King, who for once agreed to overlook the etiquette that demanded that no military uniforms be worn at court.

A mutiny had indeed been successfully crushed, but it would have repercussions, many of them adding to the chaos into which France was sliding. And it would have repercussions for Frédéric too, now marked as an enemy of the ordinary soldier. A number of reforms were indeed immediately instigated – cruel punishments and press-ganging were abolished, along with venal posts, and officers were to be selected henceforth on the basis of merit rather than lineage – but something about the bloodshed at Nancy, and the fact that it had been until that moment a garrison of model discipline, was not forgotten. Officers from across the entire French army began to emigrate, crossing the borders to join the Princes at Koblenz: in the next 18 months, the army would lose a third of its officer class in resignations or emigration. And many ordinary soldiers, lured by better pay and greater freedom, drifted away from their regiments to join the *Garde Nationale*, in whose clubs they heard much talk about the perfidy of the nobility. Soon France possessed a whole parallel army, over which M. de la Tour du Pin, as Minister for War ostensibly in charge, had no control. 'The army,' he warned, 'risks falling into the most turbulent anarchy . . . Today, the soldier has neither judges nor laws: we need to give him back both.'

What Lucie did not know, and Mme d'Hénin did not tell her, was that the first news of the defeat of the mutiny at Nancy to reach Geneva contained a rumour that Frédéric had been killed.

Lucie as a young woman: the pen suggests that she was already drawn to the idea of writing.

Comte Arthur Dillon,
Lucie's father, an impetuous and brave soldier, colonel-proprietor of the Dillon regiment.

Thérèse-Lucy de Rothe,
Lucie's charming but weak mother, lady-in-waiting to Marie Antoinette and much courted by the nobility at Versailles.

A miniature of Lucie's half-sister Fanny, daughter of Arthur Dillon and his second wife. Headstrong and moody, Fanny later married Napoleon's faithful General, Henri-Gratien Bertrand.

Arthur Richard Dillon, Archbishop of Narbonne and Lucie's great-great-uncle, who was widely known to be the lover of her tyrannical grandmother. Lucie grew up in their household.

Lucie's husband, Frédéric-Séraphim, Comte de Gouvernet and later Marquis de la Tour du Pin, with whom she spent fifty happy years.

Humbert, the first of Lucie's and Frédéric's six children.

Frédéric's imperious aunt, the Princess d'Hénin, lady-in-waiting to Marie Antoinette and a lifelong influence over Lucie.

Lucie's English aunt, the forceful Lady Jerningham, who offered the family a home in Norfolk during their second flight from France.

(*Above left*) Humbert, at around the time he joined the Black Musketeers. (*Above right*) Charlotte, born soon after Lucie's voyage home from America in 1796. Lucie, who was especially close to her, noted that, though not exactly pretty, Charlotte was quick, cheerful and always helpful.

(*Above left*) Cécile, Lucie's third daughter, born unexpectedly in the middle of a freezing winter in North Germany, while the family was returning to France during the Directoire.
(*Above right*) Aymar, Lucie's youngest son, and the only child to survive her. In the 1820s Aymar's escapades sent both his parents to prison in Bordeaux.

Cécile, Charlotte's daughter, whom Lucie brought up. Lucie loved her dearly and admired her strength of character.

Lucie's grandson Hadelin, Cécile's brother.

Lucie, dressed totally in black, which was the only colour she wore for the last thirty years of her life.

She learnt the full details of his escape only in a letter that reached her after he had recovered. His death would have been such a profound blow that she could not bear to dwell on the thought of it. In early October, he arrived in person to collect her. But he was anxious not to spend too long away from his father, whose problems at the Ministry of War were becoming more acute every day. They travelled back by way of Alsace, where Frédéric had a brief meeting with M. de Bouillé on the road near Neuf Brisach, the two men pacing up and down as they talked, while Lucie waited in the carriage. M. de Bouillé had sent Frédéric a horse, 'which I hope you will keep for affection of me', to replace the one shot under him at Nancy. As they drove on through Nancy their carriage passed below the windows of the room in which M. Désilles, the brave young officer, lay dying, a sentinel posted at the door to prevent passers-by from making too much noise. They reached Paris to find Humbert in excellent health and 'much improved in looks', lovingly watched over by Marguerite with what Lucie called 'incomparable, unfailing care'.

* * *

Lucie, returning to the Hôtel de Choiseul, resumed her life of dinners, rides, musical soirées at the Hotel de Rochechouart, and visits with Frédéric to Mme de Staël's salon where she endured, rather than enjoyed, the hours of her forceful hostess's 'masculine attitude and powerful conversation'. The abolition of noble titles had done little to curb the sense of privilege in many of the older families, and Lucie found herself acting as chaperone to the young Nathalie de Noailles, whose mother, Mme de Laborde, though pleasingly rich, was not considered sufficiently well bred to take her daughter into society. Nathalie's father-in-law was the Prince de Poix, in whose house Lucie had lived at Versailles, and she looked on Nathalie as a younger sister; the two girls often went out in matching dresses and with similarly dressed hair. Looking back on this friendship many years later, Lucie remarked complacently that she had never 'suffered from that smallness of mind which made some women jealous of the success of other young women'.

More surprising, perhaps, was how unreflective Lucie appeared to be of the revolution gathering pace around her. Even as the army was losing all its officers, as the Assembly was turning steadily more hostile towards the aristocracy and senior prelates, as friends and relations were frantically packing to leave France – all things she witnessed every day and heard about at length from Frédéric, from her father-in-law and from Arthur – Lucie continued to pay visits and to enjoy herself. If her refusal to acknowledge what was happened seems wilful, it has to be remembered that she was just 20, that she had a small baby, that she had finally escaped her terrible grandmother and that she was very much in love with her husband. She was happy.

* * *

In the winter of 1790, Paris was busier, more frenetic, than it had ever been. Leaving Versailles deserted, the several thousand courtiers and ministers, together with their relations and servants, in all some 60,000 people depending for their livelihood on the court, had followed the royal family back to the capital and were established in mansions in the faubourgs of Saint-Germain and Saint-Honoré, where their carriages added further chaos to the crowded narrow streets. Even the now daily departures of frightened noble families, for whom, noted Lucie, 'to emigrate was a point of honour', did little to diminish the bustle of the city.

In the Tuileries, the elaborate ritual of the *lever* and the *coucher* remained unchanged, the royal family dined in public on Thursdays and Sundays and the 5-year-old Dauphin, using a special small hoe and rake, gardened in a little railed-off corner of the park, guarded by two grenadiers. Though not in formal attendance at court, Lucie occasionally joined the diminishing throng of courtiers in the gloomy galleries of the Tuileries. In the Tuileries gardens, aristocratic women in the latest fashions strolled and gossiped, while at the same moment a crowd in the nearby Palais-Royal listened to a speaker haranguing against 'the perfidy of the court, the arrogance of the nobles and the cupidity of the rich'. Visitors remarked on a new air of pride and independence in ordinary men and women, who seemed to walk and stand differently.

Even the salons, where Lucie spent her evenings, were changing. Gone were Suzanne Necker's polite disquisitions on the nature of piety, or Mme du Deffand's play of words on platonic love. In their place had come gatherings of furious debate and partisanship, where the talk was all of the new Constitution, of how far liberty could be allowed to govern the workings of a modern state. The salons had lost their lightness, the Goncourt brothers would later write; they were no longer schools for manners and gallantry, but political debating societies. Manon Philipon, passionate admirer of Rousseau and married to Roland de la Platière, former inspector general of manufactures for Picardy, held a salon for the rising Jacobin stars; Mme de Staël, back in the rue du Bac, drew the liberal royalists; while Josephine de Beaumarchais, the Creole wife of a member of the lower nobility and related to Lucie through her new stepmother, gathered the centrists, in what would later be called a 'nest for the Assembly'.

In the house of the romantic novelist Adelaïde de Flahaut, Talleyrand, father of her son, held sway and here guests debated the political nuances of events that seemed to come fast one upon another. Talleyrand, noted Gouverneur Morris, was referred to as the '*monstre mitré*', the mitred monster. Morris himself was having an affair with Mme de Flahaut, and recorded in his diary that she was an 'elegant woman and a snug Party'. One of the few remaining Americans in Paris, Morris was enjoying himself. He found the city to be 'in a sort of Whirlwind which turns [one] round so fast that one can see nothing . . . as all Men and Things are in the same vertiginous Situation, you can neither fix yourself nor your Object for regular examination'. The Parisians, he added, were good, but the background was 'deeply and darkly shaded'.

The cafés, too, of which, according to Mercier there were now over 600 and more opening all the time, had taken on different political hues. In the Café Turc, on the rue Charlot, men of the centre played billiards, draughts, chess and tric-trac. The Jacobins met in the rue de Turnon, while in the Taverne Anglaise in the arcades of the Palais-Royal, surrounded by Chinese wallpapers and lanterns, entrepreneurs and soldiers discussed finance and political stability over cutlets, sweetbread and fricassees of chicken. Zoppi's was where Danton and Marat drank punch. Street life, too, had changed. Reformers, looking to nature as a source of freedom and to natural laws as the way to restore it, had succeeded in getting public animal fights banned. Posters all over the city showed birds bursting out of cages. Festivals and street theatre were to be about peace, not ferocity. In caricatures, the nobility was portrayed in the shape of carnivores; in one picture, the King appeared as a voracious pig-like animal, timid and very fat, who spent his days drinking.

Nor had fashion ever been so full of meaning. Always a mirror to French life, it now provided subtle indications of political allegiances. A tricolour cockade, worn in the buttonhole or pinned to a hat, was mandatory, but a red-and-black pierrot jacket with white feathers indicated that you supported the 'non-juror' clergy, while plain *déshabillé*, with white fichu and bonnet, were signs of democratic leanings. A military look was also in fashion; as the *Magasin des Modes Nouvelles* observed: 'We have all become soldiers now.' Plates, cups, posters, even prints from the toile-de-Jouy factory, all bore scenes of social commentary, one of the favourites being the fall of the Bastille. The 'bizarre and enchanting' spirit of Louis XV, with its soft and sentimental contours, had given way entirely to a taste for purity and antiquity, lines that were rigid, straight, 'inexorable' and 'unfriendly'. All shades of pink, citrus yellow and maroon had disappeared: everything was now red, green or dark brown, which was known as Etruscan. Beds were '*patriotique*' or '*à la Fédération*'. The bolder *ci-devant* nobles painted clouds over the coats of arms on their carriages, to suggest a mist temporarily obscuring their glory, and dressed their servants in outrageous costumes, since liveries were forbidden. When she went out visiting now, Lucie dressed more simply and used a plain, unadorned carriage.

Paris in the winter of 1790 was still full of the English, residents who had stayed on, travellers, seamstresses, grooms, soldiers of fortune who had served with the French. The new fashions, and the debates in the Assembly, were noted and reported back to London where Edmund Burke's *Reflections on the French Revolution* had been published in November. It caused a great stir. The fall of the Bastille had been greeted initially with enthusiasm, the British Ambassador to Paris, the Duke of Dorset, writing in a glowing despatch about the 'greatest revolution that we know anything of' while Wordsworth composed his memorable lines: 'Bliss was it in that dawn to be alive / But to be young was very heaven'.

However, the England of the 1790s was no longer the revolutionary England of the 17th century, but a thriving, conservative land, and stories of heads on pikes and the assault on Versailles had troubled even those who strongly opposed absolute monarchy.

Burke's view – that reform based on abstract philosophical principles was doomed to failure, and that only slow modifications, produced over time, lasted – found willing listeners in London. To make a *tabula rasa* of the past, Burke warned, was to 'insult nature'. Though he attacked the profligate spending of the nobility, and maintained that the existing order needed to be reformed, his views were somewhat tempered by his great admiration for the French court and particularly for Marie Antoinette, whom he described as 'glittering like the morning star, full of life, splendour and joy'. As for the nobility, they were 'men of high spirit and a delicate sense of honour . . . well bred, humane and hospitable'.

In France, *Reflections* sold well, particularly to the royalists and those who still believed it possible to combine reform and monarchy. For Frédéric and his father, as for Mme de Staël, Burke's words struck a chord. The Jacobins preferred to buy Tom Paine's *The Rights of Man* with its attacks on despotism and tyranny and its celebration of God-given 'rights'.

* * *

After the mutiny at Nancy, M. de la Tour du Pin appeared discouraged and depressed. He was, wrote Lucie, increasingly 'powerless before the intrigues of the Assembly'. It was becoming clear to him, as it was to the Marquis de Bouillé, that discipline in the army was collapsing and that civil war threatened. After the plans he had drawn up for the reform of the army were rejected, he had written to offer his resignation to the King. It was refused, but in the Assembly there were cries of 'Sack the ministers' from deputies who wanted the appointment of diplomats and ministers to be put in the hands of the people rather than the King. Later, M. de Bouillé wrote that it had been exceptionally fortunate to have had such an excellent and virtuous man in so essential a post at that time; he urged M. de la Tour du Pin to make whatever concessions were needed in order to remain at his post 'for you may find yourself one day in the position to render the King a great service'.

But Frédéric's father's position was becoming untenable. On 10 November, Danton accused him in the Assembly of being an enemy of the revolution, a despot and an incompetent; he called for him to be tried. Clamours for his dismissal followed; the ministers, the deputies shouted, were all 'clowns who do nothing but move their lips'. M. de la Tour du Pin again tendered his resignation; this time it was accepted. Writing to the King, he said that every day he had seen his honour compromised and been forced to witness 'every kind of sickening event'. Later, among the King's papers, was found a draft of a letter Louis had written to his Minister for War. M. de la Tour du Pin had been, it said, a good and devoted servant to him, and he would never forget it.

Frédéric himself had spent some months working on a plan for the reorganisation of the army: the King now proposed to him that he take his father's place at the ministry. Frédéric, fearing a political backlash, refused. But he did agree to go instead to Holland as Minister Plenipotentiary in the autumn, after the Constitution had been accepted. Since they were obliged to move out of the Hôtel de Choiseul, Lucie, Frédéric, Humbert, Marguerite and M. de la Tour du Pin accepted the Princesse d'Hénin's offer of her house in the rue de Varennes, taking with them Cécile and her children. Augustin was away with his regiment. For the first time, Lucie began to understand the precariousness of their financial position. With her father-in-law's salary gone, and their seigneurial dues vanished, there would be only Frédéric's wages to support a considerable number of people.

Cécile was fascinated by the deliberations in the Assembly and spent most of her days in a box at the edge of the Chamber lent to her by one of the King's equerries. Lucie occasionally accompanied her, but preferred her piano and Italian lessons. In the afternoons, she rode in the Bois de Boulogne with her cousin Dominic Sheldon, on a lively thoroughbred whose 'step and manner I enjoyed tremendously'. Since game was no longer protected as a royal privilege there was little to be seen in forests once teeming with deer, wild boar, pheasants, partridges and rabbits. The King had given up his hunt, and most of the birds and animals had been slaughtered. It was characteristic of the strange sense of calm that pervaded Paris, 18 months after the fall of the Bastille, that a young woman on a fine horse, evidently a member of the nobility, could ride in safety in the company of only her cousin or a single groom, often staying out until dusk.

But even Lucie could not be oblivious to the violence that now marked most Parisian days. One afternoon, in the spring of 1791, returning home in the early evening accompanied only by her English groom, Lucie was stopped at the end of the rue de Varennes by the *Garde Nationale.* She turned back and tried to enter the street by one of the other roads leading to it. All were blocked. Eventually, in the rue de l'Université, she was allowed in. As she rode past the Hôtel de Castries, she saw that the house had been ransacked, its furniture dragged out into the courtyard, its mirrors shattered, the windows and doors wrenched off their hinges. A riot, it turned out, had been orchestrated by Charles and Alexandre de Lameth, after Charles had been slightly injured in a duel with the Duc de Castries that morning. Lucie knew the duke and his family well, having been a visitor to the house all her life. Lafayette had been warned in advance of the attack, she wrote later, and had delayed sending the *Gardes* to the house, through 'laziness', and she hoped that it had not been through some more sinister motive.

From her windows in the rue de Varennes, Cécile had watched the de Lameths' Italian secretary urging the rioters on. Since the fall of the Bastille, the three de Lameth brothers had taken very different political paths. Augustin, Cécile's husband, remained a liberal monarchist, like Frédéric and his father, while Charles and Alexandre were emerging as powerful voices among the Jacobins, trying to become, Lucie contemptuously noted, 'the idols of the people'. Neither Cécile nor Augustin any longer spoke to his brothers, and Lucie herself cut them dead if she met them in the street.

On the eve of Epiphany, Talleyrand had written to his former mistress, Mme de Flahaut: 'Poor Kings! I think that their celebrations and their reigns will soon be a thing of the past. Mirabeau himself fears that we are moving too fast, with too large steps, towards a republic.' For some time now, Mirabeau, who, for all his booming oratorical skills, had a shrewd sense of the workings of government, had been fighting in the Assembly to preserve for the King at least some of his powers, hoping to find ways to reconcile the revolution with the monarchy, in order to safeguard liberty. 'The people,' he warned, 'have been promised more than can be promised; they have been given hopes that will be impossible to realise . . .'. When, at the end of February, the Assembly debated whether to set up a committee to determine the right of anyone to enter or leave France, Mirabeau, seeing in this the beginnings of a police state, fought hard against it, saying that it was not constitutional. Among the Jacobins, he was loudly criticised for betraying the revolution, but his standing was such that he survived: his flourishes of rhetoric and passion charmed his audiences.

Then, one morning, he was struck down with severe pains to his stomach. He struggled on, making speeches, while day by day his great voice shrank to a gravelly whisper. On 2 April, he asked to be shaved; then declared that he was dying. 'When one has come to that,' he observed, 'all one can do is be perfumed, crowned with flowers, enveloped in music and wait comfortably for the sleep from which one will never awaken.' He was 42.

The idea of transforming Soufflot's stark neo-classical Church of Sainte Geneviève into a Pantheon for patriots and heroes, stripped of all religion, a place to celebrate life and not death, well predated the revolution. As Paris mourned the turbulent, philandering Mirabeau – his sexual appetites and colossal debts obligingly forgotten – plans were made to give him the most magnificent funeral. On 4 April, at six o'clock in the evening, a lead urn containing Mirabeau's heart, where his passion and candour were said to reside, was borne through the streets of the capital by eight black horses covered in black velvet studded with silver stars, followed by a procession of Guardsmen, their rifles reversed and their drums muffled, and a band with piccolo, flutes, oboes, clarinets, French horns, bassoons and trumpets. Virtually the entire Assembly as well as most of the Jacobins followed behind. It was later said that 300,000 Parisians had turned out, in the fading dusk, to watch their dead hero carried past by the light of flaming torches. 'It seemed,' wrote Nicolas Ruault, to his brother, 'that we were travelling with him to the world of the dead.'

The terms 'left' and 'right' were being used regularly in the Assembly – the patriots sitting to the left of the President, the monarchists to the right – where debates, as the spring of 1791 wore on, were increasingly divisive and acrimonious. Out of power, but anxiously aware of the anger which seemed to consume Paris, Fréderic and his father listened, talked and felt powerless. They were sometimes joined by Arthur, still a member of the Assembly.

The lifting of censorship meant that pamphleteers and editors on all sides could join in the discussions, so that the speeches of the men who were coming to dominate the Assembly and the Jacobins – Robespierre, Danton, Marat, Desmoulins – were printed in the newspapers. The Jacobins, taking some of their rituals and symbols from the Freemasons, had won followers in provincial towns and regarded themselves as guardians of revolutionary purity. But there were many other clubs, where men came to discuss politics and the new order. Paris was alive with orators, spinning webs of revolutionary fervour.

The Declaration of the Rights of Man and the Citizen had generously promised freedom and equality to all Frenchmen. But on the subject of slaves and women it had been silent. Among the debates raging in the Assembly were calls for the abolition of France's highly lucrative slave trade and the granting of rights to the men, women and children working the sugar, coffee, indigo and cotton plantations of Martinique, Guadeloupe and Saint-Domingue. Just before the revolution, Brissot de Warville, spurred on by the new anti-slavery movements among English and American Quakers and liberals, had founded a Société des Amis des Noirs, with the goal of bringing to an end the 'horrible traffic in negroes'. Among its early members were three dukes, two princes, seven marquises, eight counts, one archbishop and any number of merchants, magistrates and financiers. For Condorcet, long an outspoken abolitionist, slavery was a crime far worse than theft.

But the Antilles were extremely important to France. The small colony of Saint-Domingue alone, with its 30,000 whites and 465,000 blacks, produced greater riches in sugar and coffee than any other colony in the world. And in the Assembly were many delegates – Lafayette and Talleyrand, both absentee owners of plantations, among them – who opposed abolition on the grounds that the slaves were necessary for the prosperity of France, and that in any case humane slavery was a better condition for black people than freedom in their own heathen African countries. As Montesquieu had written, it was hard to comprehend how God, in all his wisdom, could have put a soul, let alone a good soul, into a totally black body.

One of these anti-abolitionists was Arthur who, having come from the Antilles with a mission to safeguard the economy of the colonies, spoke passionately in the Assembly against over-hasty abolition. To counter the arguments of the Société des Amis des

Noirs, Dillon and a number of plantation owners set up their own club, the Massiac, pointing out that there was no one among the abolitionists who had actually ever been to the Antilles or understood how colonialism worked. The Massiac established branches in the French ports, to keep a close eye on the comings and goings of '*négrophiles*'. But as the tide of liberty swept on, Arthur and his friends were overruled. After many fierce debates and reports of growing revolutionary ferment in the islands, the Assembly voted to grant full legal rights to blacks, whether in the colonies or in France; but only, for the moment, to those born to free parents.

Arthur, having argued that France would end up losing its colonies and its trade to the English by pandering to egalitarian chimeras and been defeated, resigned from the Assembly. He was now without a job and, as always, without money, though his fortunes took a turn for the better when compensation was at last paid for the confiscated Dillon regiment which had been formally incorporated into the French army as the 87th regiment of the line. Lucie, visiting her father frequently, was conscious of the many years she had been without him. She had become extremely fond of him, as had Frédéric, though their views on many issues – such as slavery – were markedly different.

Women, however, fared rather less well than slaves in the Assembly. In 1789, France's for the most part illiterate female population had listened to the discourse of rights and wondered what it might achieve for them. It was women who had, after all, led the march on Versailles. Freed at last to reimagine a world made on their own terms, they began to suggest that they should have a say in their choice of husband, and even over how they wished to live. Like Frédéric's mother, any adulterous woman could be put away by her husband in a convent for two years, and forced to remain there indefinitely if he did not want her back. In Paris, groups of women now opened their own clubs, went to meet friends and talk in cafés, and sat in the public gallery of the Salle du Manège, where they heckled the delegates.

They soon found a champion, in the shape of the 42-year-old daughter of a butcher and widow of a banker called Olympe de Gouges, who came up with her own Declaration of the Rights of Women. It was time, said de Gouges, that Frenchwomen were made aware of their 'deplorable fate'. Contrary to the opinion of dozens of tendentious philosophers, she declared, there was no proper evidence to prove that women, supposedly frail creatures with poor powers of reasoning, were actually inferior to men. They should be entitled to vote, to take government jobs, to have parity with men over money and, 'especially', they should resist the 'oppression' under which they currently lived. And they also needed their own, female, national assembly. A devoted monarchist, de Gouges dedicated her Declaration to Marie Antoinette, which may have done neither her nor the Queen any good.

Another champion for women was a novelist and journalist called Louise de Kéralio, who achieved rather more when she proposed in an article in the *Mercure Nationale*, of which she was editor-in-chief, that the polite form of '*vous*' should be put in quarantine, and the more familiar '*tu*' used instead, suggesting that this would lead to more 'fraternity' and consequently to 'more equality'. Not long afterwards, *tutoiement* became obligatory, the polite form effectively vanishing for some years, along with Monsieur and Madame, who became *Citoyen* and *Citoyenne*. But neither de Gouges nor de Kéralio made much progress against 18th-century men's fears that women would cease to be women if given rights. It would not be long before the Committee of Public Safety disbanded the successful women's clubs, after so-called '*Jacobines*' were spotted in the markets wearing red trousers and red bonnets. As the committee pointed out, man, born strong, robust, full of energy, courage and boldness, destined to play a part in commerce, navigation, wars and 'everything that demands force, intelligence, capability', was alone suited to 'great mental concentration'.

Lucie, growing up in a world in which clever and articulate women wielded considerable subtle influence, would see herself, not freed, but silenced, relegated to a life of domestic obscurity, forbidden all political activity. The day of the true salon hostess, with her deft and delicate sense of power, was, for the time being, over. The role of the *citoyenne* was to be the virtuous Roman matron of David's republican paintings, not a politician.

* * *

After a winter of incessant rain, spring had arrived early, bringing blossom to the Tuileries gardens, where pear, apricot and peach all flowered in March. By May 1791, there were peas, asparagus and strawberries in the markets. Frédéric's job as Minister to Holland had been confirmed and in the rue de Varennes, where Lucie was packing cases to send ahead to The Hague, the lilac smelt very sweetly. The city was quiet enough for a visiting soldier called Desbassayns to spend a peaceful day at Versailles, where he visited the very dilapidated Ménagerie, to find the rhinoceros sharing its cage with a dog, and a strange, pale, stripy animal which he identified as a cross between a zebra and a donkey. The novelty on the streets of Paris was a horseless carriage, propelled by two men pedalling, while a third sat in front to steer.

When Voltaire became the third person to be interred in the Panthéon – the first had been Descartes, the second Mirabeau – his coffin removed from the abbey in which he had been buried, Lucie and Frédéric went to watch what she called the 'unseemly splendour' of the procession. An immense chariot, drawn by four white horses, its wheels cast in bronze, carried the porphyry sarcophagus on which an effigy of Voltaire reclined, asleep; in front of it walked 20 young girls in white robes strewing flowers from golden baskets to the martial beat of military bands. In the midst of such turmoil and uncertainty, it was a curiously innocent scene.

Mirabeau's vision of a modified monarchy, with ministers accountable to a legislative body, might perhaps have worked; but by the late spring of 1791 too many forces were stacked against it. Its chief architect, Mirabeau, was dead; the King's chronic indecisiveness was getting worse; and Marie Antoinette was in fact behaving precisely as her enemies suspected, encouraging schemes for counter-revolution. Hostility towards the woman they now called '*l'Autrichienne*' was turning more vicious all the time; in the Tuileries gardens, when Marie Antoinette went walking, her clothes were sometimes ripped by jeering passers-by. *Mesdames Tantes*, Adélaïde and Victoire, the King's elderly aunts, had signified their distaste for the way the revolution was going by leaving for Rome (having taken the precaution of ordering four '*grand tabliers de taffetas vert d'Italie*' to take with them, the current fashion being to wear little aprons in bright colours), though their departure was delayed by angry crowds.

From early 1791, there had been secret plans on all sides – from Gouverneur Morris to Lucie's father-in-law – to rescue the royal family and spirit them abroad, but the depressed and bewildered King continued to waver. In the evenings, in the rue de Varennes, Frédéric, Arthur and M. de la Tour du Pin talked urgently about what might be done to help them. An idea to take at least the Dauphin, who had recently been decreed as belonging to the nation rather than to his family, had been abandoned after the King and Queen announced that they did not want the family to be separated. Only after Easter did the King make up his mind.

Louis XVI had never reconciled himself to the Civil Constitution of the Clergy. In order to avoid taking communion from a 'juror priest', the royal family decided to spend Holy Week at Saint-Cloud, where it would be easier to smuggle in a non-juror. But their departure from the Tuileries, misconstrued by the people as an attempt to flee the country, ended in insults and threats, and even Lafayette was unable to persuade the *Gardes* to allow their carriage to pass. Something of the horror and hatred of the day, the hour and a half spent trapped in a carriage while the crowd shouted abuse all around them, suggested that flight might now be the only course left.

Lucie, having sold her saddle-horses, had gone to join Cécile at Hénencourt. She had taken with her not only Humbert's wet-nurse, but Zamore, her fashionable black manservant, one of the 400 or so black Africans working for noble families in Paris at the time of the revolution. One morning, Zamore appeared in her rooms in a state of agitation. Two strangers, he told her, had passed through Hénencourt with a story that the royal family had vanished from the Tuileries. Lucie, fearing that Frédéric had been involved and was now in great danger, sent Zamore to Paris for news. She waited, in a 'state of indescribable anxiety'. 'The days,' she wrote, 'seemed centuries long.'

It was not until the evening of the third day that Zamore returned, bringing with him a 'long and desperate' letter from Frédéric. In it, he described how at midnight on 20 June, a coach carrying the King, the Queen, the two children – the Dauphin dressed as a girl – Mme Elisabeth, and the children's governess, driven by Axel von Fersen, Marie Antoinette's supposed lover, and accompanied by a small number of outriders, had left Paris on a circuitous route to the border with the Austrian Netherlands, leaving behind an open letter full of bitterness and accusations; how, on discovering their flight, a furious mob had besieged the Tuileries; and how, through a succession of errors, misjudgements, faulty timing and simple bad luck, they had been recognised at the small village of Sainte-Ménehould by a zealous postmaster called Drouet, and returned, disgraced, exhausted and frightened, through dust and blistering heat, to virtual imprisonment in the Tuileries. As their carriage had trundled back into Paris, crowds lined the streets to watch; ordered to show respect, Parisians remained silent, but they did not take off their hats. Lucie, learning that one of the courtiers who had been with them had offered to take the Dauphin on his horse and gallop off to safety, but that the Queen had refused, observed: 'Unhappy Princess, mistrustful even of the most faithful among her servants!'

The King's brother and his wife, travelling by another route and in a faster carriage, had managed to get away; even Fersen had been able to reach the border. The entire flight, Lucie wrote, had been nothing but a succession of 'blunders and imprudences', starting with the amount of luggage the royal party had insisted on taking with them: chamber pots, a heater, a walnut travelling case, along with the King's robes and his crown, all of which served to slow down the carriage to little more than walking pace. Even so, when caught, the royal family were just 40 miles from the border. Before hearing of their capture, Gouverneur Morris wrote to Tom Paine: 'If the King escapes, it means war; if not, a republic.'

For the next two months, until the Constitution was finally drafted and ready for signature, the royal family lived mournfully as prisoners inside the Tuileries, their existence tolerated as necessary by the dwindling number of deputies who continued to favour the retention of a king and a court. Many others were now calling openly for a republic. Louis and Marie Antoinette were closely watched day and night, by men who no longer bothered to show them much deference. On the gates of the Tuileries, someone had put up a placard: '*maison à louer*', house for rent. The sacrosanct mystique and inviolability of the monarchy had effectively disappeared, never to return. Barnave, the deputy despatched by the Assembly to escort the royal prisoners from Varennes to Paris, casually ate his meals with them, an act of *lèse-majesté* that had not been seen before. 'People call him Louis the false or the Fat Pig,' wrote Mme Roland. 'It is impossible to envisage a being so totally despised on the throne.'

The Constitution, redefining citizenship and setting out the role of the monarchy and the election of a government to serve the people, was presented to the King early in September. Louis was to remain king, with limited powers: he would be allowed to veto new laws, and to choose new ministers, but little else. Frédéric drew up a long memorandum which he signed 'M. de G', urging him not to endorse it; but Louis did so. The memorandum, instead of being destroyed, went into an iron chest, together with other documents. Frédéric's position, as an aristocrat loyal to the King in a country hastening towards republicanism, was becoming very dangerous.

At the beginning of October, Lucie, Frédéric, Humbert and various servants set out for Holland, leaving behind them uncertainty, intrigues and talk of counter-revolution. On 21 September, the National Convention had proclaimed France a republic and abolished the monarchy. 'To aggravate the horrors of this place,' wrote William Augustin Miles, a pamphleteer and friend of Pitt, 'every maniac almost is an assassin either in thought or deed.' Cécile, Frédéric's sister, who was ill and coughing constantly, went with them, taking her sons and their tutor, finding the thought of a lonely winter at Hénencourt unbearable. Augustin, who had left the army in March and had ceased to call himself a marquis, stayed behind to put together a local militia.

CHAPTER SEVEN

Standing on a Vast Volcano

Lucie would remember her year in Holland as the last heedless moment of her life. Later, she would wonder at her own frivolity and insouciance; but in the autumn of 1791, a determined young woman with a small healthy son and a husband she loved, accompanied by Marguerite and Zamore and her former tutor M. Combes, who had returned from America and was now secretary to Frédéric, she set out to enjoy life as the wife of a French ambassador. She was 21; she had a large wardrobe of Parisian dresses and good jewellery; and wherever she went, she was looked on with admiration.

Frédéric, for his part, was turning into a highly competent administrator, though somewhat outspoken. When doubts had been raised in the Assembly about appointing known monarchists to the diplomatic service, M. de Montmorin, the Foreign Minister, had pointed out that, given the royalist feelings at the courts of Europe, this was not the moment to send 'people who have declared themselves in favour of revolution'. Frédéric's political credentials, liberal yet aristocratic, were precisely what were needed. Even so, Mme de Staël, who, estranged from her husband was now mistress to the Comte de Narbonne and working hard behind the scenes for her friends, advised Frédéric to 'study closely the men you meet and try rather to fascinate them with their own ideas rather than with yours'. He would do well, she said, to lie low, for the word around Paris was that he was 'not very Constitutional', and too close to the King. Try to write despatches 'that sound very patriotic', she added: to do otherwise had become foolish.

Frédéric's position at The Hague was not an easy one. He had never been a diplomat and France's relations with Holland were tense. After years of struggle between the French and the English for supremacy in Holland, with fortunes spent on spies and secret diplomacy, William V, Prince of Orange, a cautious, irresolute man with protruding eyes and bulging cheeks, had only recently agreed to recognise the new United States. At heart, he remained firmly in favour of the English. There were two main parties, the ruling House of Orange, of which William was 'stadtholder' – the chief executive and military commander – and the Patriots party, younger, more liberal and more open to France. Dutch citizens who supported William wore orange ribbons, whether as cockades in their hats or their buttonholes, or pinned to their dresses, a custom followed by Frédéric's predecessor, a timorous diplomat called M. Caillard. Caillard, noted Lucie scornfully, was 'prudent to the point of timidity' and had kept his post 'only by sending despatches exaggerating the difficulties'.

Frédéric, on the other hand, was already showing signs of the diplomatic stubbornness which would later mark many of his decisions. Though titles and liveries had been abolished in France, French diplomats abroad were allowed to use the King's livery. Frédéric ordered his household to leave off both orange ribbon and tricolour cockade; Lucie supported him, remarking tartly that it would be unseemly to clutter the Bourbon livery with the insignia of someone who was nothing other than the senior officer of a republic, even if he was of excellent family and had married a Royal Highness.

Their stand earned them a few clashes with angry crowds, but the Dutch people were soon more interested in Zamore and his elegant wardrobe. Frédéric now settled down to negotiate his way between the cross-currents of revolutionary fervour drifting across Europe, while Lucie became, as she wryly wrote later, 'the acknowledged leader of all society gatherings'. Unlike Sir James Harris, former English Minister at The Hague, who had found the city unremittingly dull, its inhabitants interested only in cards and food, Lucie was in a mood for parties. 'I was dazzled,' she would say, 'by my own success', and, she added, 'little realising how short a time it would endure'. She was also pregnant again.

By the spring of 1792, what had started as a trickle of frightened people trying to escape the uncertain temper of Paris was turning into a river even if Talleyrand, when consulted, would advise his friends to hide somewhere, stay quietly in their châteaux, rather than leave the country. The word 'émigré' was a creation of the French Revolution, Edward Gibbon, in 1791, remarking on the large numbers of 'emigrants of both sexes . . . escaping . . . the public ruin'. What had become known as '*l'émigration élégante*', the elegant émigrés, had crossed the Channel to England; '*l'émigration pauvre*', the least well-off, had gone to Soleure and Fribourg in Switzerland. There were French dukes, marquises, counts and their retinues, as well as royalist soldiers and 'non-juror' clerics, in every state and country bordering on France, their numbers growing with each violent incident. More or less tolerated by their hosts, they led lives as similar as they could make them to their former existence at Versailles. Horses with which to leave Paris were, at times of particular turbulence, in such short supply that oxen were harnessed to carriages instead, to be seen creaking at little more than a crawl out of the city gates. Until March 1792, there was no need for a passport to leave.

Koblenz, meanwhile, standing on the confluence of the Rhine and the Moselle, had become the capital of the military emigration. It was here that the King's brothers, the Comte d'Artois and the Comte de Provence, had set up their headquarters and were assembling an army, *l'Armée des Princes*, the Army of the Princes out of disillusioned, royalist soldiers. Six thousand men, half the French officer corps, had gone into exile after the royal family's disastrous attempt at escape. The 24-year-old René de Chateaubriand, recently returned from America, observed that they were a motley bunch of 'grown men, old men, children just out of their cots, speaking the jargon of Normandy, Brittany, Picardy, the Auvergne, Gascony, Provence, the Languedoc', sometimes several generations of the same family.

Viewed with extreme misgivings in France, and the cause of constant rumours of imminent invasion and counter-revolution, these soldiers, when not engaged in military tasks, spent their days intriguing. Swinging between optimism and pessimism, their mood was constantly jostled by inflammatory pamphlets smuggled over the border from France, and by the comings and goings of spies, Jacobins, recruiters and people carrying letters. Their families, mourning the *douceur de vivre* they had left behind, were already beginning to appal their hosts with their arrogance and frivolity. 'We dance and we enjoy ourselves,' wrote the Duchesse de Saulx-Tavannes about the émigrés in Brussels. But Lucie, nearby in The Hague, heard stories from her friends that the Belgian noble families, known for their modesty and simplicity, were terrified lest they become contaminated by the profligacy of their visitors.

Among the recent arrivals in Koblenz were Mme de Rothe and Archbishop Dillon, who had fled from Hautefontaine as more and more 'refractory' priests were rounded up, and there were fears that he might be taken with them and imprisoned. Though never overly concerned with piety, the Archbishop had a stubborn and courageous streak. He did not intend, at any price, to renounce what he believed to be the true vows of the Catholic Church.

At The Hague, Frédéric, closely following local events, listening to the gossip and reading the papers, both clandestine and those published openly, was soon aware that the French émigré army in Koblenz was looking for horses and weapons. His reports to Paris, either written in his own hand, or dictated to M. Combes, some of them in a code of neat numbers, gave precise accounts of the success of these ventures. Frédéric's tone was careful, but already he was filling his despatches with his own opinions, which reflected his thoughtful nature. 'The Dutch,' he wrote, having not greatly taken to them, 'are not people to give away cheaply what is theirs.' In December, hearing of 15,000 guns and ammunition for sale, he asked the Foreign Ministry in Paris whether he should make an offer for them, to keep them out of the hands of the émigré officers. 'I am as yet to find a single educated man,' he wrote, with that sense of false security in which he and his colleagues continued to live, 'who believes in the success of the French Revolution.'

In March 1792, as France's revolution was heading into a new phase, Frédéric was abruptly informed that he had been dismissed. The weak but amiable M. de Montmorin had been replaced as Minister for Foreign Affairs by General Charles-François Dumouriez, an energetic, canny, ambitious soldier who had been a close friend of Mirabeau's. Dumouriez, whose proposed reforms of the army were opposed to those of Frédéric's father, bore a grudge against M. de la Tour du Pin: he repaid it by getting rid of Frédéric.

Moving out of the splendid French Embassy in The Hague into a smaller, rented house, Lucie gave proof of her spirit and sense of humour. To embarrass Frédéric's replacement, she held a very public open-air sale of all the furniture and possessions for which they no longer had room, holding it on the promenade immediately in front of the embassy. Profiting from the lessons learnt about housekeeping and money during her childhood at Hautefontaine, she made a handsome sum, which she prudently entrusted to a Dutch banker in whom she had confidence. But she kept her lavish wardrobe and, having taught Zamore to dress her hair, continued to go to court, to dance and to pay visits. It was becoming impossible, however, even for the still carefree Lucie, to ignore the fact that the news from France, brought by visitors or carried back by Frédéric from his frequent trips to Paris, was becoming more alarming with every week that went by.

* * *

The acceptance by the King of the revised Constitution in September 1791 had been greeted with pleasure and relief. A *Te Deum* was sung in Notre Dame and a balloon, trailing tricolour ribbons, was sent up to hover above the Champs-Elysées. In the new Legislative Assembly that replaced the old Constituent Assembly – for which no former deputy was permitted to stand – the clergy and nobility had virtually disappeared. Apart from one or two revolutionary aristocrats, the Assembly was dominated by lawyers, with a sprinkling of mathematicians and historians, all well versed in revolutionary discourse, as well as journalists and editors such as Camille Desmoulins.

A new political grouping, the Feuillants, which included the two more militant de Lameth brothers, had split off from the Jacobins and was preaching a constitutional monarchy of the centre, with a stronger sense of public order to quell the frequent outbreaks of popular insurrection, and discipline imposed on the press, the political clubs and the army. But the Jacobins, though numerically smaller in the Assembly, had Marat, Danton and Robespierre, whose cold, moral earnestness and invective against royalists and aristocrats was remarked on by all visitors to Paris. Robespierre had about him 'something sardonic and demoniac', and, with his tawny-coloured eyes, looked like a brooding bird of prey. His political voice formed by the Oratorians and by his devotion to the teachings of Rousseau, Robespierre shared with his mentor a vision of an austere, virtuous and authoritarian republic, obedient to a controlling social order, views that would soon become the guiding spirit behind revolutionary absolutism.

The steady stream of émigrés leaving France to join the Army of the Princes in Koblenz had fed a growing sense of paranoia and mistrust in Paris. By the spring of 1792, the Assembly was in a fever of war rhetoric, speaker after speaker rising to denounce the legions of fanatical émigrés apparently massing on the border to attack France. Their passions were further enflamed when, on 7 February, Austria and Prussia signed a formal treaty of alliance. The Emperor Leopold, while concerned for the safety of his sister, Marie Antoinette, had been extremely reluctant to go to war, but his death had brought his more war-minded son Francis to the throne. On both sides, the mood was turning to war.

In the second week of April, 50,000 Austrian troops were moved to the Belgian frontier. On the 20th, Louis XVI, believing that a successful war might also restore him to his proper place, went to the Assembly and proposed that France should begin hostilities, though much of his family and his court were by now in exile and on the opposing side. War – which would last, on and off, for almost 23 years – was declared. In the garrison at Strasbourg, a young army engineer called Rouget de Lisle came up with a catchy, rousing song that would remain famous long after the revolutionary music of Gossec and Grétry was forgotten. Soon, men all over France were marching to war to the *Marseillaise*, with its heady appeal to valour and patriotism and its images of the tainted blood of tyrants.

For the French, the war did not start well. Despite the presence of able and seasoned military veterans in all three major theatres of war – Lafayette, Rochambeau and Luckner – the army was disorganised and still plagued by memories of the troubles at Nancy in which Frédéric had played a role. Many soldiers remained suspicious of their aristocratic officers. One of the first to fall foul of his men was Théobald Dillon, Arthur's first cousin, who had served with him in the American war. Théobald had been commanding a force sent to attack the Austrians at Tournai. Retreating after encountering unexpectedly heavy opposition, he had been mistaken for a spy, lynched and hanged from a lamp-post in Lille, before his leg was hacked off and paraded around the town. Other officers, learning of his fate, hastened to resign and crossed the border to join the Army of the Princes.

France itself, like its army, was in a troubled and volatile mood. As the value of the paper money, the *assignats*, kept falling, and news of grain shortages in the south and the south-west was reported, so there were increasing attacks on the barges and wagons carrying food. A 'famine plot' was talked about, apparently planned by scheming 'counter-revolutionary forces'. Uprisings in the French West Indies, where tens of thousands of slaves in Saint-Domingue, under the free, highly intelligent Toussaint L'Ouverture, had destroyed a large number of plantations and made coffee and sugar, two commodities which had become staples in French life, prohibitively expensive. Into this mood of paranoia about food and anger towards those responsible for the shortages was emerging a new band of revolutionaries, the *sans-culottes*, artisans, shopkeepers and functionaries of the more militant Paris sections, who saw in the silk stockings and breeches of the old court everything they most deplored. The enemy was perceived to be not just the émigré army and the Austrians, gathering against France, but the 'false patriots' and fifth columnists at home. 'To your pikes, good *sans-culottes*,' urged Jacques-René Hébert, whose acerbic *Père Duchesne* was becoming the most popular of the revolutionary papers, 'sharpen them up to exterminate aristocrats.' The true patriot now was a man who wore, along with his tricolour cockade and red bonnet, pantaloons, suspenders, clogs and a short jacket, the *carmagnole*. Writing to Jefferson, Gouverneur Morris, the last American diplomat left in Paris, remarked: 'On the whole, Sir, we stand on a vast Volcano, we feel it tremble and we hear it roar . . .'

* * *

At the end of May, as Frédéric kept a close and anxious eye on events in France from The Hague, and Lucie nursed her increasingly sick sister-in-law Cécile, the King in Paris was forced to agree to the disbanding of his own personal guard. Using his veto in the Assembly, he refused, however, to accept a decree to deport all 'refractory' priests. It was a stormy, wet summer. The Paris crowd was not in a mood to tolerate royal vetoes. On 20 June, a mob forced its way into the Tuileries, brandishing a gibbet with a doll dangling on a rope labelled '*Marie Antoinette à la lanterne*', dragged a cannon up the staircase and battered down the doors to the royal apartments. The King was cornered, forced to place a red bonnet on his head and to swear allegiance to the Constitution.

Events were moving very fast. The insults against Louis and Marie Antoinette had become such that they no longer dared to walk in the gardens. While 20,000 Guardsmen were converging on Paris for the 14 July celebrations, singing the *Marseillaise* and reducing the city to new levels of lawlessness, the Duke of Brunswick, commander-in-chief of the combined Austrian and Prussian forces, issued a memorandum promising an 'exemplary and unforgettable act of vengeance' against anyone who harmed the French royal family.

In the Assembly, where Robespierre called for unity and emergency powers, '*la patrie*', the nation, was proclaimed to be 'in danger'. An 'insurrectionary commune', made up of people from the more militant sections, and including Robespierre and Danton, was set up. The travelling soldier Desbassayns remarked that though the markets were full of peas there were very few apricots, and noted: 'We have reached a moment of crisis.'

All through the night of 9 August, the tocsin rang continuously. On the morning of the 10th, the royal family was taken for safety to the Assembly, where they were forced into a caged space usually reserved for reporters. A crowd, swollen by many *Gardes* armed with pistols, sabres, scythes, swords, pikes and knives, surrounded the Tuileries, which was lightly guarded by 900 Swiss soldiers together with a number of mounted gendarmes and about 300 aristocrats. The Swiss fired; the crowd retaliated. In the hours that followed, those defending the palace were stabbed, stoned, clubbed, bludgeoned to death. People poured into the Tuileries, killing everyone they caught, dragging them out of hiding places in the chapel, the attics or the cellars, looting and drinking as they went. The naked bodies of Swiss Guards, stripped of their distinctive red, blue and gold uniforms, were mutilated. When an English visitor to Paris, Mr Twiss, ventured out at three o'clock, he found the bridges, gardens and quays surrounding the Tuileries covered with bodies, dead, dying or drunk; *sans-culottes* and *poissardes* running about covered in blood; and bodies being loaded on to carts to be taken for burial in common lime pits. Looters, rummaging in the wardrobes of Marie Antoinette and her servants, emerged decked in feathers and pink petticoats. Marie Grosholz, later famous as Madame Tussaud, described seeing the gravel stained red, with flies buzzing around clotted pools of blood. Some of the women wore ears and noses pinned to their caps.

The violence abated; but the monarchy had fallen. That night, the royal family slept in the Convent of the Feuillants. On the 13th, they were taken to the medieval tower of the old monastery of the Templars in the Marais. They were now prisoners. In *L'Ami du Peuple*, Marat spoke of 10 August as a 'glorious day'. There would be, he added, no 'false pity' for tyrants. The Princesse de Lamballe and the children's governess were separated from the royal family and taken to the prison of La Force. Inside the Temple, life resumed some kind of normal course. The King read in the excellent library of the Knights of Malta; Mme Eloffe was allowed to deliver clean linen; there were prayers, lessons, needlework. The meals remained relatively luxurious, soups, entrées, roasts and desserts, served on silver. What would happen to the royal family had not been decided.

But the sudden outbreaks of frenzied violence were not over. Santerre, a brewer known for his revolutionary fervour, was made head of the *Gardes*. The events of 10 August were quickly recast by those now in power as a royal plot, courageously foiled. The newly formed Insurrectionary Commune called for the establishment of a military tribunal to try those accused of plotting against the republic. It had powers to arrest, question and punish, with no right of appeal. The task of tracking down the guilty was put into the hands of a new *Comité de Surveillance*; 'domiciliary visits' began, late at night or at dawn, to search for fugitives and incriminating documents. A revolutionary police state had effectively come into being. The Terror had begun.

Over a thousand people, many of them 'refractory' priests, bishops, almoners and vicars, hiding in seminaries and churches, along with aristocrats and their servants, were rounded up; editors and printers deemed 'counter-revolutionary' were seized and their publications closed down. Among those sought were former ministers. Frédéric's father, M. de la Tour du Pin, managed to make his way to Boulogne and crossed the Channel to Dover. Mme d'Hénin's lover Lally-Tollendal was caught in the net and sent to the prison of L'Abbaye, from where he wrote to a friend asking for clean shirts and a few bottles of wine.

The war was still going well for *L'Armée des Princes* and its allies, who had crossed the French border on 14 August, scored a victory at Longwy and were advancing on Verdun. Goethe, attached to the Prussian army, found time to pin down the noise made by the cannons. They sounded, he wrote, like 'the humming of tops, the gurgling of water, and the whistling of birds'.

In the Assembly, Danton rose to describe fifth columnists, traitors to the *patrie*, lying low in Paris waiting to exact revenge should the royal family be threatened. The *Orateur du Peuple* printed a rousing article about the enemies at home. 'The prisons,' it said, 'are full of conspirators . . .' Marat spelt it out. 'Good citizens', he declared, on posters that went up on walls all over Paris, should go to the prisons, seize the priests and conspirators held there and 'run a sword through them'. The city was sealed, its gates barricaded shut; drums were beaten; the tocsin was rung.

Among the first to be dragged from their cells were 24 priests, hacked to death in the gardens of the prison of L'Abbaye. Over a hundred others were slaughtered in the Convent of the Carmelites, following a brief parody of a trial. Bishops were shot as they prayed. Vicars, abbots, parish priests, canons, almoners, seminarians, men known for their piety and learning, were murdered. The massacres, which had started on 2 September, continued into the 3rd. Nothing was done to stop the violence. The prisons of Bicêtre, La Force and La Salpêtrière housed not only the newly arrested clergy and nobility, but beggars, prostitutes, old women and the insane; they, too, were hauled out and killed, the youngest being a vagrant boy of 12. By the time the killing had played itself out, 1,400 people were dead, most of them hacked and battered to pieces. One of these was M. de Montmorin, Frédéric's former employer at the Ministry for Foreign Affairs. Another was the Princesse de Lamballe, the blue-eyed, fair-haired friend of Marie Antoinette who, refusing to denounce the monarchy, had her head chopped off; it was impaled on a pike and borne off to the Temple, to be danced about under Marie Antoinette's windows, before being carried to the Palais-Royal to be shown to her lover, the Duc d'Orléans. Lally-Tollendal miraculously survived by escaping through a window.

* * *

Though the Duke of Brunswick was continuing to advance, the fortunes of the émigré army and its allies were in fact turning. The magnificent squadrons of mounted cavalry, noted by the Comte d'Espinchal in his diary of the campaign as being 'almost entirely composed of the cream of the French nobility' and filled with 'indescribable enthusiasm and zeal', were being routed. The initial force of 30,000 men, eroded to little more than 10,000 by incessant heavy rain and inadequate supplies, was halted at Valmy on 20 September. On the 30th, the Prussians sounded the retreat. It soon looked more like a stampede. Dragging carts piled high with sick or wounded men, the émigré force fell back through torrential rain and deep mud, losing men to marauding bands of French attackers, or abandoning them to die of dysentery in the ditches. The magnificent horses described by d'Espinchal were soaked and bedraggled. Returning to Koblenz, the Army of the Princes fell apart; its members began to disperse, carrying the French diaspora still further afield.

Lucie's father Arthur, to the fury of the Archbishop his uncle, who accused him of treachery, had remained in France. Many aristocratic families were now divided, their members at war with each other, the royalists outside France, plotting to bring down the revolution, the liberals inside the country, still hoping to be able to save a reformed constitutional monarchy. Promoted to lieutenant-general, Arthur had initially hoped to be sent to Martinique as colonial governor, to join his wife. But in the spring of 1792, shortly before the declaration of war, Arthur had been drafted to serve in the army. He was on the border with Flanders when he heard of the massacres in Paris. Always rash and outspoken, at heart a convinced monarchist, Arthur immediately declared that he would indeed renew his oath of allegiance: but it would be an oath to the King as well as to the Constitution, and he asked that his men do the same. His position, as a former aristocrat, with three Dillon relations serving on the other side with the Army of the Princes, was made more precarious when his friend Lafayette, judging that civil war was looming, deserted and crossed the lines to the Austrians. Arthur could, like Lafayette, have remained abroad in safety. Lucie begged him to join them in The Hague; but her father insisted on returning to Paris.

For a while, Arthur, who was an excellent and brave soldier, was protected by Dumouriez, who had been promoted to supreme control of the French armed forces. He was sent to command the Army of the Ardennes, at one point holding out against 80,000 Prussians with only 10,000 men. At Verdun, however, he made a fatal mistake. Hoping to confuse the enemy with a series of phoney letters that he was confident would be intercepted, he was himself accused of treachery and recalled to Paris to defend himself before the Assembly. There, Arthur managed to persuade the deputies of his innocence, and was even able to secure compensation for the widow of his murdered cousin, Théobald, despite Marat's accusation that Théobald was nothing but a royalist intriguer. But Arthur's manner was brusque, and he had never been diplomatic. He was relieved of his command. Offered lowly alternative positions, he refused to take them, announcing that he would remain in Paris and keep requesting military reinstatement until offered a worthy command. As a friend observed, Arthur had defended himself 'with zeal, too much zeal'.

For the revolutionary French forces, victory followed victory. In Paris, every revolutionary triumph was greeted with jubilation, the heroic deeds of the victors celebrated on posters, in paintings and on the stage. 'We cannot be calm until Europe, all Europe,' declared Jacques-Pierre Brissot, emerging as one of the leaders of the revolution, 'is in flames.' Battalions of volunteers, some as young as 14, had responded when invasion threatened. All over France every man with two pairs of shoes gave up one for the defenders of the frontiers; women made bandages and scraped the walls of cellars to extract saltpetre for explosives. By the end of October, General Custine, M. de la Tour du Pin's whiskery, red-faced friend, had occupied the Rhineland; in the south, Savoy and Nice fell to the French. The 6th November saw the winning Battle of Jemappes, after which Dumouriez and his men marched on Brussels, mercilessly requisitioning all they set eyes on as they went, and demanding vast indemnities, thereby setting a pattern for French conquest for the next 20 years.

In Brussels, the thousand or so émigrés, settled in lodgings and hotels around the Grand Place, had believed themselves safe. But one evening, just as Lucie was preparing to leave for a reception at court in The Hague – at which she was a frequent and popular guest, despite Frédéric's dismissal as ambassador – the Austrian Minister, the Prince de Starhemberg, arrived, looking 'distraught'.

'Everything is lost,' he told her. 'The French have defeated us completely. They are occupying Brussels.'

For the Dutch, and for the French émigrés who filled their cities, Dumouriez's advance was indeed disastrous. A friend arriving at The Hague from Brussels described to Lucie the flight from the city of the thousand or so émigrés who had taken refuge there. He told her of scenes of chaos, as frantic people, piling their belongings on to every waggon, dray and carriage they could get their hands on, fled the city. For many it had been their second, or even third, sudden flight. The wisest and those 'most plentifully provided with funds', made their way to England. Somewhat sternly, with the self-righteousness of youth mixed with her own tendency to take a firm moral line, Lucie noted how appallingly some of the richer émigrés behaved towards those without money. 'Many of them,' she observed, 'presented a sorry spectacle of the most shocking heartlessness towards their companions of misfortune.' Vowing that she herself would certainly never help them should they fall on hard times themselves, 'I hastened,' she wrote later, 'to offer my services to the most heavily stricken and paid very little attention to the richer.' Lucie herself had just miscarried again, at around five months into her pregnancy. She was not very well, but she did not complain.

* * *

The French émigrés, perched precariously in towns and villages over the borders in Switzerland, Italy, the German States, Belgium and the Austrian Netherlands, were indeed in an increasingly impossible position. In their absence, ever more punishing edicts had been issued against them: one by one, their rights had been curtailed, their relations threatened, their lives menaced, their houses and lands marked for sequestration and sale as *biens nationaux*, national properties. Already dispossessed of their feudal dues, desperate not to lose everything they owned, but fearing imprisonment and execution if they returned home, they were also running out of what little money they had taken with them into an exile that all assumed would only be brief. Their hosts in the border towns were, for their part, growing increasingly impatient with guests who seemed incapable of understanding that the *ancien régime* was over, and who continued to gossip, gamble and hold salons, while borrowing vast sums of money. In several places, these unwanted exiles were now being ordered to move on, and parties of desolate French counts and marquises, sinking ever deeper into debt and penury, could be encountered trailing along the roads of Europe, spending their nights in their carriages and selling their last pieces of jewellery. Outside villages could now be seen signs: 'émigrés not wanted here'. 'Our fate,' noted one mournful count, briefly finding refuge in Spa, 'is to be ceaselessly victims of events.'

Lucie and Frédéric, who had been sent to The Hague on official business, were in a somewhat different category, and Frédéric, since his dismissal by Dumouriez, had spent a considerable amount of time back in France. When, in the late autumn of 1792, fresh penalties against émigrés were voted through the Assembly, banishing them from France in perpetuity and arranging for the immediate sale of their properties, Frédéric decided that it would be sensible for Lucie to return, to try to ensure that the house in the rue du Bac would not be seized. From Paris, he wrote to tell her that too much was at stake for her to delay any longer. Frédéric's father had also decided to leave London and come back to Paris, saying that he refused to do anything that would jeopardise the remaining inheritance of his children. The full meaning of the dispossession of the Church and the nobility was at last clear to Lucie. Up until this moment, she wrote, 'I was still no campaigner, but as delicate, as much a fine lady and as spoiled as it was possible to be . . . I still thought that I had accepted the greatest sacrifice that anyone could require of me when I agreed to do without the services of my elegant maid and my footman-hairdresser.'

* * *

It was an extremely cold winter; snow lay across northern France. Taking with her Humbert, now aged 2½, Zamore, Marguerite and one other manservant, Lucie left The Hague in a carriage on 1 December 1792. Buried under a pile of pelisses and bearskins, she clutched Humbert to her, wrapped up tightly like a small Eskimo. They had taken the painful decision to leave Frédéric's sister behind, with her two young sons. Cécile, very weak and coughing blood, was too ill to travel. Many years later, Lucie would regard this journey as the defining moment in her life, the one at which she finally recognised that nothing would ever be the same again, that the pampered years of balls, dazzling clothes and jewellery, times she described as 'weakness and illusions', were over.

What she did not know, when she left The Hague, was just how far the French had already advanced. Reaching the border with the Austrian Netherlands, she learnt that Antwerp had fallen and though a room was at last found for them, the town was crammed with rowdy soldiers, celebrating their recent victory. Standing at the window, Lucie watched as a vast bonfire was lit in the main square, the flames soon 'leaping high as the rooftops'. Drunken French soldiers, reeling and swaying, fed the flames with furniture, books, clothes and portraits dragged out of the surrounding houses. Meanwhile 'dreadful-looking women, their hair loose and their dress in disorder, mingling with this gang of madmen', plied them with vintage wine looted from the cellars of rich Antwerp merchants, singing obscene songs and dancing around the flames. All night, appalled, Lucie watched from her window, 'fascinated and terrified, unable to tear myself away despite my horror'.

Leaving Antwerp next morning, accompanied by a friend of Arthur's, they drove through the French encampment, and Lucie noted how few of the men had proper uniforms. 'These conquerors,' she observed, 'who were already making the fine armies of Austria and Prussia quake in their shoes, had all the appearance of a horde of bandits.' Their cloaks, hastily made up in every conceivable colour out of material requisitioned from stores along the way, reminded her of a vast human rainbow, or a gigantic flower-bed, brilliant against the white of the thick snow. It was their red bonnets that lent them an air of menace.

Lucie's heavily laden carriage moved very slowly along roads cluttered with wagons and ammunition carts, occasionally halted by officious commanders wanting to see their papers. In Mons, drunken men tried to break into the room in which the women had barricaded themselves with Humbert, but they fell back just as Lucie was preparing to hit them with a flaming log pulled from the fire. Nearing Hénencourt, where she was hoping to find Frédéric, they came across a squadron of well-mounted black soldiers, led by the Duc d'Orléans's manservant, Edward, going to join the fighting in the north. Many of these black former footmen had been in service with the nobility and, their employers having fled, were joining the army as volunteers. Lucie feared that she would lose Zamore to them, but after spending a day with his friends, riding part of the way with them, he returned that night to her side. As they crawled their way over frozen, muddy tracks, past châteaux that had been looted and set fire to, Lucie reflected on her life. At that moment, she wrote later, she felt overwhelmingly sad; her 'carefree youth' struck her as sickeningly frivolous. Doing what she had so often done in the past, imagining situations and events that might lie in the future, she started to sketch out the dangers that awaited her, so that she might be better prepared. From that day onward, she would write, 'my life was different, my moral outlook transformed'.

Frédéric was not, as she had hoped, at Hénencourt with Augustin de Lameth. Travel around France had become extremely difficult, with detailed passports and permits – specifying age, height, length of nose, shape of mouth and face, type of chin, and colour of hair and eyebrows – needed for the shortest journeys, and he had been unable to leave Paris. Augustin was alone, very gloomy and anxious about his wife Cécile and the children and about the encircling violence. Lucie, entrusting Humbert to Marguerite's care, set out to look for Frédéric, taking with her Zamore for the 160-kilometre journey to Paris and carrying false papers, provided by a friendly official in Hénencourt, to say that she had been resident there since the spring.

After an interrupted and uneasy journey, constantly stopped at roadblocks by suspicious *Gardes*, she found Frédéric at Passy on the outskirts of Paris, living secretly in a house belonging to the Princesse de Poix. They were its only occupants and decided to keep to the back rooms, leaving the front of the house shuttered to give the impression that it was empty. Humbert would be left in safety at Hénencourt. Frédéric's father was in hiding not far away in a house belonging to his cousin at Auteuil. It could be reached from Passy by little-frequented footpaths and every day, slipping out of the house through a concealed side door, Lucie and Frédéric made their way through the wintry countryside to see him. They had discovered a ramshackle cabriolet and a very old horse, and planned to make forays into Paris to find out what could be done about salvaging the house in the rue du Bac.

* * *

Paris was much altered. In the 15 months that Lucie had been away everything that had once made the city the envy of the civilised world was either destroyed or under attack: libraries and priceless antiques chopped up for firewood, churches and convents desecrated, crowns, coats of arms, fleurs de lys all defaced or chipped away, the statues of the French kings in the places Vendôme, des Victoires and Royale toppled from their pedestals, volumes of heraldry and registers of nobility publicly burnt. In the Faubourg Saint-Germain, Lucie and Frédéric, venturing cautiously past the Église des Théatins, now a grain store, and the Convent of the Récolettes, now the Théatre des Victoires Nationales, found the once exotic and sweet-smelling gardens abandoned and overgrown. The Hôtel de Bourbon, confiscated from the Prince de Condé, had been transformed into a prison, where the few Swiss Guards who had survived the massacre in the Tuileries were locked up. Lucie and Frédéric learnt that the liberal monarchist Stanislas de Clermont-Tonnerre had been murdered outside his house in the rue du Cherche Midi, and the Duc de la Rochefoucault dragged from his carriage and lynched in front of his wife and daughter. 'In Holland,' Lucie wrote later, 'I had been spoiled, admired and flattered . . . The revolution was now all about me, dark, menacing, laden with danger.'

The signs of revolution, as Lucie and Frédéric moved around Paris, were indeed everywhere; it was the speed with which it had happened that was so striking. Many of the street names had already changed – there would be 4,000 new names by 1793 – losing their ecclesiastical and noble connotations: the rue de Condé had become rue de l'Égalité, the rue Comtesse d'Artois was now the rue Montorgueil. Citizens addressed each other as '*tu*', and signed their letters '*ton-Concitoyen*'. And just as Notre Dame was now the Temple of Reason, so playing cards had shed their kings and queens, to become '*Génie*', talent, and '*Liberté*'.

In this new world, time itself was to be thought anew. A special commission had been appointed to give the old Gregorian calendar a revolutionary twist, and the model they had come up with, disdaining the superstitions and tyrannies of the past, looked to the seasons and changing weather of the agricultural world. 'We can no longer count the years,' said Fabre d'Églantin, 'during which kings oppressed us as a time in which we lived.' The first moment of the declared Republic, 9.18 plus 30 seconds on the morning of 22 September, 1792, had been the time of the *vendange*, the grape harvest, and so became *vendémiaire*; *brumaire* was the month of mists; *frimaire*, that of cold. Each of the twelve months was divided into 10-day periods, the *décades*, periods in which people were encouraged to contemplate the fruit, flowers and produce of the moment. For *fructidor*, on the cusp between summer and autumn, the calendar prescribed eglantine, roses, crayfish, chestnuts, hops and sorghum. Since each month was to be an equal 30 days, the five days left over in the year were declared to be *sans-culottides*, to be devoted to festivals of industry, heroism and patriotic games. *Fructidor* was becoming a popular name for new babies.

And, for all the shortages, the almost total lack of coffee and soap – which meant that skin diseases were multiplying – for all the bread queues and bread rationing, there was something hopeful and bracing in the air, remarked on by the few foreign visitors left in the city. The English traveller, Mr Twiss, walking around the streets, observed how much better dressed ordinary women seemed, even wearing earrings as they served behind their stalls, and how few children were to be seen barefoot. He approved of the simpler dresses, the way lace, silk and velvet had been replaced by cotton and linen, and the new short haircuts for men, shorn like David's heroes of antiquity. David had designed a new uniform for men, with a short tunic, held together at the waist by a wide scarf, tight breeches and a three-quarter-length cape, inspired by the Romans and the Renaissance; but it was generally found to be too fanciful to be worn.

Most of the fashionable *modistes* had followed the émigrés into exile, Rose Bertin taking with her four of her famous dolls, like the one given to Lucie by de Lauzun, splendid in the full panoply of the *ancien régime*. It was no longer either right or appropriate, noted the editor of the *Journal de la Mode et du Goût*, for women to have an air of meekness or submission; on the contrary, given the new laws promising them freedom from parental and marital control, and providing for divorce on equal terms, their manner should be 'decisive', their head 'held high', their step 'resolute'. *Citoyens* and *citoyennes* were encouraged to abandon the flowery talk of the aristocrats and use coarse and vulgar oaths. Some of these bold *citoyennes* could be found at Meot's, in the Palais-Royal, where a former chef served 22 white and 27 red wines, some of them looted from the cellars of émigrés, or eating ices in the boxes at the back of the Assembly, as they heckled the speakers.

Since the revolution, the Palais-Royal had become '*un grand marché de chair*', a market place for flesh, with girls as young as 12 soliciting for custom, no longer harried by the police, who were more intent on catching refractory priests and suspicious nobles. Mr Twiss, strolling in the arcades, noted that there were canaries, white mice and linnets for sale, and what he called 'lion cats' from Syria, with ruffs and tails spread over their backs like squirrels, as well as a lively commerce in libellous books. A two-volume *Private Life of the Queen*, complete with obscene drawings, was selling well. Prevented from exploring further by the 'disturbances', he was disappointed to hear that 'aristocratic plants' had been banned, though just what those were he did not say. At Versailles, he discovered that most of the famous stable of horses had gone off to the army, and that the magnificent furniture, tapestries and pictures that had once made Versailles one of the most beautiful courts of Europe, were soon to be auctioned in 17,182 lots in the Cour des Princes. Lucie, taking stock of the extraordinary changes brought about in the 15 months of her absence from Paris, reproached herself for the 'futility' of her past life. 'A bitter sadness filled my heart,' she wrote later, adding that it made her determined to face the future with 'very great courage'.

On 1 January, Mme de Staël, who in November had given birth to a second boy, Matthias Albert, arrived in Passy to stay with Lucie and Frédéric. Nicknamed by her enemies as the '*Bacchante de la Révolution*', having 'Girondins for dinner, Jacobins for supper, and at night everyone', and 'derided and lampooned by every journal in the capital', she had left her children with her parents in Switzerland. She was frantic with anxiety about the Comte de Narbonne, her son's father, whom she had earlier helped to escape to London, but who was now proposing to return to Paris rather than risk permanent exile and the forfeiture of all his property under the new laws.

Lucie had not seen her father since her departure for The Hague. Returning into Paris early in January 1793, she found him living in a furnished house in the Chaussée d'Antin. He was running up further debts in wine, trying to avoid creditors pressing with unpaid bills for lace, porcelain, carpets and shoes, and unable to leave the city until a trusted messenger he had despatched to Martinique returned with money from his wife. While waiting, he continued to drink, and to gamble, particularly at cards. Arthur was now 42, but appeared considerably younger. 'I looked on him more as a brother than a father,' wrote Lucie. 'No one was ever more noble in manner or more aristocratic in bearing.' They often dined together, when she and Frédéric came into Paris from Passy, and she was delighted by how attached to each other her husband and her father seemed to be.

Arthur, always heedless of his own safety, was taking enormous risks, and spent his days lobbying the Assembly on behalf of the King. He was trying to convince the deputies that it was in their own political interests to remove the royal family to somewhere that they could be protected, where they would then be unable to communicate either with foreign or hidden royalists. But Arthur had enemies, particularly Dumouriez, who had turned against him just as he had once turned against Frédéric.

* * *

Just how possible it was to try Louis XVI was a question that had divided the National Convention for some months. In theory, under the Constitution of 1791, he could not be tried at all, for no court had jurisdiction over a king. But the endless debates had brought to the fore a young protégé of Robespierre, Louis-Antoine Saint-Just, who wore a single gold earring and let his black hair fall loose down to his shoulders. 'One cannot reign innocently,' Saint-Just told the Convention. His case was greatly strengthened by the discovery of an iron chest hidden in the Tuileries, containing damning evidence of royal duplicity. This was the chest in which Frédéric's incriminating memorandum had been put. For a few alarming days, Frédéric and Lucie feared arrest. But the initials on the document – M. de G – were believed to be those of a M. de Gouvion, who had since conveniently died.

On 11 December, the man now known to the French as 'Louis Capet' – the name of an earlier French dynasty – was taken before the Convention and indicted for 'shedding French blood'. Louis had been tubercular in his youth, but his years of passionate hunting had made him physically robust. Locked up in the Temple, he spent his days reading, later saying that he had read 250 books, most of them history or geography, in four months of captivity; he also translated Horace Walpole's *Richard III* into French. The royal family lived quietly and simply, both the King and the Queen giving the children lessons and playing skittles with them. The once plump Marie Antoinette had grown thin and haggard, and her hair had turned completely white; her sumptuous wardrobe had gone, to be replaced by a brown linen gown and lawn cap.

On 26 December 1792, the trial of the King opened before the National Convention. An account of the trial of another king, the English Charles I, was being read all around the hall. Saint-Just rose to argue that Louis should die, not for what he did, but for who he was; Robespierre declared that he had condemned himself by his actions; Danton observed that revolutions could not be made with rosewater. In his diary, Marat noted: 'I shall never believe in the Republic, until the head of Louis Capet is no longer on his shoulders.' Since the trial itself was a violation of France's new criminal code, Louis could, perhaps, have emerged innocent. In fact his guilt had already been decided.

On 4 January, voting began. There were three questions. Was the King guilty or innocent? If guilty, what should the sentence be? And should there be a popular referendum?

There were 749 deputies present when, on 15 January, the first and third questions were settled: a verdict of guilty was passed by 693 of them, the few others abstaining, or being absent, and the decision was taken not to hold a referendum. In the corridors, Arthur, going from friend to friend, still believed that the King could be saved. At nine o'clock on the evening of 16 January, one by one, the deputies went up to the tribune to decide on the King's fate. The hall was full of a whispering, almost festive, crowd. Tom Paine, who had been made a deputy despite his poor French, suggested through an interpreter that Louis should be deported to America; it was wrong, he said, that the 'man who helped my much loved America to burst her fetters' should die on a scaffold. Condorcet voted that the King should be sent to the galleys. Twelve hours later, at nine on the morning of the 17th, there was a majority of 70 for execution. Among those voting for death were the King's cousin, Philippe-Egalité, the

The execution of Louis XVI.

former Duc d'Orléans, who looked miserable and distracted. Lucie and Frédéric had spent the night in the Chaussée d'Antin with Arthur, 'suffering,' as she wrote later, 'an anxiety it is quite impossible to convey'. When the news of the verdict came, they began to discuss ways in which the King might still be rescued. Arthur was convinced that some kind of 'revolt' or popular uprising would take place.

The execution had been set for the morning of 21 January. The King was allowed a brief, miserable farewell from his family. Early that morning, Lucie and Frédéric, back in their house in Passy, stood at the open window, hoping to hear the 'rattle of musketry' that would tell them that 'so great a crime' had not gone unchallenged. They heard nothing. 'The deepest silence,' wrote Lucie later, 'lay like a pall over the regicide city.' The gates to Paris had been barricaded shut. It was a bitterly cold, damp, foggy morning. The King, escorted by 1,200 guards, was brought from the Temple to the Place de la Révolution in a closed carriage along streets lined four-deep with soldiers.The procession lasted two hours. People had been ordered to keep their shutters and doors closed.

'One might have thought that the freezing atmosphere of the day had benumbed every tongue,' observed J. G. Millingen, one of the very few foreigners left in Paris. The drums beat. Louis, composed and dignified, spent a few minutes in prayer before climbing the steps to the scaffold and raising his arm: '*Peuple, je meurs innocent.*' He was 38. Later, Mercier, who had abstained from voting in the Convention, recorded that the executioner gathered up little parcels of hair and fragments of the King's clothing to sell.

At 10.30, still standing shocked and silent at their open window, Lucie and Frédéric heard the familiar noises of Paris resume. Trying to appear calm and composed, they set out on foot for the city. Avoiding the Place de la Révolution, they went in search first of Arthur, then of their elderly friends, Mme de Montesson and Mme de Poix. 'People scarcely spoke,' noted Lucie, 'so terrified were they.' It was as if, she thought, each one of them felt a measure of personal responsibility for the King's death. Frédéric blamed himself bitterly for not having believed that the execution would actually go ahead.

Returning that night to Passy, they found two monarchist friends, Mathieu de Montmorency and the Abbé de Damas, waiting for them. Both men had witnessed the execution, and both now feared that their angry exclamations might have been overheard, which could bring instant arrest. Lucie and Frédéric agreed to hide them until they could get away. In Paris the mayor had decreed that the city would be illuminated for the next few days, despite a full moon, saying that these 'unsettled times' were not the moment to practise economies.

An English visitor, who had attended the execution, was later said to have dipped his handkerchief in the King's blood and sent it to London, where it was prominently displayed. The English reacted with horror and disbelief to the news of Louis's death. The theatres closed and at a memorial service the King's last written words were read aloud to a weeping congregation. 'I leave my soul to my creator . . .' Louis had written. 'I advise my son, should he have the misfortune to become king, to give himself up entirely to his subjects . . .' Until now determinedly neutral, Pitt, calling the execution the 'foulest and most atrocious' act the world had ever seen, ordered Chauvelin, the French Ambassador, to leave England.

The war was spreading, both against enemies coming from across the borders, and within France itself, where counter-revolutionary groups were joining forces. In the Vendée, an isolated, rural area in the south-west, an eruption of popular anger, triggered by the arrival of recruiting agents, and fanned by energetic, popular, 'non-juring' priests, burst out into rebellion against Paris.

Towards the middle of March, as the Convention announced that it intended to hunt down enemies of the state and try them before special revolutionary courts, M. de la Tour du Pin was arrested, together with a cousin, the Marquis de Gouvernet, mistaken by the police for Frédéric. Taken before the Paris Commune, they were repeatedly questioned about the events at Nancy in which Frédéric had played a part and blamed for the death of so many 'good patriots'. Though they were soon released, M. de la Tour du Pin begged Lucie and Frédéric to leave Paris for Bordeaux. He thought that they should hide on his estate at Le Bouilh before crossing the border into Spain or even joining up with the rebels in the Vendée.

Lucie was extremely reluctant to leave her father, whose careless ways and outspoken views terrified her. 'It grieved me profoundly,' she wrote. 'His originality of mind and evenness of temper made him a most agreeable companion. He was my friend, and a comrade also to my husband.' Arthur had been offered and refused the Army of the Rhine as second-in-command under General Custine. He had since been struck off the list of general officers and been warned that he was under suspicion for 'anti-civic activities'. France was at war on all fronts: against possible invasion, against angry Vendéens, against counter-revolutionaries. It was an anxious and sad small party that now set off south, for news had just come that Cécile, alone with her two children in The Hague, had died. And Lucie was once again pregnant, though she was still weak from her miscarriage in Holland.

Taking with them Humbert, Marguerite and Zamore, she and Frédéric set out by carriage through countryside plagued by brigands and hungry and sick deserters from the army. They left behind them a city increasingly short of food and riven by rumours of conspiracy, black marketeering and treachery, with the *sans-culottes* and the new extreme left-wing *enragés* closing in on anyone suspected of uncertain loyalty, and a new tribunal to judge all those said to violate the security of the state; and the Jacobins, day by day, acquiring ever greater power.

CHAPTER EIGHT

Heads Falling like Tiles

Le Bouilh was to have been the finest château of the Bordeaux nobility, a mansion fit for the King, as M. de la Tour du Pin had assured Louis XVI when he complained that there was nowhere of sufficient grandeur in the region for a royal visit. Making the most of the celebrated Victor Louis's presence in Bordeaux to design the city's new theatre in the late 1780s, M. de la Tour du Pin had asked the architect to build him a château consisting of two wings, joined by an arcaded gallery and a cupola, with rows of Doric columns and a terrace above. It was to be vast, and extremely grand. But when, in 1790, he was summoned to Paris to take up the post of Minister for War, all work on Le Bouilh stopped, leaving only one wing finished, the columns unadorned and just 90 of the planned 180 rooms completed. The property had been left with an awkward,

The original designs for the château of Le Bouilh.

somewhat top-heavy air, not so much unfinished as oddly proportioned.

Even so, the building that had risen on the ruins of an earlier fortified medieval mansion was extremely imposing, with high ceilings, drawing rooms decorated with stuccoed garlands, shells and musical instruments, great marble fireplaces and chandeliers. The rooms were filled with tapestries and there was a renowned collection of maps and globes which M. de la Tour du Pin planned to spend his old age studying. Approached along an avenue of poplars, through a curved courtyard with stables and a forge, Le Bouilh also had a neo-classical '*château d'eau*' or pump house. There was a magnificent walled kitchen garden, with pools fed by the hydraulic water system. The French windows in the drawing room opened on to a formal garden and lawns with views over vineyards to the Garonne beyond. It was a tranquil place.

By 1793, the journey from Paris to Bordeaux by the rapid new *turgotine* had come down to five and a half days, but Lucie's carriage moved very slowly, on account of the atrocious roads and her anxiety not to lose her baby. It was mid-April by the time the party reached Le Bouilh, to find the great house shuttered. There were wheelbarrows and scaffolding, abandoned from the day work had ceased. Lucie was nonetheless delighted with the house. There was an excellent library and while Frédéric read aloud to her from books of history and French literature, she spent the evenings sewing clothes for the coming child. She had done well, she reflected, to pay such close attention to her sewing lessons at Hautefontaine. Below, in a semi-basement, were the vaulted kitchens, with rows of copper pots and a flagstone floor, soon taken over by Marguerite and Zamore. Later, Lucie would remember the four months they stayed at Le Bouilh as a time of perfect love for Frédéric, when the feelings they had for each other seemed to shelter them from the surrounding troubles. 'There was,' she would write, 'no flaw in our domestic happiness. None of the disasters that threatened had the power to alarm us as long as we could bear them together.'

Every day, however, it was becoming harder to keep these dangers at bay. The news from Paris was increasingly grim. On 1 July, Arthur was arrested and sent to the Luxembourg, one of some 50 different schools, convents, barracks and former hospitals that had been turned into prisons. Charged with being a ring-leader in a counter-revolutionary plot, Arthur was questioned about how he spent his days. He replied that, since his dismissal from the army, he had led a retired life, with no intimate friends; that he ate at home, sometimes in the company of Condorcet or Camille Desmoulins, and that he occasionally went to a club to play billiards. Among his few possessions seized by the police were maps and atlases, several volumes of travel and history, and a file of letters from General Dumouriez, who, not long before, had deserted and crossed over behind enemy lines.

In the national archives in Paris, written on crumbling paper in faint ink, are the accounts of Arthur's expenditure on food while a prisoner, in orderly columns giving date, item and cost. They make poignant reading. On 15 August, he ordered in soup, a dish of braised cabbage, a roast, a bowl of apricots, coffee, bread, three bottles of wine and four of beer. On the 16th, he asked for soup, roast pork, a 'fresh egg', some more apricots, salad, a dessert, coffee and bread, as well as two bottles of wine and three of beer. Most days, there was wine, beer and brandy; occasionally an omelette, artichokes or spinach. In among the accounts are letters from Arthur's wife in Martinique, begging him to come home, and to stop gambling, as well as his own petition to the Ministry of War for a pension, in recompense for 38 years of service, including a 'harsh' tour of duty in the colonies.

Soon after hearing of Arthur's arrest, Frédéric and Lucie learnt that M. de la Tour du Pin had also been detained, and that orders had been issued to sequestrate his property in Saintes and his château at Tesson. Papers relating to his case gave his height at 5 feet 2 inches, his nose as aquiline, his eyes as blue and his face as 'full'; they stated that he wore a wig. Since a 'rascally' local lawyer in nearby Saint-André-en-Cubzac had spread a rumour that Le Bouilh itself was subject to a lien, Frédéric feared that the château would also be impounded. Anxious about the coming baby, they accepted an invitation from their friend M. de Brouquens, provisioner of the army in the south, to move into a small isolated house he owned called Canoles, set in the middle of the Haut Brion vineyards on the edge of Bordeaux, with useful tracks leading off in every direction, and no village nearby. They moved in September, taking with them Zamore and Marguerite.

* * *

It was not an altogether prudent move: the Terror was coming closer, and Bordeaux was in the eye of a new storm. All during the spring and summer of 1793, a battle had raged in the Convention in Paris between the Montagnards, the fanatical men of the Mountain, and the milder Girondins sitting below them, their disagreements fuelled by accusations of corruption and speculation, by plots and counter-plots, the protagonists tearing each other apart in furious speeches that lasted long into the nights. By late July, the Girondins were on the run, many in hiding in and around Paris. A new nine-man Committee of Public Safety under Robespierre and Saint-Just was turning itself into the most concentrated and powerful state machine France had ever known, with a programme of economic regulation, military mobilisation and an official ideology to replace the haphazard and anarchical politics of the street. 'Let us be terrible,' declared Danton, 'so that the people will not have to be.' Networks of spies and informers fanned out to provide information about 'non-juror' priests, hidden nobles and hoarders of food, while revolutionary mobs of *sans-culottes* ransacked the countryside for concealed sacks of grain.

With the murder of the Jacobin Marat in his bath by the young Charlotte Corday, and the sense of panic created by the news that Toulon had opened its harbour to the British fleet, the Jacobins were able to rouse French citizens to war against the 'enemies of the people', both within France and across its borders. It was enough, now, simply to be a priest, a nobleman, a royalist, a would-be émigré or even a 'false republican' to become a 'suspect'. 'Those who want to do good in this world,' Saint-Just told the Convention, 'must sleep only in the tomb.' The *levée* in March had brought 300,000 men under arms; metallurgical factories were turned over to producing cannons, rifles, balls and shot; church bells were melted down; the Tuileries gardens were dug up and planted with potatoes. Under thc Tree of Liberty in Paris, the Committee of Public Safety made a bonfire of newspapers and 'counter-revolutionary speeches' before debating a proposal to burn all libraries.

Reading the Paris papers, when they were able to get them at Saint-André-en-Cubzac, and questioning travellers arriving from the north, Frédéric and Lucie learnt that the guillotine in the Place de la République, placed by the side of the statue of Liberty, cap on head, spear in hand, shield by her side, was working at such speed that the *tricoteuses* were splattered with blood as they knitted, and came away with their feet wet. During a 47-day period, citizens lost their heads at the rate of 30 a day, sped on their way by the implacable Public Prosecutor, Fouquier-Tinville, borne to the guillotine in carts through crowds of jeering onlookers. On 13 September 1793, there were, according to *Le Moniteur*, 1,877 people awaiting trial. There was very little mercy. To process the guilty faster, Robespierre longed for a blade which could cut several heads at a single go.

For the first time in the history of France, the prisons contained not petty thieves and murderers but counts, seamstresses, lawyers, tanners, maids, priests, wig-makers, marquises, schoolteachers, all held together in a way they had never been before. Once the daily list of names for the guillotine had appeared, and people knew that they were not on that day's, they then threw themselves into a frenzy of games, cards, madrigals and charades. There had been an expectation that the women would cry and collapse: for the most part they remained stoical and calm. Frédéric's father was reported to be insisting on keeping his wig and dressing every day as for his former life.

In August, while Lucie and Frédéric were still at Le Bouilh, Marie Antoinette, '*veuve de Louis Capet, ci-devant roi des Français*', gaunt and white-haired, had been moved to the Conciergerie, the 'anti-chamber of the revolutionary tribunal' from which few people emerged alive. She was held in a small damp cell, whose only light at night came from a lantern hanging outside her barred window. She spent the days crocheting and reading. The case against the 'Austrian she-wolf' was that she was immoral, corrupt, greedy, lustful and treacherous; she had, said her accusers, squandered France's fortune and held orgies at Versailles and committed incest with her son. Her trial opened on 14 August.

The following day, the 15th, M. de la Tour du Pin was called as a witness. Asked if he knew the Queen, he bowed low before her, and said that he had indeed had the honour of knowing her for many years; but, he added, he knew no ill of her. Challenged to admit that it was on Marie Antoinette's orders that he had sent his son to massacre the 'brave soldiers' of Nancy, he denied it. Accused of running down the army on her orders, he replied that he was not aware that he had run it down at all. Frédéric's father was questioned for several hours; his manner remained calm and dignified. Not everyone called to give evidence was as loyal.

Right up until the moment of the verdict, Marie Antoinette herself seemed to believe that she would still be ransomed and sent to her nephew in Austria. When informed that she was to go to the guillotine, she was asked whether she wanted to say anything. She shook her head. At eleven o'clock next day, her hands bound, wearing a white piqué dress, white bonnet and black stockings, she was taken to the Place de la République in an open cart, staring straight ahead of her, silent and calm. On Marie Antoinette's feet were a pair of kid shoes, carefully preserved through 76 days of captivity. She stumbled as she mounted the steps to the scaffold, but remained composed.

After the execution, Hébert, in the *Père Duchesne*, wrote of his joy at having seen with his own eyes, 'the head . . . separated from the fucking tart's neck'. In his memoirs, the Comte de Saint-Priest would say that though the Queen had been weak and not very bright, 'she was never evil or cruel . . . she never betrayed France and . . . at moments of great danger she showed a kind of magnanimity'.

Marie Antoinette on her way to the guillotine. Drawn from the life by David.

In Bordeaux, lists of those who had gone to the guillotine were passed from hand to hand. Every day that Frédéric ventured into the city he returned with the names of friends and relations who were either already dead or in prison, awaiting their trials. There was Mme du Barry, Louis XV's last mistress, betrayed by her black servant, who had so often accompanied her to balls dressed as a hussar with a plumed cap. In the cart carrying her to her death, Mme du Barry was reported to have wept and called for help. There was M. de Genlis, husband of the royal tutor, father of Lucie's friend Pulchérie, who was now in prison herself; the Prince d'Hénin, Frédéric's aunt's wayward husband; and Philippe-Egalité, whose betrayal of his cousin the King had not been enough to save him. Philippe-Egalité's servant, Edward, whom Lucie had encountered leading his troop of black volunteers near Hénencourt, followed his master to the guillotine.

Titled soldiers, many of them men with whom Frédéric had served in America, were being rounded up and sentenced for treachery and incompetence. General Custine, M. de la Tour du Pin's good friend, went to his death with the words 'I die calm and innocent'; the Duc de Lauzun, who had given Lucie her court doll, insisted on finishing a bottle of Burgundy and a plate of oysters in the tumbril. 'I am dying at a moment when the people have lost their senses,' Larousse, one of 21 leading Girondins sentenced to death, told his judges, 'you will die the day that they recover them.' In prison, Condorcet, the celebrated author of *Mathématique Sociale*, took poison. On hearing her sentence of death, Olympe de Gouges, who had dreamt of a female Convention and a world in which women would be equal, cried out: 'I wanted so much to be someone.' For her own troublesome political fervour, the strong and articulate Mme Roland followed her soon after; her husband, hearing that she had been guillotined, slashed his own throat with a dagger. It was becoming harder every day for Lucie and Frédéric to believe that they would ever escape alive.

* * *

When the Terror had begun, Bordeaux, of all French cities, was perhaps the most open to reform. Grown rich on wine, slaves and commerce, its ships had carried eau-de-vie, lard, beef, linen and the famed red wines of Haut Brion and Lafite to America, bringing back coffee from Martinique, indigo from Saint-Domingue, and sugar, to be refined in warehouses around the city. Its real fortune, however, lay in wine. Foreign merchants most often came to collect their cargos themselves, the English taking the reds, the Dutch some of the whites, and the Bordelais keeping the best Graves and Sauternes for themselves. At Le Bouilh, M. de la Tour du Pin had planned extensive vineyards.

Drawn by legislation that was both liberal and protectionist, hundreds of foreigners had settled in the city, bankers, shipbuilders and merchants, who spent lavishly on handsome, unadorned buildings in the creamy local stone along the quays. From their balconies, they looked out across the city to the surrounding hills, where many had also built châteaux or country mansions to escape the summer heat. The skies were blue, the river was majestic, and on fine evenings the Bordelais strolled along avenues planted with limes and poplars. It was in his Château de La Brède near Bordeaux that Montesquieu wrote *Les Lettres Persanes* and *De l'Esprit des lois*, and where, inspired by him, *belles-lettres* and scientific and artistic academies flourished throughout the 18th century. In Bordeaux, 'as in Paris', noted one visitor, 'it was *le bon ton* which counted'. Because of the lively overseas trade, most Bordelais knew a great deal about what was going on in other parts of the world.

In Bordeaux, after the fall of the Bastille, a revolt, inspired by and paid for by the Girondin municipality, had flared up against over-zealous local Jacobins, and a federalist army was mobilised to defy the dictatorship of Paris. Young men drilled on the slopes of the Château-Trompette, where jugglers, cock fights and exotic animals had once entertained the crowds. But the federalists were weak and there were too few of them, and Bordeaux soon set up its own *Garde Nationale*. But it took a milder form than in Paris. Ruled in an orderly and fair way by a kind of spontaneous council and a moderate city administration, it had at first remained largely impervious to the revolutionary violence stirring the north. By the end of 1791, the old *noblesse de robe*, which had dominated Bordeaux for the previous three centuries, had for the most part retired to their properties in the surrounding hills, where, like M. de Brouquens, they hoped to sit out the troubles. Their places had been taken by rich merchants who kept up the tradition of orderliness and dealt calmly with the overthrow of guilds and corporations. On the tracts of land confiscated from the Church, they built new roads and planned imposing new buildings. Right up until the end of 1792, a cult of revolution, efficient and even somewhat aristocratic, seemed to have taken over smoothly from that of religion.

But the tranquillity could not last. Bit by bit, a city driven for centuries by work, by trade, by its vineyards and intellectual circles, began to alter. Merchants, their fortunes hit by the revolt of the slaves of Saint-Domingue, by the war with England and by the closure of the Chamber of Commerce, ceased to trade. Victor Louis's colonnaded theatre was turned over to plays of a patriotic, exhortatory nature, in which kings were portrayed as rapacious and tyrannical, nobles as frivolous, priests as artful and hypocritical. And a rival authority, incoherent and domineering, began to emerge in opposition to the municipality through Jacobin clubs and popular societies. The day came when two refractory abbots, seized in a nearby town, were brought to Bordeaux and had their throats cut in the courtyard of the old archbishop's palace. The statue of Louis XV was pulled off its pedestal in the Place Royale. In January 1793, not long before Lucie and Frédéric fled Paris, battles raged between the Jacobins, raised from among the discontented and the disoccupied, and the more moderate Girondins, who were forging links with the federalists across the south, resolving, like the rebels in the Vendée, to defy the dictatorial rule of Paris.

Bordeaux would not experience the same atrocities as federalist Lyon, where the ex-Oratorian Joseph Fouché and Jean-Marie Collot d'Herbois sent 1,700 men, women and children to their deaths, many of them mown down by cannon and shovelled into mass graves. But the temerity of the Bordelais would not be allowed to go unpunished. As Fouché remarked, 'Terror, salutary terror, is the order of the day.'

The first agents of the terror, the *représentants en mission*, sent from Paris to quell the rebels, bring orthodoxy to disaffected areas, and complete the violent de-christianisation of France by smashing altars, images of saints and crucifixes, arrived soon after Lucie and Frédéric moved to Canoles. These agents were heckled, threatened and forced to retire to the safety of a nearby town.

But on 16 October 1793, the day of Marie Antoinette's execution, a revolutionary army of 2,000 *sans-culottes*, with 200 cavalry, under the command of General Brume, escorting three men from the Convention, marched into the city. Not one of the three was vicious in the way of Fouché or Collot d'Herbois: Claude Ysabeau, a former vicar, was ponderous but could be generous; Marc-Antoine Baudot, a doctor, though an ardent republican, was optimistic by nature; while Jean-Lambert Tallien, volatile and given to violence, was also capable of clemency. What made them lethal, however, was their choice of men to carry out their orders: a clock-maker called Bertrand, better known for his thieving than for his clocks, was appointed mayor, and Jean-Baptiste Lacombe, a crafty and corrupt teacher and militant Jacobin with a long thin face, was named President of the Revolutionary Tribunal. It would later be said that between them, and the scores of spies, fanatics and criminals they recruited, justice in Bordeaux would be turned into a 'tribunal of rapine and blood'. It would also be remarked that neither Bertrand nor Lacombe was actually a native of Bordeaux, and that the Bordelais had needed such men to sully their hands, so that their own might stay clean. From Canoles on the outskirts of the city, Lucie heard of 'this army of butchers, dragging a guillotine behind them', getting greedier every day.

On 23 October Bordeaux woke to the sounds of the guillotine going up in the former Place Dauphine, recently renamed the Place Nationale, and to the shouts of town criers beating their drums and calling out that the time had come to cut off the heads of traitors. What surprised everyone was the ease, the lack of all opposition, with which the Terror had arrived. It was as if, by being submissive, many Bordelais believed they might yet escape retribution. Lucie, observing the speed with which they capitulated, remarked with scorn that a little courage might well have been enough to keep the 'fanatics' at bay. When an order went out for all citizens to surrender their weapons, and people hastened to obey, she noted that it did not 'seem to have occurred to anyone that it would have been far braver to use them in self-defence'.

Lucie and Frédéric discussed leaving for Spain, but Lucie was now in the ninth month of her pregnancy, and to reach Spain they would have had to pass through the French lines. Once the baby was born, they decided, Frédéric would seek shelter somewhere else.

Late one night, just at the moment that Lucie realised that she was going into labour, news reached Canoles that people in nearby country houses – 'bad citizens' – were being rounded up and taken to one of the makeshift detention centres in former seminaries, convents and public buildings, to await hearings before Lacombe. Dr Dupouy, come to attend the birth, was said to be on the list of wanted men. Their host de Brouquens sent a trusted servant to stand on the road leading to Canoles, in order to be able to warn of approaching danger so that Frédéric could escape from the house and slip away by another path through the vineyards. The birth was fortunately both quick and easy. No sooner was the baby born, a girl, and given the name Séraphine, than Frédéric left for Le Bouilh, where he intended to stay secretly while they worked out what to do. Dr Dupouy, not daring to go home, was found a hiding place in an alcove in the baby's room. 'I suffered more from fear,' wrote Lucie later, 'than from the actual pain of my daughter's birth.' Not knowing what 'fate held in store' or when she would see Frédéric again, all she could think about was recovering her strength 'so that I might deal with whatever might arise'.

Three days later, de Brouquens returned to Bordeaux, where the revolutionary committee was already at work rounding up suspects in the middle of the night to send before Lacombe, whose phrase 'we are fascinated by your case' had already become synonymous with execution. The Terror had taken its first victims, borne away to the Place Nationale by tumbril at ten each morning, the mounted guards of the escort, the '*cavaliers de la mort*' clattering over the cobblestones. The terrified Bordelais kept their windows shuttered. One of the first to die was a good friend of de Brouquens's, the much-loved and respected former mayor, François Armand de Saige, accused of having incited the Girondins to rise against the Jacobins. Four other prominent Girondins, who had come to take refuge in the city, were hunted down; two managed to commit suicide before they were caught, but one, Barbaroux, succeeded only in wounding himself and was dragged dying to the guillotine.

Bordeaux, like Paris and the other cities in which the executioners were at work, had its own *tricoteuses*, knitting and singing at the foot of the guillotine. Actors, accused of being subversive, merchants suspected of hoarding, priests caught hiding in attics or cellars, or behind altars, were dragged before the tribunal. De Brouquens himself was arrested on his return from Canoles, but managed to convince Lacombe that it was not in anyone's interests to do away with the official provisioner of food for the army in the south.

Three nights later, just as de Brouquens was going to bed in his house in Bordeaux, guards arrived with orders to escort him immediately to Canoles, insisting that they needed to examine any papers and documents held there. In vain, he protested that he seldom used the house. Lucie was asleep when a terrified servant shook her awake to say that soldiers were on the way, a friendly guard, loyal to de Brouquens, having found a way to send word ahead. Before disappearing into the night, the servant slipped a packet under Lucie's pillow, containing money that de Brouquens had put on one side for emergencies such as this. Taking care to conceal the packet from Séraphine's nurse, a girl Lucie deeply mistrusted, she pushed Humbert's bed across the alcove in which the doctor crouched, and got back into her own bed, taking the baby with her. In the drawing room, Zamore hastily laid out a meal of pâté, wine and liqueurs. And then, Lucie wrote, 'we all waited resolutely for the arrival of the enemy'.

Half an hour later, she heard the clatter of clogs on the flagstone floors. She lay stiff with fear, clasping Séraphine to her, listening as the men combed the house for incriminating documents. At last, she heard a voice ask: 'Who is in the bedroom?' And then there was silence. For the next two hours, the men searched the house, pausing only to cat and drink; but they did not open Lucie's door and nor did they discover Dr Dupouy. When they left, de Brouquens told her that he had said that he was looking after the sick and delicate daughter of an old friend, who had just given birth to a baby.

Once Dr Dupouy had recovered from the fright of the nocturnal visit, he was happy to spend his days instructing Lucie in the rudiments of surgery and midwifery. In return, she taught him embroidery, dressmaking and knitting, accomplishments learnt from the servants with whom she had spent so much of her lonely childhood. Many years later Dr Dupouy told her that his new skills had saved him from acute boredom when he had been forced to hide for many months with a peasant family in the Landes, whose kindness he was able to repay by making them shirts and stockings. In the evenings in Canoles, while Lucie sewed, the doctor read aloud from the newspapers, though what they contained often terrified and appalled them.

It was in the local paper that one evening Lucie and Dr Dupouy came across a full account of Marie Antoinette's trial. It brought home to them the true extent of their danger. The callous nature of the proceedings dismayed them – when a juror called Antonelle had heard the sentence, he had ordered a celebration dinner of foie gras, thrushes, quails *au gratin*, sweetbreads, a pullet, Sancerre and Champagne – but they were more alarmed when they read further. The Public Prosecutor, according to the report, had asked M. de la Tour du Pin where his son was living. Frédéric's father, for whom lying was repugnant, replied that he was staying on his estates near Bordeaux. A warrant for Frédéric's arrest had apparently already been despatched to Saint-André-en-Cubzac.

Frédéric had taken the precaution of keeping a strong horse stabled and ready at Le Bouilh. As soon as he learnt of his father's indiscretion, he set out, in thunder and driving rain, disguised as a merchant in search of supplies of grain, planning to make his way to his father's other estate at Tesson. There a trustworthy servant called Grégoire and his wife were still living in the house, though it had been sequestrated after orders that the 'nation be put in possession of the property of traitors'. Shortly before dawn, as he rode past a posting stage not far from Tesson, he was offered shelter by a villager. Sitting by the fire, he found an elderly man: this turned out to be the local mayor. They discussed the high price of grain and cattle. Frédéric was then asked to produce his passport. Finding no visa on it for any forward journey, the mayor ordered Frédéric to remain where he was until morning, when the local Municipal Council could be informed. Frédéric kept his head. Very calmly, he walked to the door, as if to check on the weather. Surreptitiously stretching out his hand, he silently unhooked his horse's bridle from the door-post, leapt into the saddle and galloped off before the mayor had time to get to his feet.

It was clearly too dangerous for him to go to Tesson. At nearby Mirabeau, he went in search of a former groom of his father's, a man called Tétard well known to the family. Tétard now kept an inn. But he also had a wife and small children, and feared for their lives should Frédéric be discovered in their house. However, he had a brother-in-law, a locksmith called Potier who, though married, had no children, and was willing to hide Frédéric in a small windowless room, little bigger than a cupboard, in return for a generous sum of money. All day, while men worked in the adjacent forge, separated from Frédéric's room only by a plank of wood, he had to stay absolutely still and in the dark. Only once the men had gone home was he able to join Potier and his wife for dinner. From time to time, Tétard came with the local gazette and news of Lucie and the children. Frédéric thought of making his way to join the rebels in the Vendée, but such was the strength of their royalist fervour that he doubted they would welcome a man who had remained for so long working for the Constituent Assembly. In any case, once it became known that Frédéric was in the Vendée, all chances of either his wife or his father surviving the rage of the revolutionaries would be lost.

The months passed very slowly. Bit by bit, Lucie regained her strength and with it her determination. On her walks around Canoles, she occasionally came across traces of hiding places, clearly occupied by others like her, on the run from the guards scouring the countryside for suspects. She would leave out food, to find it gone next day. By late winter she began to worry that her continuing presence at Canoles might endanger de Brouquens, who had been placed under permanent house arrest in the city. One of the few people who was occasionally able to visit her was a young relation of de Brouquens's, a M. de Chambeau, who was also in hiding, having found refuge in the house of a man called Bonie in the centre of Bordeaux. Bonie was, to all appearances, the perfect Jacobin, an ardent member of a political 'section' and never to be seen without his *sans-culotte* jacket, clogs and sabre. But he was also a good-hearted and generous man, and when de Chambeau told him of Lucie's predicament, with her father and father-in-law both in prison in Paris, her husband in hiding, and two small children to care for, he offered her rooms in an empty, dilapidated apartment overlooking a small garden and hidden away behind a wood store in the heart of the old city.

As a hiding place, the rooms in the Place Puy-Paulin were excellent. Lucie moved in, taking with her Marguerite, constantly ill with malaria, Zamore, who as a free slave ostensibly waiting to join the army could come and go as he wished in Bordeaux's large community of blacks, Humbert and Séraphine and their nurse. She also took with her the cook, who was able to get a job with the *représentants en mission* and bring back not only news but food to cook for the household. Restless and energetic, it was not long before Lucie was looking for ways to keep occupied. Hearing that an Italian singer called Ferrari was in Bordeaux and that he was a fervent royalist, she invited him to give her singing lessons, which had been interrupted by the years of revolution. It served to pass the long hours of idleness and fear. Even in the shadow of the guillotine, some semblance of normal everyday life went on. And it was indeed in its shadow: from her room behind the wood store, Lucie could hear the clatter of hooves from the horses pulling the tumbril to the Place Nationale, the rattle and crash of the plunging blade, the shouts and songs of the crowd, as heads and trunks were thrown into baskets ready to be taken for burial in communal pits in the cemetery of Saint-Seurin.

Much of her time was spent trying to find food to feed her household. A tough new form of rationing had been introduced, with taxes imposed on all goods and punishment by death for all who sought to evade them. Since farmers now received less money for their grain, meat and vegetables than they could produce them for, many preferred not to sell them at all. Famine and malnutrition spread around the city. Ration cards were issued for bread that was black and glutinous, and lists were posted on every house giving the names of all their inhabitants and what they were entitled to. They were written on paper with tricolour edges under the words 'Liberty, Equality, Fraternity or Death'. Lucie wrote their own names as faintly and illegibly as possible, to be washed away by rain, and was able to boost Zamore's meagre ration by obtaining double for herself as a wet-nurse since she was still breast-feeding Séraphine.

In the streets, carefully dressed as a peasant woman wearing a waistcoat, with kerchiefs on her head and round her throat, a disguise she rather enjoyed, she marvelled at the docility of the Bordelais, standing in endless queues for inferior food, doing nothing to protest when the cooks of those in power pushed their way to the front to collect freshly baked white rolls and choice cuts of meat for their masters' tables. There was something about the perilousness of their situation that seemed to galvanise her, and she would later speak of these months in hiding as times when it was still possible to laugh. Lucie herself was in a fortunate position. A local farmer's wife from Le Bouilh, devoted to the family, brought provisions in the baskets strapped on to her donkey when she came twice a week to sell potatoes and cabbages in the nearby market. Frédéric kept in touch by sending letters concealed in a loaf of bread, brought to Bordeaux each week by a young boy. Le Bouilh had been sequestered but before it had been sealed shut, the caretaker had managed to remove all the best linen and small objects, and these she sent, a few at a time, by trusted courier to Bordeaux, along with a supply of logs for firewood.

* * *

Jeanne-Marie-Ignace-Thérésia Cabarrus was regarded by many as the most beautiful woman of the revolution. She was small, not much over 5 feet, but she had silky black hair, very white skin, a slightly turned-up nose, perfect teeth and a ravishing smile. Even Gouverneur Morris, never lavish with his praise, remarked approvingly on her sprightliness. Born in July 1773 to a French merchant father and a Spanish mother, she had arrived in France at the age of 5 to be educated by the nuns. Added to her many other charms was that of her voice, which retained a barely perceptible foreign lilt. When only 14 she made a good marriage to the Marquis de Fontenay, a councillor at the Paris parliament; the following year, not yet 16, she gave birth to a son, Théodore. Lucie, who was three years older than Thérésia, had been introduced to her when they were both girls. 'Nothing,' she wrote later 'could dim the radiance of her wonderfully fair skin.'

It was in the studio of the fashionable portrait painter Mme Vigée-Lebrun that Thérésia met Jean-Lambert Tallien, the son of a modest maître d'hôtel in an aristocratic household. Tallien was tall, with curly fair hair and regular, pleasant features, and he was then working for the printer Pancoucke. By their second meeting, not long afterwards, he had become secretary to Alexandre de Lameth and was moving rapidly up the political ladder. By then Thérésia too was causing a stir, both for her looks and for her salon, frequented by Lafayette and La Rochefoucault. With the revolution, Tallien found his true vocation. Though not as brutal as Danton, nor as outrageous as Marat, nor as cold as Robespierre, he shared most of their ideals. By the summer of 1792, he was clerk of the court to the Paris Commune, with the job of signing warrants for arrest.

At 25 one of the youngest deputies elected to the Convention, sitting on the Mountain with Robespierre, Tallien was sent, in the spring of 1793, to Tours to speed up recruitment for the army and to set up committees of security and surveillance. The city was in chaos: Tallien proved competent and not overly cruel. When *représentants en mission* were needed for rebellious Bordeaux, he seemed an obvious choice. In the special *représentant* uniform of long, tight-fitting blue redingote, tricolour sash, plumed hat, cross-belt and sabre, he cut a glamorous figure as he rode into the city behind General Brume.

The summer of 1793 was extremely hot, and Tallien took a few days' holiday in the Pyrenees. Here he again met Thérésia, on vacation with her brothers and uncle from their house in Bordeaux, where she had taken refuge from the violence in Paris. She was divorced – she had been one of the first to take advantage of the new laws – and after falling out with her uncle she moved into a large apartment in the Cour Franklin where she filled her study with musical instruments, books, an easel and paints, and planted her terrace with orange trees. At the end of November, Thérésia was arrested and taken to the Fort du Hâ, a grim, humid 15th-century fortress; Tallien was able to have her freed. After this, the two were inseparable, and though Thérésia kept her rooms, in which lived Théodore and her maid Frenelle, she spent much of her time in the sumptuous apartments of the *représentants en mission* in the former Grand Séminaire, enjoying their unrationed food and the excellent wine confiscated from unhappy Bordeaux merchants. Though by the middle of the winter of 1793 Bordeaux was a ghost city, many of its inhabitants reduced to penury and malnutrition, those in power lived well.

And there were also occasions of jollity, even if of a forced nature. Virtue and terror, Robespierre maintained, were both aspects of the self-improvement and moral regeneration he believed should govern every part of life, both private and public. The festivals marking the revolutionary calendar were growing ever more theatrically absurd and opulent, with specially commissioned revolutionary oratorios, regiments of choristers, triumphal arches and the releasing of hundreds of white doves. For the *Fête de la Raison* on 10 December, a cortège of Bordeaux dignitaries, followed by 40 women feeding their babies, 100 girls dressed in white, soldiers, schoolchildren and the members of Jacobin clubs, processed slowly through the centre of the city. Behind followed mummers, grotesque figures parodying the Church and the nobility, a dwarf sitting on a donkey and shouting benedictions, criminals and prostitutes decked out in church vestments, later burnt, along with books of 'chicanery and superstition' in a gigantic auto-da-fe in the Place de la Comédie. Soon after, encouraged by Tallien, Thérésia gave a rousing speech on education, for which she wore an 'amazon's' dress of bright blue cashmere, with yellow buttons and red velvet facings, her hair cut *à la Titus* under a fur-edged scarlet bonnet. Thérésia did not always take the revolution very seriously. From her hiding place, Lucie followed her exploits.

Whether because of the great prosperity of Bordeaux, or because Lacombe and Bertrand were exceptionally venal, Bordeaux's Terror had been from the start as much about money as about survival. The revolutionary committees were more than happy to cut deals, accepting fortunes in exchange for acquittals. The Hôtel de Ville was already filled with works of art plundered from Bordelais churches and the nobility. 'Tallien,' the historian Michelet would write later, 'sold lives, while his mistress served behind the counter.' Tallien and Thérésia were soon tainted by charges of extortion and greed, and stories would later circulate about Thérésia's great fondness for certain kinds of jewellery. On returning to Paris after the revolution, the Marquise de Lage de Volude would recount a particularly damning incident.

Mme de Lage de Volude had been a lady-in-waiting to Marie Antoinette. She was in Bordeaux, nursing her dying mother, when she heard of the September massacres. Rather than return to Paris, she had gone into hiding in Bordeaux. One day, a friend of hers who had been to call on Thérésia returned saying that she had spotted a pile of unsigned passports on Thérésia's desk. Learning that Thérésia had a particular fondness for antique cameos, Mme de Lage de Volude bought an exceptionally fine piece and had it mounted with some diamonds that she still had left. She secured an introduction to Thérésia and told her that she desperately wanted to join her husband, who had managed to escape to America. Before leaving, she presented her cameo. A few days later, a passport arrived; Mme de Lage de Volude was able to leave France.

But there was another side to Thérésia. She had a generous heart and she was determined to save as many people as she could from the guillotine, whether they paid her or not. The Comte de Paroy, whom she counselled and protected, was only one of many who later stated that they had remained alive only through her help. By early 1794, it was widely reported that she was having a softening influence on her lover Tallien; and that she was not afraid.

The Terror, early in 1794, was reaching a peak in Bordeaux, *ci-devant* nobles, priests, peasants, merchants following one another rapidly to the guillotine in the Place Nationale, where people were beginning to complain of the persistent smell of blood, despite constant washing of the scaffold. On the lists of the accused in the Gironde were some of the charges that sent them to their deaths: 'suspicious opinions', 'nostalgia for the *ancien régime*', 'weak nature, concerned only with the abolition of the monasteries', or simply 'consorting with the nobility'. As in Paris, a few young women won reprieves by stating that they were pregnant, but once they had given birth they went to the scaffold. From her hidden room, Lucie listened in terror to the roll of drums before each beheading, and the crash of the blade that followed. 'I could count,' she wrote later, 'the victims before seeing their names in the evening papers.' When arrests of English and American merchants and their staff began, Lucie, light-skinned and fair-haired, feared for her own life each time she left the house.

Frédéric's own position was also becoming ever more precarious. When Potier, in whose house he was still hiding, came one day to Bordeaux to buy iron for his foundry, he happened to pass through the Place Nationale as a woman mounted the scaffold. Enquiring as to her crime, Potier learnt that she was a '*ci-devant noble*'; and this he found reasonable. But when the next to climb the steps was a peasant, whose crime was that of sheltering a nobleman, the full horror of his own position hit him. 'In that poor man's fate,' wrote Lucie, 'he saw his own.' He hastened back to Mirabeau and told Frédéric that he was to leave instantly.

For a while, Frédéric was able to hide in his father's house, Tesson, with the faithful Grégoire and his wife, but this came to an end when the local municipality sent men to draw up an inventory of the house's valuable contents. Grégoire found him temporary shelter with the sister of another trusted groom. When her nerve went, Frédéric returned to Tesson, where, each day, there were rumours that the municipality would soon be taking over the château for its own use. Though barely recognisable in his rough peasant clothes, Frédéric was too well known locally to believe that he could survive for very long without hiding.

Lucie, listening to the talk around Bordeaux, was well aware of Thérésia's relationship with Tallien. Sensing that the time was fast running out for her family, she decided to take a bold and terrifying step. She sent a letter to Thérésia. Without giving her name, she wrote: 'A woman, who met Mme de Fontenay in Paris and knows that she is as good as she is beautiful, asks her to accord a moment's interview.' A friendly reply came back, with an invitation to call. 'In a state of agitation difficult to convey', Lucie went to Thérésia's apartment, where she was greeted with great affection. Lucie told her that she wished to have the order of sequestration on Le Bouilh lifted, in order that she and her family could live there quietly. The only way to achieve this, said Thérésia, was for her to ask Tallien in person.

That evening, accompanied by the equally nervous de Chambeau, Lucie went to see Tallien, listening in terror to his approaching carriage, unmistakable in the silence of Bordeaux's dark and shuttered streets. When Thérésia announced that Tallien was ready to receive her, Lucie was unable to move her feet forwards until Thérésia gave her a little push. Tallien, she was sure, would be her 'executioner'. The meeting did not go well. As soon as Tallien realised that Lucie was the daughter of General Dillon and the daughter-in-law of the man who had addressed the 'widow Capet' as 'Your Majesty', he made a gesture with his hand indicating beheading, saying: 'All the enemies of the Republic will have to go.' Even in extreme adversity, Lucie was never undignified. 'I have not come here, citizen,' she replied, 'to hear the death warrant of my relatives . . . I will not importune you further.' Walking home, she reflected that if Tallicn would not protect them, then 'death seemed inevitable'.

The net, bit by bit, was drawing tighter. Sweeps, sudden descents by the revolutionary soldiers on houses at unexpected times, were gathering in people ever closer to Lucie. She now left the house only under the cover of darkness. M. de Chambeau, picked up while dining with friends, was taken off to the Fort du Hâ, where he spent a terrible 28 days, believing that each day was his last, before finding himself mysteriously released, the tribunal apparently ignorant of his aristocratic connections. Opposite the Fort du Hâ lived a teacher called Saint-Sernon and his daughter, who wrote the names of those arrested and sent to the guillotine on a blackboard and held it up for the prisoners to read from their barred windows. There was a community of men and women in hiding all over the city, seizing moments of gaiety, meeting fleetingly to talk, play cards and exchange food and news, but it was shrinking all the time. Every day now, there would be friends among those going to the guillotine, people who had been certain that they would never be touched. A young royalist, M. de Morin, to whom Lucie had fed omelettes made from eggs smuggled to her from Le Bouilh and truffles stolen by her cook from the stores of the *représentants en mission,* and who changed his hiding place every night, was picked up along with others in the small monarchist association they had formed and taken straight to the guillotine.

Then, one day, Lucie herself was recognised. She had discovered that unless she had a certificate to prove ownership of her house in Paris and her state bonds, she would lose them all. Taking Bonie, her Jacobin protector, with her, she went to the municipality, in which sat a dozen clerks, all wearing, to her 'extreme distaste', red bonnets. Bonie urged them to hurry, saying that his young friend was nursing a baby and could not be kept waiting. To Lucie's horror, when her case was heard, the officer in charge insisted on reading out loud every line of the certificate. When he came to the word Dillon, one of the clerks looked up and asked her whether she was by any chance a sister or niece of 'all those émigrés of that name who are on our list'? She was about to confess that she was, when a man who happened to be in the room interrupted, insisting loudly that she belonged to quite another family. As she left with her certificate, he whispered in her ear that he had known her great-uncle, the Archbishop.

Ever more desperately, Lucie was trying to think up ways to escape. For a while, she wondered whether she might not pass herself off as her Italian singing master's daughter, taking her children back to Italy, with Frédéric disguised as her servant. As her agitation and fears for Frédéric and the children grew, so Lucie's milk began to dry up.

* * *

M. de Brouquens was still under house arrest, but from time to time Lucie was able to pay him a visit. One day, she happened to be standing by his table when her eye fell on an item in the morning paper. A ship, the *Diana* of Boston, was to leave Bordeaux in eight days' time. It was indeed news. For the whole of the past year, 80 American ships had been anchored in the Garonne, trapped there by France's state of war with its neighbours. As Lucie hurried from the room, M. de Brouquens asked where she was going. 'I am going to America,' she replied.

Now that she had a plan in mind, Lucie's sense of anguish and uncertainty seemed to leave her. She was still only 24, but it was she rather than Frédéric who was emerging as the more resolute and forceful of the two. Rarely buffeted by self-doubt or tormented, as he was, by the greyer shades of meaning, it was as if her solitary and loveless childhood had equipped her for such a challenge. She now had just eight days in which to have Frédéric fetched back from Tesson and kept hidden and safe, and to obtain passports and visas for them all, while keeping everything secret from the spying nurse. Marguerite, unable to shake off the malaria which had kept her shivering and feverish for many months, would stay behind, Lucie fearing that the long sea journey might prove fatal to her. Nor would Zamore accompany them. The question as to who could be trusted to go for Frédéric was solved by Bonie, who insisted that he would go himself. Thérésia, whom Lucie asked for help in securing the passports, pressed her to hurry, for Tallien had been denounced for 'excessive moderation' towards the Bordelais, and was likely to be recalled to Paris at any moment. He was to be replaced by an austere and pitiless 18-year-old called Marc-Antoine Jullien, who was known as the 'shadow of Robespierre'.

All seemed to be going well when Lucie, accompanying de Brouquens – who had unexpectedly been released from house arrest – to Canoles, encountered Tallien on his way to visit the Swedish Consul on the outskirts of the city. With all the grace and courtesy of a noble of the *ancien régime*, Tallien accepted her story that she and her family needed to visit Martinique, where she had financial matters to attend to, and promised to have papers prepared for her. 'I can today,' he told her, 'make amends for the wrongs I have done you.' Two hours later, his secretary Alexander brought her the documents requesting the municipality of Bordeaux to issue passports in the name of Citizen Latour and his family. (Thérésia told her that she had threatened never to see Tallien again unless he produced them.)

The first hurdle had been crossed. Next day Bonie set off to find Frédéric, taking with him a spare set of identity papers, borrowed from an unsuspecting friend, telling everyone that he was going in search of grain, more plentiful in the Charente-Inférieure than in the Gironde. The journey was uneventful but, dressed in his full *sans-culotte* costume, Bonie had difficulty in convincing the terrified Grégoire and his wife of his loyalty until he thought to show them a slip of paper, written in Lucie's hand, which he had fortunately stitched into his jacket. The return journey was slow, for Frédéric was not strong, and the months of anxiety and concealment had taken their toll. Lucie, having calculated the hour and the exact time she expected to see him, went down to the quay where the travellers were to arrive. But Frédéric did not appear.

The river grew dark, the curfew was sounded and she could think only of her stupidity at having entrusted the life of the person she loved most in the world to a revolutionary she barely knew. Desolate, she spent a sleepless night listening for sounds that he might have arrived by some other means, imagining Frédéric arrested, recognised and dragged to the scaffold. The house had never seemed to her so still. 'If I had an enemy,' Lucie wrote later, 'I could not wish for him any worse punishment than the mortal agony I endured.'

When, early next morning, the sly nurse arrived to dress Séraphine, she brought with her a message. Bonie was at home and wished to see her. Very casually, keeping her excitement under control, Lucie finished dressing and then, saying that she was going out, flew along the corridors to another part of the house, where Bonie had a secret room. Bonie, it turned out, had hired a fishing boat, and brought Frédéric to Bordeaux by a different route. Though they had exchanged letters, Lucie had not seen Frédéric for six months. 'In every lifetime,' she wrote later, 'there are a few luminous memories that shine like stars in the darkness of night . . . We were happy, and death, which we felt so very close to us, no longer frightened us, for it was possible again to hope that if it struck it would strike us down together.'

Time, however, was pressing and there was still much to do.

A friend of Lucie's father, a ship-broker, agreed to arrange with the captain of the *Diana* for their passages. Surreptitiously, so as not to alert the nurse, Zamore was given the task of packing the linen and silver, brought from Le Bouilh and hidden locked in cupboards, into crates, ready for removal to the *Diana*. With them went 50 bottles of Burgundy, a few jars of potted goose, a small case of bottled jam and some potatoes, the only food they had managed to gather for the journey. A piano went too, no one imagining that Lucie could survive in the New World without one, or indeed that she might find one there.

Then one morning, to their great alarm, the whole family had to present themselves to the clerk at the municipality for their permits to be exchanged for passports. Dressed in elegant but well-used clothes – for Lucie was to pass herself off as an Englishwoman fallen on hard times – they shrank back into the dimly lit corner of the busy passport office, relying on Bonie, who accompanied them, to smooth over all awkwardness. There was a terrible moment later that evening when Lucie took the passports for their final visas to Thérésia, who was to give them to Tallien, only to find her in tears, Tallien having already departed in disgrace for Paris, and her own immediate future far from safe. It was now up to his colleague, the cold-blooded Ysabeau to provide the indispensable visas; and these were only obtained because Tallien's secretary Alexander was a kind man and was able adroitly to hurry the process through by presenting the passports late at night, when Ysabeau was about to dine and did not bother to scrutinise what he was signing.

* * *

Lucie, Frédéric and the two children were now free to depart. The timing could not have been more fortunate. The next morning the harsh young Jullien, little more than a boy, reached Bordeaux. Announcing that Ysabeau and Tallien had both been far too lenient, too lacking in true revolutionary ardour, he set about hunting down the remaining Girondins and nobles, hidden away in attics, in concealed vaults behind wine vats, in cellars, in cupboards and on their country estates. In the coming weeks, he would send 71 people to the guillotine.

In Paris, where Arthur and Frédéric's father remained in

prison, the Terror was approaching its most murderous phase. Robespierre and Fouqier-Tinville, dreaming of a commonwealth of virtue, were despatching 30 to 40 people to the guillotine each day. Many of them were taken away in groups: magistrates one day, tax inspectors the next, or even several generations of one family. 'Heads,' remarked Fouquier-Tinville, 'are falling like tiles.' One by one, the men who had shaped the revolution and sent others to the guillotine were becoming victims themselves of stronger and cannier political forces.

There was Hébert, who went to the scaffold with 19 of his friends. After Hébert came Danton, fighting to the bitter end with great booming tirades of rhetoric, who left behind him a 16-year-old wife; and Camille Desmoulins who, when asked his age by the Tribunal, replied: 'The same age as the *sans-culotte* Jesus when he died, the critical age for patriots.' Accused of spending his time with aristocrats and of praising his friend Arthur Dillon with the words that the General was 'neither Royalist, nor Republican, nor Jacobin, nor Aristocrat, nor Democrat: he was simply a Soldier', Desmoulins could have gone back on his words. Instead, he spoke out. 'I am proud,' he declared, 'even if I am alone to oppose the injustice of Rome for the services of Coriolanus.' There are, observed Robespierre, 'only a few serpents left for us to crush'.

Paris's places of detention were filled to overflowing, some 7,000 people crammed into humid and unhealthy barracks, convents and former palaces, prey to typhus, dysentery and influenza. The surviving *ci-devant* nobles continued to keep fear at bay by charades, cards and rhyming couplets, and by singing, even on the way to the scaffold, faithful to *bon ton* until the end. As the historian Taine would later observe, the nobility went to their deaths with dignity, ease and serenity, all the savoir-faire that was second nature to them.

By March, M. de la Tour du Pin was imprisoned in the Conciergerie, on the Quai de l'Horloge, next to the Palais de Justice, where he continued to receive visitors in formal dress, wearing his wig. Arthur had also been moved to the Conciergerie, accused, in the wake of his friend Camille Desmoulins's death, of a conspiracy involving Lucile, Desmoulins's 23-year-old wife. Dillon, said his accusers, had never 'ceased to conspire against the republic'. Arthur knew himself to be guiltless. 'If,' he had written to Desmoulins earlier, having handed over all his papers willingly to the authorities, 'a single suspect line is found, then I will accept the harshest treatment.' Later, an anonymous writer would describe Lucie's father during his weeks in prison as drinking heavily 'and when he was not drunk, he played backgammon'. Lucile Desmoulins, arrested soon after her husband's death and charged with the same conspiracy, had been to plead with Robespierre for Desmoulins's life; Robespierre had been witness at their wedding and was the godfather of her 18-month-old son, Horace.

In Bordeaux, there was one last gathering with the friends who had saved Lucie and Frédéric's lives: M. de Brouquens, described by the Comte de Paroy as 'the best man who ever lived', M. Meyer, the Dutch Consul in whose house Frédéric had been hidden, and Thérésia, who wept as they left. Lucie found the parting from Marguerite particularly hard, and both she and Frédéric dreaded the thought of what might befall their fathers, shut up in Paris's revolutionary prisons. And then the moment came when, feigning an afternoon stroll with the children in the public gardens, they climbed into the dinghy sent by the captain of the *Diana* to collect them at the end of the Quai des Chartrons. At the very last minute, M. de Chambeau, who had just learnt that his father had been denounced by a servant who had worked for him for 30 years and had been arrested with papers revealing that his son was in hiding in Bordeaux, was able to get a passport as their legal representative and went with them. 'There is no doubt,' wrote Lucie, 'that the heave of the oar with which the sailor pushed us off from the shore was the happiest moment in my life.'

CHAPTER NINE

Mrs Latour from the Old Country

Sturdy, with a broad beam and a single, tall mast, the 150-ton *Diana* was one of hundreds of boats built in dockyards dotted along America's eastern seaboard to transport cargo between Europe and the New World in the 18th century. It was small and uncomfortable, with just four sailors under a young and inexperienced captain and an older mate, both from Nantucket. Lucie, Frédéric and the children shared one cramped cabin, M de Chambeau had another. But in the first relief and euphoria of escape, none of this mattered. They were in any case immediately distracted by being stopped, on three separate occasions, by guards doing sentinel duty along the Garonne, each time risking discovery and fearing for their lives. On the last inspection, Lucie was forced to hand over a lamb, secured at the last minute by M. de Chambeau as food for the crossing, to a covetous official.

It was when the *Diana* finally pulled out into the open sea that the full discomfort of their journey hit them. The Atlantic passage westward, particularly during the equinoctial spring gales, was renowned for its battering winds and immense seas. It was not unusual for a boat to lose its mainmast and to have its sails shredded. Soon after they left the estuary, one of the sailors slipped from the masthead to the deck below and though he survived, his injuries kept him confined to his hammock. Frédéric was an appalling sailor. With the first big waves, he took to his bunk and for the next 30 days seldom left it, able to keep down only weak tea and biscuits soaked in wine. Apart from Lucie's 50 packing cases and the piano, the *Diana* carried no cargo, which made the rolling more pronounced.

What chiefly afflicted Lucie, however, was hunger. The captain had also had trouble finding food for the voyage and as the days

The *Diana*. 1794.

went by and their rations grew smaller she worried that her milk was again drying up: Séraphine, who slept happily all day, soothed by the swaying of her bed, was beginning to look shrivelled and pinched. 'I could see,' wrote Lucie later, 'my daughter shrinking visibly.' At night, fearing that as the boat rolled she might squash her, Lucie fashioned a strap with which to tie herself firmly to the bed-frame. Having slept all day, Séraphine spent the nights awake.

The young captain, for his part, was terrified of Algerian pirates, of privateers, the semi-legal pirates encouraged in wartime to seize the ships of enemy nations, and of the French navy, known to be patrolling the seas off the French coast in search of royalists escaping revolutionary justice. Determined to avoid a life of slavery in Algeria, or summary justice in a French court, he insisted on heading due north to Ireland, into pitching seas.

One black, rolling day, what they most dreaded happened: the *Atlante*, a French man-of-war, drew alongside and only the rough seas prevented its sailors from boarding the *Diana*. Instead, the French captain ordered the *Diana* to follow him back to Brest, making his point by sending a couple of cannon shots across their bows. The *Atlante*, as Lucie knew, had recently escorted a party of escaping French back to Brest, where they had all been guillotined. But a dense fog came down, the captain of the *Diana* dawdled, and as soon as the *Atlante* was out of sight, he hoisted all sails, and they scuttled away towards the north-west. The *Diana* was now blown to the Azores, where Lucie and Frédéric begged the captain to put them ashore, thinking that they would pick up a boat for England. Though angry at his refusal, Lucie would later feel nothing but gratitude. Just at the time that they would have reached England, an expedition of French royalists set out for Quiberon Bay on the coast of Brittany, planning to confront the revolutionary army. Frédéric, she reasoned, would have felt bound to join the ill-fated raid and would certainly have died with the other aristocrats in the ensuing massacre.

For Lucie, life on board was taken up with looking after the seasick Frédéric, breast-feeding Séraphine and keeping Humbert from falling overboard, with the help of the obliging M. de Chambeau, and trying to learn something about life in America from the entertaining young cook, Boyd. Finding her long hair unmanageable, she took the kitchen scissors and, to Frédéric's displeasure, chopped it off and threw it overboard, and with it 'all the frivolous ideas which my pretty fair curls had encouraged'. She spent her days propped up in the galley, open to the winds but warmed by the stove, listening to Boyd's descriptions of his Boston childhood, and trying to stretch out the ever-dwindling supplies of dried haricot beans and ship's biscuit. She was disgusted to see weevils squirm out the moment it was mixed with water. Her gums began to bleed. Humbert, constantly in tears, begged her for food that had been eaten days before. 'I could not rid myself of the fear,' Lucie wrote later, 'that I would see my children die of hunger.' Day after day, strong westerly winds beat the *Diana* back so that often they seemed merely to be marking time. For ten days, an impenetrable fog blanketed out even the riggings.

The captain had a little terrier, Black, to which he was much attached. On 12 May, after 60 days at sea, on a morning of warm weather and calmer seas when Lucie and Frédéric had taken the children up on deck, Black was seen to behave very strangely, racing up and down, barking and licking their hands. Suddenly, out of the mist, a pilot boat came in sight. Unbeknownst to the captain, the *Diana* lay just outside Boston. Even better, the pilot had with him a large, recently caught fish, some fresh white bread and a jar of butter. As they devoured the food, the *Diana* was towed out of the choppy seas and into a passage of flat water, surrounded by green fields with dense, flowering vegetation. After so many days of blue and grey light, of salty air and the sucking sound of the waves, the colours and the smell of land from the pine, spruce and balsam that grew along the shores were overwhelming. 'Like the changing decor on a theatre stage,' Lucie wrote, 'the friendly land appeared, waiting to welcome us.' Four-year-old Humbert, who had suffered acutely from the terror of their escape and the loss of 'the fine bread and the good milk of earlier days', was beside himself 'with transports of joy'.

Just outside the harbour, the captain dropped anchor and had himself rowed to shore, promising to find them lodgings. While they waited, supplies of fresh fruit and vegetables were brought out to the boat; these were soon followed by parties of Frenchmen, clamouring for news from home. Most were small tradesmen and artisans, ruined by the revolution, who had left France several years earlier in search of work, but whose sympathies remained firmly with the revolutionaries. Lucie's first human encounters in the New World, after almost two months of solitude and pitching seas, were far from pleasant. She found herself harangued by belligerent Frenchmen, angry that she had so little news to give them, and angry too that another aristocratic family had escaped the guillotine. After a few heated exchanges, the Frenchmen pulled away for shore. Lucie and Frédéric were left to contemplate their new home. Boston's fine port lay spread out before them, with its tidy rows of ships and its church spires rising behind tall buildings, and the hills beyond.

* * *

Long before dawn the next morning, Lucie was awake and dressed. As soon as the dinghy was ready, they had themselves rowed to shore, sad to leave the crew with whom they had spent so many close and anxious weeks. The long wharf in Boston, where passengers were landed from their boats, was a confused, noisy mass of people. There were porters heaving crates, stevedores pushing carts and loading bales, all speaking a dozen different languages as they jostled their way along the quays, between horses pulling drays, pigs and cows being herded on to boats, carriages carrying merchants, passers-by come to watch. Above screeched gulls.

Behind were narrow, crooked roads, most of them unpaved, lined by brick or pine clapboard houses painted in different colours, two or three storeys high, with small roof terraces and gardens with apple and cherry trees in blossom. Between the low green hills, cows and sheep grazed. In the evenings, and on Sundays, people strolled up and down a public walk they called the mall, half a mile of turf and gravel, shaded by elm trees. It was very ordinary, and welcoming, in ways that Lucie had forgotten existed. Only someone who had been through the anguish she had lived through, she noted, would ever 'be able to fully appreciate my joy when I set foot on that friendly shore'.

By 1794, America's Atlantic seaboard had a population of 2 million people. The promise of the New World, and the conflicts of the old, had between them brought English, Scots, Welsh and Irish, Germans and Dutch, Swiss, Swedes, Jews from all over Europe and slaves from Africa. Driven from their homes by dispossession and penury, they had come to settle and to trade in ports and cities along a thousand miles of coastline, importing manufactured goods, olive oil, salt, wine and brandy from Europe, cloth, tea and spices from India, sugar, coffee, indigo and rum from the West Indies, trading them on to the different states, often by river or canal.

The first thing Lucie and Frédéric did on landing was to be taken by the young captain to the best inn in town, where he had ordered the sort of food of which they had been so long deprived. Lucie would later write of this meal that it gave her 'a pleasure so vivid that it surpassed any pleasures I had known until then'. After this, he took them to a lodging house run by a Mrs Pierce, with the help of her mother and her daughter, where their drawing room looked out over Market Square, the liveliest part of town, and their bedroom across to the shipyards. The three Bostonian women took the family to their hearts, the grandmother taking charge of the precocious Humbert, who was already speaking some English, while the daughter looked after Séraphine. That first night the household was woken by scratching at the door. The captain's small terrier Black was discovered shivering outside, having escaped from the *Diana* and swum to shore. Despite his fondness for his dog, the captain agreed to part with her.

Lucie and Frédéric were taken to meet Mr Geyer, one of Boston's richest merchants; Mr Geyer spoke good French, but his wife and daughter did not, and they were delighted to find that Lucie's English was excellent. Having followed the news from France closely – the French papers, brought over by packet boat and merchant vessels, arrived with a delay of about six weeks, when their contents were picked up and printed by papers across the country – they were extremely sympathetic towards their visitors, insisting on believing that Lucie's cropped hair had been cut by the executioner in preparation for the guillotine. 'By the evening of the first day,' wrote Lucie, 'we felt as settled there as if no grief or anxiety had ever troubled our lives.'

Though Lucie did not again encounter the unfriendly French who had rowed out to the *Diana*, Boston was full of French émigrés, come over in two waves, the first as voluntary exiles in 1789 after the fall of the Bastille, the second those escaping like Lucie and Frédéric from the revolutionary guillotine. More recently had come French colonial settlers from Saint-Domingue, fleeing the slave uprising of Toussaint L'Ouverture. By 1794, some 20,000 French men and women had reached America and were eking out precarious livings as dancing masters, chefs, teachers of French and music or on small farms. On his travels around America in 1791, Chateaubriand, who wrote lyrically about the primitive natural life, described coming across a former chef, M. Violet, teaching Indians to dance near Albany, half-naked men with rings in their noses and feathers in their hair, prancing while he fiddled, in full dress of apple-green coat, his hair powdered, taking his pay in beaver skins and bear ham.

Some of these French exiles, preceding Lucie and Frédéric and landing on shores they had imagined hospitable and full of possibilities for work and happiness, had quickly become disillusioned. In place of a libertarian arcadia, they had found a hard land, energetic and tolerant, but vulgar and materialistic in ways they had not expected. They felt alien, sick with longing for France and for the graceful, polished society they had known; they were poor, ill-adapted to hard work, and fearful of the moral and cultural dangers that might lie in all this rugged prosperity. As Gouverneur Morris noted, they had chosen to disregard Benjamin Franklin's warning that the point about America was that it was neither rich nor poor, but 'rather a general happy mediocrity'.

And some of the émigrés had been swindled. From 1788, American speculators had been sending their agents to Paris to sell frontier lands, spinning tales of 'smiling farms', rivers full of fish, forests teeming in game, and vegetables of unimaginable size growing wild. As the revolution took hold, the offers had begun to look more enticing. A month after the fall of the Bastille, a poet called Joel Barlow, working for the Scioto company and helped by a crooked Englishman called Playfair, had succeeded in selling title deeds to 100-acre plots for the planned new town of Gallipoli on the Ohio, a river known, they claimed, for its enormous fish and the way that the trees along its banks spewed sugar from their trunks. A thousand émigrés set sail, to find the company on the verge of bankruptcy and in any case without rights to the land it had promised. Though half of them did indeed spend the first winter crouched on the banks of the Ohio, they were 'peruke'-makers, artists and shopkeepers, utterly unsuited to life in the wilderness. They found the Indians menacing and the cold appalling. A visitor to the colony wrote that the inhabitants had a 'wild appearance and sallow complexions, thin visages and sickly looks'. Within six months the project collapsed, leaving them destitute and reviving all the Abbé Raynal's fears that the New World was no place for civilised Europeans, and that America's peculiar climate had caused its people to degenerate, morally, physically and intellectually. All through the terrifying Parisian winter of 1793 and spring of 1794 other Frenchmen, fleeing the Terror, had been duped by similar schemes, lured by similarly chimerical promises, most of them doomed to bankruptcy.

* * *

Mr Geyer had offered Frédéric and Lucie a farm that he owned 18 miles from Boston. But Frédéric, who spoke very little English, had dreamt of settling nearer to French-speaking Canada. Soon after reaching Boston, they received a letter from the Princesse d'Hénin, who had taken refuge outside London with a group of aristocratic émigrés, with letters of introduction to General Schuyler, one of the heroes of the American War of Independence. Schuyler lived on his estates in Albany in upper New York State. The general was married to Catherine van Rennsselaer, descendant of the Dutch 'patroon', a diamond merchant who had first colonised the area and owned over 700,000 acres of land; they were rich farmers known for their hospitality towards earlier French travellers. In reply to a letter, the Schuylers wrote back, pressing Frédéric and Lucie to visit Albany and saying that it would not be hard to find them a farm. Before leaving Boston, Lucie held a sale of some of the possessions Zamore had hastily crammed into crates and which she believed would find no place in their new lives. Elegant dresses, laces and fabrics went to fashionable Bostonian ladies, who still looked to Paris for new designs, as did Lucie's piano, which had made the journey safely. The money raised was converted into bills of exchange. Having been so careful to bring all these things with her, Lucie now parted with them without regret, reflecting that the life that awaited her would involve 'conditions comparable with those of peasants in Europe'. She appeared little bothered by their loss.

Early in June, with the weather good and the trees in flower, Frédéric, Lucie and the children, taking the devoted M. de Chambeau and the terrier with them, left for the 165-mile trip to Albany. They had decided to travel by road rather than by sloop up the Hudson, hoping to get a clearer picture of their new country. Much of their route lay through forest of oak and pine, the road little more than a cutting opened by felling trees and leaving them lying along the side to mark the way. They saw squirrels the size of small cats and 'thickets of flowering rhododendrons, some of them purple, others pale lilac, and roses of every kind'. In the many creeks and rivers grew water-lilies in full flower. 'This unspoilt nature enchanted me,' wrote Lucie. From time to time they came across clearings where farmers had put up wooden frame houses and were growing crops of peas, beans, potatoes and turnips. One night, stopping in an inn, Lucie heard 'a stream of French oaths' coming from the adjoining room. M. de Chambeau, not having been warned about this local custom, had been woken by another traveller climbing into his bed.

It was only once they had left Boston that Frédéric told Lucie the news that had just arrived on a boat from France. He had waited, hoping that the journey might distract her.

On 13 April, not long after they had set sail from Bordeaux, Arthur had been brought before the Revolutionary Tribunal in Paris and accused of conspiring against the republic, of being in league with Pitt in trying to put Louis back on the throne and of having helped France's enemies to escape. 'It is Dillon,' said *Le Moniteur* in its coverage of the trial, 'who is the spirit behind all the counter-revolutionary plots.' In court, Arthur had answered each absurd accusation clearly and with courage. Not long before, he had written to Robespierre: 'I have always fled before even the shadow of any plot, and I have despised and detested plotters . . . An incorruptible patriot like yourself should love and respect the truth.' Arthur was one of 27 people tried that day, most of them accused with him of the 'Luxembourg conspiracy', a fanciful and convoluted tale of intrigue and spying. Arthur and Desmoulins's widow, the 23-year-old Lucile, were, so their accusers claimed, guilty of trying to raise money to finance a movement to assassinate true patriots. The jury had retired for three hours. They returned with 19 verdicts of death. Arthur's and Lucile's were among them.

At six o'clock that same day, their hands tied behind them, they had been put into tumbrils and taken to the Place de la République. The frail 77-year-old former Bishop of Paris went with them, even though he had loudly renounced his faith. When the carts arrived to take them on their last journey, Lucile asked Arthur to forgive her for being the cause of his death. 'You were only the pretext,' he replied. At the foot of the scaffold, her nerve gave way: she shrank back from the executioner's grasp and asked Arthur to go first. Arthur climbed the steps, calmly removed his cravat and shouted '*Vive le Roi*' before the blade fell. Lucie had been expecting his death; but it did not lessen the grief she felt.

This was not the only painful news to reach the travellers. On 28 April, Frédéric's father, who had so enraged the Public Prosecutor by referring to Marie Antoinette as '*Sa Majesté*', followed Arthur to the guillotine. He was executed on the same day as his first cousin, the distinguished Marquis de la Tour du Pin, who had been a member of the Assembly of Notables. With them went three carts of 'nobles, foreigners, indolent men and hired orators of the combined powers', as the French papers described them.

Paris, according to the newspapers reaching America, was running out of food, its prisons were overflowing, its inhabitants cowed by terror. 'The very paving stones smell of blood,' wrote one reporter, 'and the river itself seems to run red.' The city, once indescribably noisy, now lived in 'the silence of the grave'. Lucie and Frédéric learnt that in Bordeaux, in order to make heads fall faster, a guillotine had been erected to sever seven heads at a single stroke. It was enough, now, to 'slander patriotism' to be found guilty. When the victims died before they reached the scaffold, as did some who had tried to commit suicide, their corpses were guillotined. The Duc de Châtelet was unfortunate: he had tried to pierce his heart with a broken bottle and arrived at the guillotine drenched in blood, but alive.

With her mother, father, father-in-law, sister-in-law all dead, her great-uncle and grandmother in England, Versailles deserted, the nobility murdered or scattered, the world in which Lucie had spent the first 19 years of her life had vanished.*

* * *

The busy trading port of Albany, known to the Indians as Muhattoes and to the early Dutch settlers as Oranienburgh, was a pleasant, small town of some 1,500 houses. It had 6,000 inhabitants, 2,000 of whom were slaves. One of the oldest settlements in North America, Albany lay on the banks of the Hudson, and many of its houses had gardens down to the water. There was a market place, a prison, a town hall, a new bathing establishment and a number of English and Dutch churches. Its streets were regular, straight and paved with cobblestones, and in front of their Dutch-style shingled and gabled houses, people had planted lime trees, often marking the birth of each new baby with a sapling. In summer, families sat out on their porches, the Dutch fathers smoking pipes, watching the cows returning from their communal pastures to be milked. On the sandy hills surrounding the town, and all along the banks of the Hudson, raspberries and blueberries grew in abundance among the willows and wild roses. The land herc, wrote an early missionary, was very like Germany, and the wine made from local grapes was good; but there were rattlesnakes, 'variegated like spotted dogs', and they made a noise like crickets. The forests that lay all around, of maple, poplar, chestnut and oak, were full of skunks.

Albany had begun as a frontier town, a starting-off point for journeys into the uncharted interior, and a depot for the furs

* With M. de la Tour's death, Frédéric assumed his title.

brought from the interior to be shipped down the Hudson. While the 1783 Treaty of Paris had ended hostilities between the British Crown and the American secessionist states, leaving Canada in British hands but accepting a new border through the Great Lakes, it had no direct bearing on the standing of the Indian tribes who had been placed on the American side of the border. Congress had quickly set about entering into treaties with the Mohawks, the Senecas, the Cayugas and the Onondagas.

In October 1784, anxious to protect their frontiers from the British in Canada, it negotiated with representatives of the Iroquois Six Nations Confederacy, the 10,000 or so Indians whose ancestral hunting grounds stretched from the Hudson Valley to Lake Erie, persuading them to relinquish lands on which military forts were built in return for guarantees that they could continue to occupy a portion of their homelands. But only in theory: in practice, revolutionary soldiers, returning from the north with tales of fertile alluvial soil and well-watered fields, abandoned their stony and debt-ridden New England farms and migrated up the Hudson, arriving in groups by river in summer and overland by sled in winter, to occupy lands 'subleased' in deals to last 'while water runs and grass grows', which effectively appropriated most of the native territory.

The Iroquois gathered, talked, complained; but to no avail. By the summer of 1794 the forests where Indians had once hunted plentiful elk, deer, bear, raccoon, porcupine and wild mink were largely in the hands of ever-growing numbers of settlers, who chopped down the trees and planted potatoes, peas and corn, let their pigs roam to eat the wild plants favoured by the Iroquois and dammed their streams for sawmills. The settlers, noted Jefferson's friend, Elkanah Watson, a man interested in the development of the land, 'are swarming into these fertile regions in shoals, like the ancient Israelites seeking the land of promise'. An early casualty had been the rare white beaver, which had totally disappeared. In the luggage of one group of settlers, migrating north through Albany, came a printing press, destined, noted *The Albany Register*, 'to shed its light abroad over the western wilds'.

A key figure in the negotiations with the Iroquois was General Schuyler, who for almost 40 years had known and dealt with the Indians, and whose ancestor, Pieter Schuyler, had taken a Mohawk chief to Queen Anne's court in London and brought him home wearing a light blue suit, trimmed with silver lace. Though loved and respected by the settlers for his generosity in restoring the town after a terrible fire in 1792, Schuyler was a controversial figure, having long taken the view that it was Albany, and not Congress, that should have the final say on all matters dealing with the Indian lands. Exploiting the trust the Iroquois had placed in him, Schuyler had flouted Congressional laws, dispossessing the Indians in favour of the farmers and merchants. His dream was to open a canal from the Hudson river to the Great Lakes, to make the flow of goods faster and cheaper. Like the van Rensselaers, to whom the Schuylers were closely related, the General was enormously rich, owning over 10,000 acres of land, along with sawmills, hemp and flax mills, and a stately Georgian mansion in the centre of the town. His daughter Elizabeth was married to Alexander Hamilton, currently Secretary of the Treasury in Philadelphia.

Both the Schuylers and the van Rensselaers welcomed Lucie and Frédéric warmly. The General, a tall, commanding, somewhat austere man, declared that he would regard her as his sixth daughter. Mrs Rensselaer, one of his own five daughters, was not much older than Lucie and spoke good French; she was an invalid, confined for months on end to an armchair, where, clever and well informed, she liked to sit talking to visitors. Like Lucie, she had been reared on the Enlightenment. She immediately impressed Lucie with her grasp of the intricacies of the French troubles, much of it learnt from a close reading of the newspapers. Her insights, remarked Lucie, into the 'vices of the upper classes and the follies of the middle classes' were far more interesting and acute than those of most of her French acquaintances. In the drawing room of the Schuyler mansion, with its hand-painted wallpapers and stucco garlands so beloved of French 18th-century decorators, its Brussels carpets, yellow damask coverings and silver and glasses brought back from Europe, where young girls did their embroidery and played the pianoforte, Lucie rediscovered some of the conversation and the music of her early childhood. After five years of uncertainty and fear, it all seemed very reassuring.

As they did not want to live in Albany itself, Lucie, Frédéric and the children went to stay with a farmer not far away, hoping to learn from him the rudiments of their new life. M. de Chambeau apprenticed himself to a carpenter in nearby Troy, returning each week to spend Sundays with them. He had just heard that his own father had been guillotined. While Frédéric searched for possible properties to buy with the money they had managed to bring from France, Lucie, determined to acquire all possible skills, rose before daylight each day to help the farmer's wife, making clothes for the family as well as black mourning dresses for herself. Opposite the farmhouse, on the other side of the river, lay the road to Canada, and on it stood a large inn, where news, gazettes and sale notices were posted. In the early mornings, Lucie explored her surroundings, marvelling at the speed with which the local vegetation seemed to grow, riding on horseback through fields of Indian corn that stood taller than her horse, and forcing herself to cross the river over a floating bridge of logs tied together, which, until she mastered them, greatly unnerved her. Abhorring cowardice, 'I was careful not to tell anyone my fears', she practised making the crossing until she could do so without hesitation.

Albany had two weekly papers, *The Gazette* and *The Register*, both full of advertisements for Irish linens, women's gloves, Malaga raisins, all the luxury items newly arrived from Europe, together with notices about escaped slaves and missing horses. Early in August, a 'composition' mounted by a travelling Italian artist reached Albany. It depicted Louis XVI taking leave of Marie Antoinette in the Temple, with, as *The Register* put it, 'a countenance very expressive of his feelings', and the actual guillotining of the King, his head falling, 'the lips which are first red, turning blue', the whole performance orchestrated by an invisible machine. Even so far from France, it was impossible to escape the revolution. It was in the local paper that Lucie read about the execution of the King's sister, Mme Elisabeth. 'Blood flowed everywhere,' she wrote. 'Nor could we see an end to it.'

In September, Frédéric found them a home. It was a newly built farmhouse, 4 miles outside Albany on the road between the new settlement of Troy and Schenectady, down which passed a constant stream of wagons loaded with pelts. The lands surrounding Troy were the former hunting grounds of the Mohawk Indians, and it was at Troy that the wide, sluggish Hudson met the greatest of its tributaries, the Mohawk. The farm was exactly what they wanted. It had 150 acres of crops, a quarter of an acre of vegetable garden, an orchard with 10-year-old apple trees producing excellent cider, and extensive woods and pastures. Since the owners did not wish to move until the first snows, Lucie and Frédéric rented a log cabin in Troy itself, only recently incorporated as a village but growing rapidly, with new potash works, paper mills and tanning yards. Lucie was in the courtyard, chopping the bone of a leg of mutton ready for the spit, when she heard a familiar deep voice behind her saying that he had never seen mutton so magnificently speared. It was Talleyrand. With him was the Chevalier de Beaumetz, jurist and reviser of the French penal code and a former deputy to the National Assembly, who after the attack on the Tuileries had escaped Paris disguised as a travelling merchant. Both men had exchanged their silk and ruffles for clothes more suited to the American outback, the elegant and fastidious Talleyrand barely recognisable in rugged hunting shirt and waistcoat. They had brought with them a present of Stilton cheese.

Talleyrand had initially left France for England with Danton's help in September 1792 on the pretext that he would write a report on France's relations with the European powers, intending to return if it seemed safe for him to do so. But when Louis XVI's iron chest was discovered, it was found to include letters from Talleyrand to the King, assuring him of his loyalty: next day, in the Convention, Talleyrand was accused of treachery and his name added to the list of proscribed émigrés. His position in London, however, was precarious. Lord Grenville, Minister for Foreign Affairs, disliked his manner, saying that it was cold and arrogant. Shunned by both the court and the government, Talleyrand spent his time with Fox, Sheridan and other members of the opposition.

And after Parliament passed the Aliens Bill in 1793, putting the French émigrés under police surveillance, Talleyrand, perceived as an influential and dangerous intriguer, received orders to leave. His first thought had been to remain somewhere in Europe, but one by one Russia, Prussia, Tuscany and Denmark closed their doors to him; even his good friend Mme de Staël, agitating on his behalf with the Swiss cantons, was unable to persuade any one of them to take him in. That left America. In March 1794, just as Lucie and Frédéric were leaving Bordeaux, Talleyrand set sail from England in the *Penn*, bound for Philadelphia, taking with him his friend de Beaumetz and a valet. He told Mme de Staël that he was not sorry to leave a country where 'life was so royally disagreeable'.

Though his first impressions of America were far from favourable – he wrote to Mme de Staël that 'if I have to stay in this country a year I shall die' – and though George Washington, committed to a policy of neutrality towards the new French government, refused to receive him, Talleyrand found friends in Philadelphia. He was soon paying calls, risking the disapproval of the town when he was rumoured to have taken a mulatto mistress. With the heat of May, and fears of a possible epidemic of yellow fever, he had decided to make a journey through Maine and Upper New York state, in search of land to invest in and hoping to meet Lucie and Frédéric, having heard that they had escaped Bordeaux successfully. In February, one of General Schuyler's other daughters, living in London, had written to her sister Elizabeth, introducing Talleyrand and de Beaumetz. 'To your care, dear Elizabeth, I commit these interesting Strangers . . . who left their country when Anarchy and Cruelty prevailed.'

Though genuinely impressed by the vast empty landscape, Talleyrand was a man of cities. The long report he wrote of his travels was full of poetical descriptions of forests 'as old as the world itself'. But the farmers he met along the way struck him as 'lazy and grasping . . . without the slightest trace of delicacy'. Revealingly, as he rode along, his mind turned to building 'cities, villages and hamlets . . .'. The Indians, he added, were smelly and useless.

The travellers had picked up news of Lucie as they passed through Boston, where they found that she had been much admired for her excellent English and unaffected manners and the fact that 'she sleeps every night with her husband'. As Talleyrand told Mme de Staël ruefully, such devotion was necessary in America, where 'illicitness' did not find favour. Tracking her down to the wooden cabin in Troy, the travellers had brought with them an invitation to dinner from the Schuylers, and promised to return next day to taste her roast mutton.

Reaching Albany later that afternoon, they found the General waiting for them in the garden, brandishing a newspaper and calling out: 'Come quickly, there is exciting news from France.' The local paper, reporting events that now lay many weeks in the past, carried an account of the fall of Robespierre and the abrupt end of the Terror in Paris: how, in June, Tallien, ousted from power and desperate to save Thérésia, who had been arrested and was about to be tried, had mounted a challenge to Robespierre's dictatorial powers over the Convention; how he had won sufficient backing to have Robespierre arrested, together with Saint-Just; how Robespierre had shot himself but survived, though half his jaw had been crushed; and how, on 10 thermidor, 11 July, Paris had woken to find the guillotine back in the Place de la Révolution – moved some time earlier after complaints about the blood from the headless corpses – and to witness Robespierre and 17 of his followers mount the scaffold. Soon after, the Public Prosecutor, Fouquier-Tinville, had followed him to the guillotine. In Bordeaux, the last to die had been the infamous Lacombe.

Around the Schuylers' table, wrote Lucie, 'we all rejoiced together'. The talk was of the end of the revolution and a quick return to France. Talleyrand was particularly pleased that his sister-in-law, the wife of his brother Archambauld, who had stayed in France in the hopes of saving the family fortune only to find herself arrested, would now be freed. It was only later that night, reading the papers more closely, that he discovered her name on the list of those who had died in the frenetic final killings of June and July. Among the last to be executed, they discovered, were the wife, daughter-in-law, granddaughter, brother and sister-in-law of the Duc de Noailles – all close friends to Lucie and Frédéric – three generations in a single day, their hands tied behind them, the Maréchale in a bonnet and black taffeta, borne to their deaths through a thunderstorm and heavy rain. Hidden in the crowd, but making his way alongside the tumbril and seen by the women, had walked the family priest, pronouncing absolution.

Lucie was struck by how very upset Talleyrand appeared at the news of his sister-in-law's death, and how 'amiable and graceful' he was when he came to visit them the next day, with his intelligent conversation and old-world courtesy. He spoke to her with an 'almost paternal kindliness which was delightful', and took great pains, with an 'exquisite sense of propriety', not to say things that might shock her. Many years later, reflecting in her shrewd way on Talleyrand's subtle and scheming nature, Lucie wrote: 'One might, in one's inmost mind, regret having so many reasons for not holding him in respect, but memories of his wrong-doing were always dispelled by an hour of his conversation. Worthless himself, he had, oddly enough, a horror of wrong-doing in others. Listening to him, and not knowing him, one thought him a virtuous man.'

For the days that Talleyrand and de Beaumetz stayed in Albany, where Alexander Hamilton had come to join them, there were long evenings at the Schuylers' and much interesting talk about the need for a liberal constitution and at least an appearance of democracy in both America and France. Talleyrand and Hamilton shared the view that real power should always reside in the hands of 'gentlemen of standing'. Hamilton, Talleyrand would later say, was one of the three great men of his age, along with Fox and Napoleon. Hamilton and Talleyrand were somewhat alike, being slender and not very tall, but while Talleyrand was pale and watchful, Hamilton had a ruddy complexion and reddish hair, and he was energetic and quick in his movements. What puzzled Talleyrand was Hamilton's decision to give up his job in the Treasury in order to earn more as a lawyer and to spend time with his eight children, and he remarked how strange it was that anyone so talented would yield power so readily.

When soon afterwards he went to Troy to dine with Lucie and Frédéric, Talleyrand took with him a medicine chest and presents of a side-saddle and bridle for Lucie, a perfectly timed gift for she had just acquired a mare to ride. He was accompanied by a tall, fair-haired Englishman called Mr Law, a former colonial governor in India, recently widowed by a rich Brahmin who had left him a considerable fortune. Lucie enjoyed Law's wit and his clever remarks, but found him highly nervous and eccentric, as all Englishmen were, 'to a greater or lesser extent'. Since she had become skilled at milking her cow, she was able to give them cream for dinner. Before leaving, Talleyrand told her that Law had been so moved by the spectacle of such a well bred woman milking her cow and doing her own washing that he had been unable to sleep, and now wished to make them a gift of some money to make their lives easier. Lucie and Frédéric, though much touched, refused: they would call on him, they said, if ever they found themselves in serious difficulties.

* * *

Lucie and Frédéric waited impatiently for the moment they could take over their farm. The snows arrived suddenly, with a force and abruptness that startled them. In the space of a few days, thick black clouds, driven by bitterly cold north-west winds, sent everyone hurrying to put under cover boats, outside furniture and tools. The winds were followed by the rapid freezing of the wide river, along which pine branches were hastily laid to mark out a path so that travellers would find their way, before heavy snow descended, reducing all visibility to just a few yards. The Hudson would now stay frozen for several months.

As soon as the snow stopped falling, Frédéric loaded their two working sledges and their one 'pleasure' sled, which was shaped like a large box and had room for six people, and harnessed them to the four horses they had bought during the autumn. Together with M. de Chambeau, who had become a competent carpenter, they set out for the farm. The settlers had left the house in a very bad state. Constructed as a single-storey house, with a dairy and cellar tucked underneath, its wooden frame was filled in with bricks dried in the sun which Frédéric finished with plaster and painted, while M. de Chambeau mended the beams and Lucie cleaned. Lucie was a ferocious worker, saying later that as long as they lived in Troy she never remained in bed beyond sunrise, and admitting that for M. de Chambeau and Frédéric a little idleness might have been preferable.

With 150 acres of crops and a large orchard, help was needed and slaves, their new friends told them, were the people to provide it. Though in his original draft of the Declaration of Independence Jefferson had included an attack on slavery, by the time the final document appeared slave-owners in the south had forced him to reconsider, and all that remained was an ambiguous reference to bondage. Even Jefferson, however, did not personally consider the blacks inherently equal, saying that though as brave and indeed more adventurous than whites, 'in imagination they are dull, tasteless and anomalous'. But the fighting against the English in the north had seriously disrupted patterns of slavery, and many slaves had fled their owners – some 30,000 were said to have escaped from plantations in Virginia alone – finding sanctuary in the confusion of armies on the move. Several thousand had fought – and many had died – on both sides. With the peace, and the heady rhetoric of the new nation, many former slaves had protested that they too had a 'natural and inalienable' right to life, liberty and happiness. A number of Northern slave-owners had gone along with the spirit of the revolution, and freed their slaves; other, less fortunate slaves had found themselves abandoned by the departing English, to be rounded up and re-enslaved. After all, as the popular English physician Dr Charles White had pointed out, blacks occupied a different 'station' from whites, being an intermediary species somewhere between white men and apes.

Lucie and Frédéric were deeply uneasy about the idea of owning a human being, but when they learnt that a slave dissatisfied with his master could officially request to be sold, and heard of just such a young man, they climbed into their red and yellow sled and set off to find his Dutch owner, a Mr Henry Lansing. The slave was called Minck, and he was eager to escape not just the harshness of his employer, but the severity of his own parents, both of them slaves in the same household. A deal was quickly struck, Mr Lansing being much impressed when he learnt that Frédéric had represented the French government in Holland. Before night fell they were on their way home, Minck driving the sledge, bringing with him only the best suit which he was wearing, everything else, down to his moccasins, belonging to his former owner and remaining behind. Lucie felt appalled, as she recorded later, by the ease with which a man could be bought and sold. For all her father's support of the slavers, her views had remained firmly with the abolitionists. In the *Albany Gazette*, where 'very lively Negro wenches' were advertised, the reward for finding an escaped slave was the same as that for a lost horse.

The next slave to arrive was Minck's father, Prime, a man well known to the Schuylers and the van Rensselaers for his farming skills, though Minck himself was not pleased to find himself again under his father's yoke. Lucie also wanted a woman to help her in the dairy, and heard of a slave called Judith, in her 30s and the mother of a small girl, who had been forcibly separated from her husband. Taking a bag of money with her, Lucie visited her owner, a Mr Wilbeck, and said to him that it was well known locally that Judith wished to leave and that he had treated her with great brutality. A truculent Mr Wilbeck took the money and handed over the woman and her young daughter; when Judith learnt that she was to be reunited with her husband, whom Lucie and Frédéric also bought, she fainted. She too was now carried back by sledge to the farm, where M. de Chambeau had prepared a room for them in the granary, something they had never had before.

Before long, Lucie and Judith were making cream and excellent yellow butter, on which they stamped the de La Tour du Pin crest and which Prime took to market to sell 'arranged daintily in a very clean basket on a fine cloth'. Though unable to read or write, Prime kept meticulous accounts in his head. Lucie made herself popular locally by adopting the dress worn by farmers' wives: a blue-and-black-striped woollen skirt, full but not too long, a dark calico bodice, with her hair parted and piled up in a coil, held in place with a comb. No one around Albany wore jewellery, fans, patches or ribbons. Only when visiting the Schuylers or van Rensselaers did she put on a gown and stays, or one of the riding habits brought over in the *Diana*.

Spring arrived with the same abruptness as the snow. Towards the end of February, the north-westerly wind suddenly dropped, and a southerly wind began to blow. The snow melted so quickly that for several days people were trapped indoors by raging torrents along the roads. The spectacle of the ice as it broke drew everyone to the river banks. As the water below the thick ice began to stir, swollen by the snow melt, a first crack appeared along the middle of the river, bursting with a roar like thunder as shoals of ice broke away and gigantic blocks of ice rose into the air, the light refracted into rainbows. In less than a week, the meadows around the farm were green and covered in wild flowers.

The savages, as Lucie called them, had not been seen all winter. With the better weather, the Iroquois began to reappear around Troy and Albany, bringing deerskin moccasins to sell, stuffed with buffalo hair or moss for warmth, along with carved wooden implements and long gaiter-like leggings, worn by the farmers to protect their calves and shins. Lucie had been startled when she had first encountered two naked Indians, walking slowly up the street, but she had become accustomed to their ways. They were, she said, as 'sensitive to good manners and a friendly reception as any Court gentleman'.

For their part, the Indians around Albany seemed to like her, finding her different from the dour Dutch farmers' wives, and referred to her as 'Mrs Latour from the old country'. Before the snows had begun, an Indian man had asked her if he might cut branches from a particular kind of willow that grew on their property, promising to weave baskets for her during the winter. Being, as she said, rather doubtful as to whether savages kept their promises, she was agreeably surprised when, a week after the snows had melted, he reappeared, carrying a neat pile of six baskets, all fitting inside each other, so closely woven that they held water as well as any earthenware jug. Refusing to take money, he accepted a jar of buttermilk. Lucie had grown fond of a very ugly elderly Indian woman, who had matted grey hair and wore nothing but a tattered shawl and the remains of an apron, to whom she gave old remnants of feathers and ends of ribbons, once part of her fashionable wardrobe. What the old squaw, as Lucie called her, really liked was to be allowed to look at herself in Lucie's mirror, after which she would cast benign spells over their chickens and cows.

In the autumn of 1794, the Jay Treaty had been signed with Britain, finally resolving the last of the border crises. The USA now had the right to build forts wherever it chose on the remaining Indian territories, and the Indians themselves no longer had much sway with either side. Knox, the Secretary of War, favoured presents along with diplomacy, but in the long run wanted to see all frontier lands transferred from natives to settlers, believing that in any case the Indians would probably die out, especially if they did not embrace Christianity. The skirmishes that had long marked life in the less charted north, with soldiers returning to describe warriors painted and tattooed, their earlobes weighed down with ornaments and uttering piercing screams, were few and far between. If around Albany the Iroquois men continued to hunt beaver and bear for meat, roasting the fatty bear on top of venison, turkey and mallard in order to baste them as they cooked, and to collect maple sap to turn into brown sugar by slashing the tree trunks with their tomahawks, it was as objects of curiosity rather than as intimidating strangers that they were increasingly regarded. Fear of intoxicated Indians causing trouble had led legislators to outlaw the trade in alcohol, but inebriated Indians, having bartered their furs for liquor, were occasionally to be seen wandering around the streets of Troy and Albany. It was a very long time since Columbus, arriving in America, had been impressed by the resourcefulness of the Indians.

An officer in Frédéric's regiment, M. de Novion, arrived one day on a visit, hoping to buy a farm and learn from the La Tour du Pins the skills that he would need. As he spoke no English, had neither wife nor children, and had never been near a farm,

Lucie had doubts as to his suitability. She took him out riding to show him the land. After a few miles, realising that she had forgotten her whip and had no knife with which to cut one, she caught sight of an Indian she knew. To M. de Novion's evident disgust and horror, she called out to him and the man approached, wearing only a very thin strip of blue cloth between his legs, and went off to cut her a switch with his tomahawk. M. de Novion asked Lucie what she would have done had she been alone. Nothing, she replied, adding mischievously that had she asked her Indian friend to do so, he would willingly have felled the Frenchman with his tomahawk. That evening, M. de Novion told Frédéric that he had decided to live in New York, where, noted Lucie mockingly, 'civilisation seemed slightly more further advanced'.

For all her reserve about many of the Dutch settlers, whom Lucie found overly keen on money, she was happy in Troy. In the spring, the flocks of geese, duck and passenger pigeons, flying low between the coastal swamps where they wintered and the Great Lakes where they bred, were so dense that they cast a shadow over the streets below. Towards May, sharp-nosed sturgeon, up to 8 feet long and weighing up to 200 pounds, known as 'Albany beef' to the locals, appeared in the river and were fished from canoes, to be dried and pickled, the oil used for bruises and cuts. The surrounding forests were full of edible mushrooms and nuts, and in the summer months the fields were red with strawberries. In 1796, the first genuinely American cookbook had been published, a pocket-sized treatise 'adapted to this country and all grades of life' by Amelia Simmons, who won much acclaim by calling herself 'an orphan', with recipes geared to settler life – Indian pudding, cranberry sauce and cornmeal cake. (For the perfect syllabub, Simmons recommended sweetening a quart of cider with 'double refined sugar', before 'milking the cow into your liquor'.)

The farm was prospering, there were enough French émigrés in the area to provide company, and they spent many evenings with the Schuylers and van Rensselaers, people with whom Lucie could talk and play music. With the invalid Mrs Rensselaer, whose knowledge of French literature surpassed that of many of her Parisian friends, she passed agreeable hours of conversation. It was at the Rensselaer house that she met the wife of the Anglican minister, Mrs Ellison, a middle-aged woman whose great sorrow in life was to be childless. Mrs Ellison was much taken by Humbert, now 5, tall for his age and speaking English better than French, and she begged Lucie to let him spend the summer with her in Albany. Lucie worried about the dangers of the farm and the amount of time Humbert spent with the horses and the slaves, and also that he had developed a habit of wandering off to look for the Indians, who occasionally, she had been told, kidnapped children; so she agreed to let him go.

Early in the summer of 1795, they received a visit from the Duc de Rochefoucault de Liancourt, who had escaped from France and was travelling through upper New York state with a former naval officer called Aristide du Petit-Thouars. Talleyrand had given the travellers an introduction to the Schuylers and van Rensselaers, but before taking them into Albany Lucie insisted that de Liancourt change out of his filthy clothes into something more appropriate. Even then she complained that he looked more like a shipwrecked sailor than the former First Gentleman of the Bedchamber. For his part, de Liancourt appeared astonished to see Lucie emerge in an elegant dress and a fine hat.

The visit to Mrs Rensselaer was not a success. She declared afterwards that she found de Liancourt 'a very mediocre man', as did Lucie, who was profoundly irritated by his opinionated manners and his ignorant criticisms of America. Later, in his immensely long account of his travels in North America – which were indeed full of 'sickly trees', apathetic, covetous Dutchmen, and cowardly bears and wolves – de Liancourt was fulsome in his approval of Lucie. Of all stalwart Frenchwomen, he wrote, she was the most admirable and but for her devoted support, Frédéric might have 'sunk under their misfortunes'. Reading his words, Lucie would feel guilty about not having liked him better, but cross when she read his verdict on the inhabitants of Albany, 'a set of people remarkable neither for activity nor politeness . . . the most detestable beings I have hitherto met'. Only a rattlesnake emerged well: though awake, noted de Liancourt, 'she showed no sort of malignity'.

M. du Petit-Thouars, on the other hand, charmed his hosts; he was, noted Lucie, exceptionally 'witty and gay'. Even de Liancourt admitted that his travelling companion 'conversed with exquisite sense'. He made Lucie and Frédéric laugh with stories about the colony of French émigrés at Asylum on the banks of the Susquehanna where he had just spent some months. Planned by the Vicomte de Noailles, Lafayette's brother-in-law, and the Marquis de Talon, Asylum had included piazzas, summer houses and shutters of a kind never seen before in America, and a special house for Marie Antoinette, designed when it was still thought that she might escape. After her death there had been talk of the Dauphin finding refuge there. But the land had turned out too rugged and mountainous for the fruit trees, while the canon, archdeacon, two abbés and various countesses had proved inept farmers; and Asylum, like Gallipoli, was foundering, its inhabitants reduced to eating robins and boiled tadpoles. Du Petit-Thouars was sad to leave Troy, where he had found his hosts less prone to the general contempt for the Americans that seemed to afflict so many of the French émigrés; and Lucie, in particular, he was enchanted by, saying that she was not only very pretty, but graceful, cheerful, a model of elegance for the local ladies, and that she possessed 'outstanding cool-headedness and endurance'. There was something about the wholesomeness of their life that did indeed suit Lucie perfectly.

But she had not been well. Ever since the spring, when she had almost had a serious accident in the river, crossing with her highly strung mare on a ferry carrying four enormous oxen, she had had recurrent bouts of fever, which annoyed her principally because they left her unable to work. She attributed them to the shock, and characterisically blamed herself for being weak. Though the existence of malaria – '*mal'aria*' – and its connection to swamps and marshes had long been known in Europe, where it flared annually to epidemic proportions, no one had yet worked out that it was transmitted by mosquitoes. Believed to come from 'poisonous miasmas', exhalations rising from decomposing vegetation – and even, on occasion, from putrid cabbages, potatoes, onions, coffee, chocolate, old books, locusts, beached whales and the entrails of fish – these spring and autumn fevers were a constant menace along the Ohio, Mississippi, Hudson and their tributaries.*

* A French traveller observed that American women in particular were susceptible to diseases like consumption and fevers, because they took so little exercise and suffered from an infirmity of will 'which acts on them like chains which compress their limbs, gnaw at their flesh, cause obstructions, deaden their vitality, and impede circulation'.

Lucie's attacks lasted five to six hours, bringing headaches, fever and exhaustion; she felt hot and sweated, then cold and shook. Even the quinqina from Peruvian bark, sent to her by Talleyrand from Philadelphia, and already known to act as a cure, did little to shake it off. (A remedy she seems not to have tried is one described in the Boston *Gazette* the previous September: 'Just before going to bed, let the patient take off his or her clothes and stand under a sieve suspended by a line; let the person pour into the sieve a pail filled with water just drawn from the well. The shock will be considerable . . . but the effect will be pleasing.')

Soon after de Liancourt's visit, a letter arrived from Talleyrand saying that he had, by good fortune, learnt that the firm holding Lucie and Frédéric's money in Philadelphia was on the verge of bankruptcy, and that he had been able to withdraw their Dutch bills of exchange just in time. However, the bank needed Frédéric to sign the release papers. As the harvest was still a month away, Humbert was happy with Mrs Ellison and a neighbour offered to care for Séraphine, who was now almost 2, they decided to accept Mr Law's invitation to be his guests in New York, where Lucie could see a doctor before travelling on to Philadelphia.

Before they left, this time by boat, the *Albany Register* carried an article about the death in his dank prison in Paris of the 10-year-old Louis Charles, Louis XVII, a sad, forgotten child, on 8 June.

* * *

Steamships were yet to be invented, though steam was already in use to power factories, and Lucie herself had a steam spit for the mutton and turkey she roasted on Sundays. With Albany set to become the new state capital, and Troy growing rapidly into a centre for ironworks, the river did a considerable trade, schooners, brigs and sloops all competing for a place by the quays, along with 'battoes', the flat-bottomed pine boats designed by the French to carry furs and wheat. Nearly all the larger sailing boats were equipped to carry passengers. The 164-mile trip to New York took between three days and a week, depending on the wind, the currents, and the expertise of the pilot whose job it was to know exactly how far each tide, whether flood or ebb, would carry the craft through the Hudson's many reaches.

Lucie was enchanted by their gradual journey, travelling so slowly on windless days that the boat seemed to glide. She sat on deck watching villages and farms slip by, tracing the smoke from the fires started by settlers to clear the land as it rose into the still air, looking out across dense forests of pine, oak, ash and elm, and at the clearings with their orderly rows of cherry and apple trees. She was particularly struck by the deep dark water of the gorges, and the mountains that climbed steeply behind them. Never, she wrote later, had she seen anything to compare with the stretch of river at West Point, its ancient trees seeming to hang far out over the water, and in later life she hoped that they had been spared the 'soulless, frenzied clearing of land'.

In 1795, New York was a town of some 45,000 people, with 2,500 slaves; founded by the Dutch in 1614, it had a busy port, churches of all denominations, numbered houses in brick, and named and paved streets packed with stores and constantly blocked by hackneys, diligences and wagons. Broadway, its widest street, cut the city from north to south. Rents were considered exorbitant, drinking water was sold in barrels – the local water being full of gnats and said to be undrinkable unless laced with brandy – and New York's pickled oysters, regarded as the finest in America, were shipped as far afield as the West Indies, preserved in a mixture of allspice, vinegar and mace. Merchants and brokers met at Tontine's coffee house, which kept the gazettes of both Paris and London. After Albany, Lucie was delighted by New York, though Talleyrand and de Chaumetz, meeting the sloop, were appalled by how pale she was and how much weight she seemed to have lost.

In the afternoons, Alexander Hamilton joined the party on the terrace of Law's house on Broadway. From then until after midnight, under a starry sky and in great heat, they would sit talking. Law described his elephants and palanquins in Patna, while Frédéric and Talleyrand argued about the 'absurd theories' of the French Constituent Assembly, and how too great a love of money had a paralysing effect on intellectual and artistic endeavour. As Talleyrand said, the USA was a young nation, with all the arrogance of adolescence, and he was yet to meet a Frenchman who did not 'feel a stranger'. When Hamilton described the beginnings of the war of independence Lucie remarked, with one of her occasional flashes of sharpness, that it was considerably more interesting than the 'insipid memoirs of that simpleton Lafayette'. The talk never ceased; the party gathered night after night and continued their conversation. Mr Law, saddened to think that these evenings would ever end, said to his manservant about his guests: 'If they leave me, I am a dead man.'

When the moment came to move on to Philadelphia, Lucie was not well enough to travel. So it was Frédéric who was introduced to George Washington and saw the city described by the Comte de Ségur as a 'noble temple put up to liberty'. On reaching Philadelphia he quickly joined the group of liberal, royalist French émigrés, men who had known each other for many years, had fled to America in ones and twos and now gathered every evening in the bookshop started at 84 First Street by Moreau de Saint Méry. There was the Vicomte de Noailles, the historian and philosopher, with whom Frédéric had served in America, and the Comte de Volney, recently escaped from a Parisian prison, who would spend the next three years gloomily scouring America for minerals, rocks and geological specimens, harassed by gnats, complaining that the evergreens were 'dwarfish', and that Indians had mouths shaped like those of sharks. At night, when Moreau closed up shop, the friends, each of whom in his own way had helped to bring about the revolution that led to their exile, and each of whom saw his own future only back in France, moved upstairs to continue talking far into the night, until asked to leave by Moreau's exhausted wife.

From France had just come news that the exiled Louis XVIII had issued a manifesto promising punishment for the regicides and a return to power of both nobles and clergy.* From Paris too came further accounts of Robespierre's death, his body torn apart by delirious crowds, the bleeding limbs paraded triumphantly around the city. Week after week, the *Courrier Français* ran stories about those who had been tried and guillotined, and how each had conducted himself at the end. In Paris, 2,639 people had died under the guillotine, half of them in the 47 days leading up to Robespierre's death; surprisingly few – 485 – had been members of the nobility. A further 14,000 had been guillotined in other parts of France. The revolution, claimed the reports reaching

* On the death of the last son of Louis XVI, the 10-year-old Louis XVII, the crown had passed to Louis XVI's brother, the Comte de Provence.

Philadelphia, had taken 35,000 to 40,000 lives, including those who had died in prison or been executed without trial.

At the *Fête des Victoires* in July, Thérésia Cabarrus, now married to Tallien, had been called '*Notre Dame de Septembre*', because Tallien was blamed for the September massacres in the prisons; but that was before his triumphant defeat of Robespierre, after which she became '*Notre Dame de Thermidor*', on account of all the lives she was said to have saved in Bordeaux.

Though George Washington had already chosen the site for the new national capital on the Potomac, and Major L'Enfant was at work designing streets, drives and avenues radiating out from Capitol Hill, Philadelphia in 1795 was still the centre of American political and cultural life. It was here that the greatest number of French émigrés had settled, earning meagre livings as teachers of French, music, dancing and fencing, as clock-menders, dress- and wig-makers and chefs. Brillat-Savarin, one of the most renowned chefs of his day, was teaching Americans how to cook wings of partridge '*en papillotte*', scramble eggs with cheese, and stew squirrels in Madeira, reluctantly admitting that there was much flavour in a good turkey. Jefferson, who was extremely interested in food and had returned from Europe with a waffle iron from Holland and a spaghetti machine from Italy, had turned his vegetable garden at Monticello over to experiments with eggplants, Savoy cabbage, endives and tomatoes, hitherto regarded as too poisonous to eat. Jefferson was said to like his ice cream enveloped in a crust of warm pastry.

Philadelphia ate lavishly. In the markets, noted one French émigré approvingly, were to be found 52 different kinds of meat and game and dozens of varieties of fish, though lobsters, reported by early Dutch settlers to grow to 6 feet in length, were not just smaller but scarce, having apparently been frightened away by cannon fire during the revolution. The French, their Philadelphian hosts reluctantly admitted, had brought a transformation to the American kitchen, even if Amelia Simmons's successor, Lucy Emerson, noted disdainfully in her popular New England cook-book that 'Gar-lics, though much used by the French, are better adapted to medicine than cookery'. The querulous Volney admitted to enjoying '*le pie de pumkine*' but marvelled at the way Americans washed down their meat and vegetables with coffee and milk. When Volney's comments on America were published not long afterwards, Samuel Beck, a Bostonian, described their author as a 'timid, peevish, sour-tempered man'.

The Philadelphians also drank well. In smart society, wine, especially the 'sombre and somewhat heavy' reds from Hermitage and Côtes du Rhone, were increasingly replacing cider and Madeira at meals. Jefferson, who called wine 'a necessity of life', said that it stimulated intelligent conversation, whereas spirits only led to drunkenness.

With the arrival of ships from France came jams, syrups, chocolates, wines, brandies, raisins, almonds, hats and the latest French dresses, advertised by merchants in the daily *Courrier Français*. But the news from Paris was mostly bleak, of outbreaks of violence between Jacobins and counter-revolutionaries, of the settling of old scores and growing disorder and scarcity in the capital. While the French Revolution had originally been greeted by the Americans with approval, the New York *Daily Gazette* observing that 'the flame of liberty expands from city to city', its turbulent course had soured the views of many. As Hamilton observed, it was hard to go on supporting a revolution that had plainly substituted 'to the mild and beneficent religion of the Gospel a gloomy, persecuting and desolating atheism'. Pleas by the new French Consul in Philadelphia, Fauchet, that the revolution in France, as in America, had simply been an attempt to embody in fundamental laws safeguards for people's political and social liberties, counted for less than the reports of bloodshed arriving on every vessel. Even Fauchet now admitted that the execution of the King and Marie Antoinette, and the fall of America's idol Lafayette, had between them stamped out most feelings of warmth towards the French. Talleyrand, writing to his friend Lord Lansdowne shortly before Frédéric's visit to Philadelphia, observed that in the long run, America's ties to England were so deep and so natural that they were the obvious allies, and not the French, with whom relations never rose above the superficial and the sentimental.

Now that the Terror was over, the French were starting to think about going home; the Americans were beginning to long for them to do so. Increasingly, the émigrés detected an 'epidemic of animosity' directed against them.

* * *

After three weeks in New York, restored to health by Law's housekeeper, who insisted that she rest and brought her endless small cups of broth, Lucie was ready to return to Albany and the children. Frédéric came back from Philadelphia, tantalising her with his descriptions of George Washington. The long evenings of talk under the stars on Law's veranda resumed. Later, Lucie would remember these weeks in New York as ones of exceptional 'contentment'.

Then a sudden rumour spread that yellow fever had broken out in a street close to Mr Law's house on Broadway. Yellow fever, the 'malignant' or 'putrid' fever, thought to have been carried to the USA from Barbados in the 1690s, was one of the most feared of all infectious diseases; people were still haunted by the epidemic that had struck Philadelphia two summers before, which had killed a tenth of the population, despite the ministrations of the 'miasmanists' and the 'contagionists' and their various remedies of purging and bleeding. Health, many doctors still maintained, was restored by eliminating morbid matter: if the patient died, it had not been sufficiently eliminated. Smallpox, for which vaccination now existed, held fewer terrors; when yellow fever struck, however, people fled.

That night both Lucie and Frédéric woke feeling extremely ill. Not knowing whether they too had been infected, or whether they had simply overeaten on the bananas, pineapples and other exotic fruits recently arrived from the Caribbean, they decided to leave New York immediately, before being trapped in the city by quarantine restrictions. By daylight they had packed and were at the port, looking for a sloop to carry them back up the Hudson. They quickly recovered: their malaise had been fruit, not fever.

This time their boat went aground on a sandbank and rather than wait to be towed off, they had themselves rowed a little way up one of the Hudson's smaller tributaries, between climbing plants and wild vines that met above their heads in garlands, where they had heard that some French émigrés were farming. The two young Frenchmen, who were delighted to see visitors, appeared to live in a mixture of penury and luxury. Having found that the farming methods brought with them from a sugar plantation in Saint-Domingue were utterly unsuited to the American climate and soil, they were fast running out of money; but they continued to eat off magnificent, if chipped and unmatching, porcelain, salvaged from earlier splendour in pre-revolutionary France.

Lucie and Frédéric reached Troy to find Séraphine well and the farm preparing for an exceptional crop of apples and grain. Lucie settled back with pleasure to the routine of her days. Neighbours dropped by to remark on the improvements they had made to the farm and to praise M. de Chambeau's elegant new 'noble hog sty'. She was, as she noted, very happy. But then she was dealt 'the most unexpected and what seemed to me then the most cruel blow that any mortal could endure'.

* * *

One morning Séraphine, a bright, cheerful, affectionate child of 2, woke with what Lucie described as a 'sudden paralysis of the stomach and intestines'. The Albany doctor, fetched by M. de Chambeau, told them there was nothing he could do. A few hours later Séraphine was dead. One of the small Schuyler boys, with whom she had been playing the previous day, died shortly after.

There was no Catholic priest in Albany. Frédéric and Lucie buried Séraphine in a small enclosure near the farm, Frédéric conducting the funeral service himself. 'This cruel loss,' Lucie would write many years later, 'threw us all into the deepest sadness and despondency.' She had always maintained that, surrounded as a child by the spectacle of corruption and scandal in the Church, she retained nothing but bad memories of religion; it was as if, she would say, 'all concern with morality had been stifled out of my heart'. But now, kneeling by Séraphine's grave, she felt herself change, drawn towards a God who would give her the courage to endure such terrible grief. 'God bided His time to work a change of heart in me,' she wrote. 'The hour had come when I was forced to recognise the hand that had stricken me.' And ever afterwards, she would write, divine will was to find her 'submissive and resigned', though there was something in her nature that was profoundly alien to submission.

To combat her sadness, Lucie set about her work with even greater frenzy. She collected 5½-year-old Humbert from Albany and brought him back to the farm. She decided to teach him herself, rather than send him to school; he was a clever and biddable pupil. At all other moments of the day, which for her started before dawn, she threw herself into the harvest and pressing of the apple crop, the cider squeezed out by an old mill driven by a horse, on which Humbert rode round and round, then put into brandy casks from Bordeaux. So delicious was the result that their cider fetched double the usual price in Albany. The maize crop was no less good, the final shucking of the ears of corn done, according to local custom, all through one night in a barn lit by candles, with the help of both their white and their black neighbours, working together over great pitchers of hot milk, laced with cider and flavoured with brown sugar, cinnamon and nutmeg. Then came the ploughing, the stacking of wood, the repairing and repainting of the sledges in preparation for the coming winter.

Lucie bought blue and white checked flannel to make shirts for their slaves and engaged a tailor to come and sew waistcoats and thick lined capes. No minute of her long day was left unfilled. Every week, she washed the family's clothes, did all the ironing, remarking how well her childhood with the servants at Montfermeil had equipped her for these tasks, and made the butter and cream, taking pleasure in the fact that her dairy was considered to be not just the cleanest but the most elegant in the neighbourhood. Slowly, very slowly, the days passed.

Winter came early in 1795. By early November, the black clouds bringing snow began to gather in the west. Within a week, the Hudson and the Mohawk were frozen solid, and the surrounding country lay under deep banks of snow. The short dark days were broken only by occasional visits to the Schuylers and the van Rennselaers. One morning a parcel arrived from Talleyrand, containing a valuable cameo of Marie Antoinette, a gold watch and a casket, all of which Lucie had left behind in Brussels with her banker friend, and which had been appropriated by the diplomat meant to be bringing them out to her. Talleyrand, spotting the cameo in a friend's house in Philadelphia, confronted the young man and retrieved her possessions.

Lucie was once again pregnant. This was her sixth pregnancy; but of all the children she might have had growing by her side, there was only Humbert. Towards the end of the winter, she came down with measles and was for a while very ill. But neither Humbert, who slept in her room, nor the others caught the disease, and by spring she had recovered.

With the end of the snows came an unexpected letter from France. Their *sans-culottes* friend Bonie wrote from Bordeaux with the news that the property sequestrated from the émigrés and the nobility during the revolution was to be restored to its rightful owners. Frédéric's mother had already taken possession of her house and lands at Tesson and Saintes. The seal on Le Bouilh could now be lifted, but only, it seemed, if its owners returned within 12 months to sign the papers. At the farm in Troy it was as if, wrote Lucie, a firebrand had exploded, each seeing in the news a different story. Frédéric and M. de Chambeau were overjoyed, released at last from all pretence that they might wish to spend the rest of their lives in America.

Despite Séraphine's death, Lucie felt differently. She had felt safe on the farm, and what she remembered most clearly about France was that it was the country where she had lost her youth, 'crushed out of being by numberless, unforgettable terrors'. She dreaded going back; the idea filled her with foreboding. But she said nothing of this to Frédéric, beyond asking that she be allowed to give the four black slaves, to whom she had become much attached, their freedom. When she gathered them in the drawing room and gave them the news, they wept. Never in her life, she would write, had she felt such pleasure in an act of her own. At the ceremony of manumission, which took place in public, in the presence of the disapproving Justice of the Peace, who was not in favour of abolition, and all the assembled black slaves of Albany, Minck, Prime, Judith and her husband knelt in turn in front of Frédéric. One by one, he laid his hand on their heads in token of liberation, following the practice of ancient Rome.

The end of their American adventure came quickly. Having distributed the smaller pieces of porcelain and some of Lucie's gowns as presents, and sold the horses, furniture and stock for good prices, Frédéric, Lucie, Humbert and the ever faithful M. de Chambeau boarded a sloop for New York. Mr Law was away, so they stayed with a French banker, M. Olive, his wife and eight children in their house on the banks of the Hudson. They found Talleyrand in residence, also preparing to leave for France, Mme de Staël having written to urge him to return as quickly as possible in order to serve the new government. For a while it looked as if they might travel together, but Lucie and Frédéric wanted to approach Bordeaux from Spain, so as to be sure that France was indeed at peace, while Talleyrand feared that Spain's Most Catholic Majesty might take against a cleric who had not been a 'sufficiently edifying bishop'.

All over America, the French were preparing to go home, selling up their properties and booking passages on all available ships, intensely relieved that they could now abandon their precarious lives as dancing teachers and pastry cooks. Though apprehensive about what awaited them in France, where they had lost relations, friends and possessions, they felt a sense of urgency to leave. Few of them had enjoyed their exile. They had found America too rugged, its food too peculiar, its women too virtuous and assertive. They had missed the culture, the subtlety and the many nuances of language and behaviour that were so familiar and pleasing to them. Lucie was among the very few who had felt herself to be genuinely content. As for the Americans, they had found their French visitors perplexing, if elegant.

On 6 May 1796, almost exactly two years after their terrifying escape from Bordeaux, Frédéric, Lucie, M. de Chambeau and Humbert embarked on the *Maria Josepha*, a 400-ton English vessel bound for Cadiz with a cargo of corn. The weather was fine. One of the other passengers, Mme Tisserandot, was also expecting a baby. Lucie spent the 40-day crossing sewing for the captain and the crew, as well as making over her own clothes. By the time Cape St Vincent was sighted, the entire wardrobe of the ship was in good order.

CHAPTER TEN

Incroyables and *Merveilleuses*

Lucie spent the first ten days of their return to Europe in quarantine in the harbour of Cadiz, a condition imposed by the Spanish to control the spread of epidemics. She was in her sixth month of pregnancy; Mme Tisserandot, also bound for France, was in her eighth. The two women, delighted after so many days at sea to see fresh fruit, pulled up baskets of figs, oranges and strawberries from fruit sellers who rowed out to the *Maria Josepha*. When finally permitted ashore, the Frenchwomen found themselves the objects of considerable curiosity, a large crowd gathering to stare at their coloured gowns and straw hats as if they were two exotic beasts. To escape the ceaseless scrutiny, they bought black skirts and mantillas. Lucie, who had dysentery, felt very ill. Accustomed to the 'exquisite cleanliness' of America, she found their Spanish inn 'disgustingly' dirty.

All her life, Lucie would be captivated by new sights and places; she was naturally curious, observant about small details, and for the most part easy-going, though she was capable of being both prim and censorious. Direct herself, she disliked evasiveness in others. She could be firm, but not unkind; she appreciated kindness in others and often remarked on acts of generosity and thoughtfulness. The French Consul General in Cadiz, M. de Roquesante, a *ci-devant* noble turned ardent republican, to whom Frédéric had to apply for their visas, struck her as extremely unpleasant, particularly after he and Frédéric had a sharp exchange over the recent death of a hero of the Vendée. Frédéric's unwillingness to compromise or feign was becoming more marked. But an English merchant in Cadiz, Mr Langton, married to the sister of Théobald Dillon, the officer murdered by his men in Lille, charmed her by his solicitousness.

In the evenings, when the temperature at last dropped to below 35° C, they strolled up and down the Boulevard de l'Alameda, looking out across the sea and enjoying the spectacle of small Spanish grandee children in their full finery of brocade and powdered wigs. One afternoon, on the Feast of St John, they went to watch the leading matador of Cadiz fighting bulls so magnificent that Lucie thought that each would have made the fortune of any American farmer. The moment of the bull's death, with the matador poised gracefully for the kill, left a lasting impression on her, as did the Spanish method of keeping the bull-ring cool by spraying cold water over the awning shading the spectators.

Since Spain had concluded a peace treaty with France, most of its previously considerable army had been disbanded. Without pay or work, destitute soldiers had taken to the mountains from where they attacked travellers labouring their way up and down the steep passes. For safety, people now journeyed in convoys of 15 to 20 covered wagons, pulled by mules, the men of the party heavily armed. Though Lucie was still not very well, she was anxious to reach Le Bouilh before the birth of her baby, due in early November; she noted, with her customary robustness, that it was not in her nature to be defeated by small obstacles. Hiring their own wagon, with a wicker and tarpaulin hood, in which a mattress for her to lie on had been laid over their baggage and packed in with straw, they duly set out for Madrid. Frédéric, Humbert and M. de Chambeau perched at the front, their frisky mule trotting rapidly over good roads. Whenever they could, they did a leg of the journey by water, rejoining the convoy at a later spot. At Xeres, they spent the night in the wagon, repelled by the dirty beds of the inn.

The convoy was approaching Córdoba, crossing a long, empty plain, when Mme Tisserandot went into labour. A halt was called and Lucie helped deliver a baby girl. Since all the clothes for the baby were packed in her wagon, and the convoy was reluctant to pause for long in such a desolate spot, Lucie and the muleteer, having washed the baby in wine, wrapped her in shawls and then hastened to catch up with the rest of the party. It was night when they finally reached Córdoba. The servants at the inn, seeing what they took to be an injured woman in the wagon, apparently the victim of an attack by bandits, fled, not wishing to be called later as witnesses. In total darkness, Lucie unpacked the bags to find clothes for Mme Tisserandot and her child.

Next morning, the convoy waited an hour for the baby to be baptised in Córdoba cathedral. The convoy travelled slowly on across Spain, through groves of lemon trees, across parched plains and up into the Sierra Morena, Lucie loving the flowering pomegranate trees around the wells and the green valleys full of weeping willows, deliciously shady after the bleaching sun of the plains. It took them almost three weeks to reach Madrid.

Once again, they had letters of introduction and soon found themselves well looked after; they were received cordially by the French Ambassador, who had once served under Lucie's father. Lucie was anxious to press on, but it was not until six weeks later that letters at last arrived from Bonie and de Brouquens, reassuring them that it was safe for them to return to France. The last stage of their slow journey, in an enormous ancient carriage pulled by six mules, with a seventh, known as the *generala*, out in front to guide the way, took them past the Escorial, where Lucie reflected severely that to collect so much splendour in a place of such solitude could serve no other purpose than to 'make us aware of the futility and vanity of the works of man'. They were all much saddened by their recent parting from M. de Chambeau who, his name not yet removed from the list of proscribed French émigrés, could not go home.

* * *

'I felt no pleasure,' wrote Lucie, many years later, 'returning to France.' She was haunted by the horror of her last six months in Bordeaux, and dreaded seeing Le Bouilh, which they knew to have been taken over by the municipality. Frédéric's mother, back in control of most of her own properties, she remembered as suspicious and obstinate. Archbishop Dillon's house at Hautefontaine had been sequestrated. The status of her own house in the rue du Bac was unknown. And, now that the lands of the former nobility and clergy had been split up and sold, the revenues on which the family had once lived had disappeared. It was to vast debts, rather than to a comfortable income, that she and Frédéric were returning. When, stopping in Bayonne for yet more papers, Frédéric was summoned before the President of the Department to answer questions, even the presence of the reassuring Bonie could not calm her fears. Though Frédéric's name had definitively been removed from all lists of émigrés, as had that of his dead sister Cécile, which meant that in theory their property might legitimately be returned to them, there were many hurdles still to cross.

Their first sight of Le Bouilh, on an autumn day when mists and rains traditionally cast the Languedoc in soft grey light, was grim. When they had last seen the château, in the autumn of 1793, it was a well-furnished, comfortable home, with an excellent library, fine pictures, silver and crystal chandeliers, a well-equipped kitchen and a vast amount of the finest linen. Now Le Bouilh appeared vast, gaunt, unwelcoming. The garden was overgrown, the outbuildings derelict; the immense high-ceilinged drawing rooms with their tile and stone floors echoed at every step. Not long after their escape to America, they learnt, men from the local municipality had arrived to conduct a sale of the château's entire contents, moving from room to room like locusts until it was stripped bare, a screen, commode and *secrétaire*, together with three walnut chairs covered in black leather, going one day, bed-hangings, napkins, sheets, aprons, tablecloths, cushions, covers, the next. No item, however small, remained, though pouring rain and cold winds during the sales had apparently kept many buyers away. All copper, iron and lead items had gone to the mint to be melted down for the army. The last object to go, lot 359, had been an orange tree in a pot. Surveying the desolate scene, Lucie and Frédéric set about unpacking the crates that had been sent on ahead by boat from their farm.

Much of the furniture of Le Bouilh had, it soon turned out, been bought by neighbours, either, as Lucie noted crossly, out of envy, or out of greed. By the next day, shamefaced people began appearing at the door carrying tables, chairs and mirrors, offering to sell them back at the prices they had paid for them. The priceless copper kitchen utensils, taken away to be melted down, had been overlooked at the mint and were now returned, as was Frédéric's father's excellent library, which was discovered unharmed in a nearby storehouse. Bit by bit, the unfinished, ungainly château began to take on something of its former appearance.

And there was an enormous, unexpected pleasure when Marguerite, having heard of their return, appeared from Paris.

She had been living with Lucie's friend Pulchérie de Valance, who had survived and been released from prison after Robespierre's fall. Marguerite herself had been saved when a stranger in the street had pulled her into an alley-way and warned her to remove the starched white apron she always wore, saying that it was bound to inflame revolutionary tempers; she had spent the Terror looking after two small children whose parents had been arrested. Marguerite reached Le Bouilh in time to help Dr Dupouy, the timorous doctor who had looked after Lucie during Séraphine's birth, deliver her new baby on 1 November. It was a girl and they named her Charlotte, after their dear friend Charles de Chambeau. There was once again a second child in the family.

* * *

The de la Tour du Pins were not alone in their misfortune, nor indeed among the hardest hit. The Gironde was one of the areas of France where the losses of the nobility had been the heaviest, and where the revolutionary committees had been most zealous in dividing and selling off vineyards. Some 16,400 members of the French aristocracy had emigrated during the revolution, 624 of them from Bordeaux alone. Though relatively few had perished under the guillotine – 79 nobles out of the 301 condemned to death in Bordeaux, along with 24 nuns, 30 priests, 28 seamstresses, washerwomen and cooks executed for harbouring refratory priests, 2 miniature painters and 6 scribes – those who now came home found their properties stripped down to the locks and wooden window-frames. Bordeaux itself, where the population had shrunk by almost a quarter, was silent and deserted after dark, the lamps no longer lit. The once long alleys of poplars leading up to handsome mansions of yellow stone, with white shutters and slate roofs, had been chopped down for firewood. It was clear to none of those returning how they would survive.

Nor was anywhere very safe. With the end of the Terror had come a surge in brigandage and lawlessness, as former Jacobins and revolutionaries found themselves without employment and those they had persecuted sought revenge. The counter-revolutionary White Terror had seen the Trees of Liberty chopped down, while royalist 'companies of Jesus' scoured the countryside for 'terrorist Jacobins'.

One of the first acts of the new government in Paris had been to add a seventh ministry, that of the Police, to the six agreed under the Constitution of Year III; but the new police were ineffective, not least because there was no money with which to pay them, and judges were in any case too afraid to pass sentences and witnesses too afraid to speak. The assault on France's magnificent forests for firewood, begun during the hard winters of 1793, 1794 and 1795, had opened great tracts of land. Up and down the country, along roads that had got much worse during seven years of neglect, diligences were regularly stopped by brigands, their bands swollen by deserters from the army.

What terrified Lucie most were the gangs of so-called '*chauffeurs*', 'heaters'. These outlaws broke into houses where they suspected they might find money and burnt the feet of those inside over the open fire until they revealed their hiding places. Not far from Le Bouilh, *chauffeurs* had crippled a shopkeeper. When, in December, Frédéric set off to inspect the family properties at Tesson, Ambleville and La Roche Chalais, leaving her and the children with Marguerite, two maids and a drunken manservant, Lucie spent the nights sitting up in bed trembling with fear, listening to the creaks in the vast, echoing house, worrying that the few dilapidated planks securing the ground-floor windows would be too flimsy to protect them against intruders. 'I longed,' wrote Lucie later, 'for my farm, my good negroes and the peace of those months.' She feared for Frédéric, travelling the countryside alone. She had started having nightmares, in which he was being hunted, and from which she woke up covered in sweat, her heart 'beating heavily and painfully'.

When Frédéric arrived back, he brought more bad news: the château of Tesson, which his father had filled with valuable pictures and furniture, had been stripped bare, its contents sold, the wall-papers torn from the walls, its shutters ripped off. The house at Saintes had been reduced to a worthless ruin. Lucie felt saddened not just for her family but for France. 'No one,' she wrote later, 'would see again the room in which he had been born, or the bed in which his father had died.'

In July 1797, when Charlotte was 8 months old, Frédéric decided that he needed to go to Paris to try to put the family affairs in better order. Humbert was now 7, and both children were thriving. Frédéric's brother-in-law Augustin de Lameth, who had spent some months in prison after Cécile's death, had become mayor of Hénencourt, but he had lost almost all of his property in the revolution. All four de Lameth brothers had survived; the family was trying to mend the political differences that had divided them during the Terror. Mme de Montesson, Frédéric's elderly friend, released from her own time in prison, wrote proposing that they should stay in her house. Mme de Staël was back in Paris, working hard on Talleyrand's behalf to find him a position in the new government, as was the Princesse d'Hénin, who was staying with Mme de Poix.

Taking Marguerite and the two children, and planning to be away no longer than six weeks and to return in time for the grape harvest, Lucie and Frédéric took very little luggage with them. Lucie was still breast-feeding Charlotte. They travelled via Tesson, where Lucie saw for herself the complete destruction of the house, and where she was able to thank the elderly Grégoire and his wife for having saved Frédéric's life during the Terror. Though as many as half a million previously landless people had become individual proprietors as a result of the revolution, having bought land at a hundredth of its true value, the countryside they saw as their hired coach rumbled its way slowly north was extremely poor, the villages filthy, the children barefoot, the inns without linen or cutlery.

* * *

Paris was very altered. The city had been overtaken by a heedless longing for happiness. No longer subdued or diminished by fear, people wanted to forget, to celebrate the fact that they were alive. Though the streets were filthy, full of wild dogs and pigs and the carcasses of dead animals, though the city had lost almost a fifth of its inhabitants to the wars, emigration and the revolution, though there had never been so many abandoned babies left on the streets, people arriving in Paris in 1797 were amazed by its air of gaiety. At least 14 of the 23 theatres had a play every night, often a tale of tyranny defeated by valiant men. A new language had been invented for the stage, a debased '*langue poissarde*', a fishwife's language, enriched by the vocabulary of the revolution, with new words like 'terrorism' and '*decadi*' heard in the popular plays about a character called *Madame Argot*, 'Madam Jargon'.

In the streets were puppet shows, ventriloquists, conjurors; balloons of every colour and size floated above the city.

And people were dancing, as they had never danced before, to forget, to pretend that nothing had happened, in cabarets, in cellars, in halls and in empty former mansions, the women dressed as nymphs and Greek goddesses, scented and floating like Venuses under the chipped gold cornices. 'There is dancing in Les Carmes, where throats were cut,' wrote Mercier, once again chronicling the everyday life of the city, 'in the Jesuites, in the seminary of Saint Sulpice, in the Daughters of Sainte Marie, in three or four churches, in the Hôtel Maubert and the Hôtel Richelieu . . .' In the Tivoli pleasure gardens, 8 acres of flowers, artificial mountains and lakes, once the property of Boutin, treasurer to the navy and executed early in the revolution, they danced the waltz, newly arrived from central Europe. It was, observed one disapproving traveller, 'an intimate dance, requiring the two dancers to join closely together, and to glide, like oil across polished marble'. To add buoyancy to their steps, men sometimes wore lead soles by day so that their evening pumps seemed weightless.

At a subscription dance for the 'victims of the Terror' at which all the guests had lost someone to the guillotine, dancers wore red ribbons tied around their necks to suggest a severed head, a 'depravation' noted Helen Maria Williams who passed through Paris that winter, 'not merely of manners, but of the heart'. At balls, lit by a profusion of candles and smelling strongly of flowers, ladies' maids hovered, waiting to replace gloves, soiled shoes and ribbons of guests who danced all night. Parisians now not only looked different, and behaved differently: they also had different tastes.

When they were not dancing, they were gambling, in spite of frequent attempts at prohibition, at lotto, *trente et un*, faro, backgammon, picquet, whist and all forms of cards and dice, people who had lost their fortunes risking what little they had left. The new game of roulette had become so popular that wheels were set up in public squares and in the foyers of theatres. After the claustrophobia of the revolution, Parisians longed to be outside, to walk in the gardens of the Tuileries, to ride in the Bois de Boulogne, to play '*diable*', a form of badminton with racket and balls, among the orange trees on the Terrasse des Feuillants. After years of military bands and the beat of drums, they wanted concerts, romantic songs, comic operas. The quarrels between melodic Piccinnistes and stern Gluckistes had resumed. For a while, the harp became the preferred instrument, favoured by women after a doctor claimed its reverberations through the chest were a protection against heart problems. 'I want Italian singers to settle in Paris,' announced the composer Grétry, 'Italian music is the antidote to the evil we must put behind us.'

In the absence of religion, magnificent displays celebrating patriotism, with mythical gods and goddesses, chariots and statues of Apollo, brought crowds onto the streets. In the underground Café du Sauvage, a man dressed up as an Indian, said to be Robespierre's coachman, played the timbals and made terrible faces and noises. In the Jardin des Plantes, Baba the elephant fired a pistol with his trunk. Along the Promenade des Tilleuls, two dromedaries grazed. Parisians were reading, too, in ways that they had not read before, particularly novels, sitting in *cabarets de lecture*, special reading rooms, or taking books out of the new lending libraries. Three poets had gone to the guillotine – Fabre d'Églantine, Jean-Antoine Roucher and André Chénier – but others were taking their place in an explosion of works in verse. Never had there been a time more dedicated to public pleasure.

Lucie and Frédéric had last seen Paris in March 1793, as the Terror was taking hold. Following the death of Robespierre and his followers, a five-member executive committee, the Directoire, had been established on 1 November 1795, elected by a new Council of Five Hundred and a second Council of Elders. Along with the monarchy and arbitrary and despotic rule, the three Estates had vanished. Having inherited an empty treasury, unpaid taxes, public institutions without funds and a near worthless paper currency, the *assignats*, the Directoire had nonetheless succeeded in bringing some measure of confidence and calm, introducing compulsory and free public education – though there were no teachers to teach it – and a more coherent system of taxation. France was still at war with Britain and Austria, but peace treaties had been signed with Prussia, Holland and Spain.

Every day, at one o'clock, one of the five Directors, in their ceremonial costumes of red, white, blue and gold, held public audiences in the Luxembourg. Not one of these men, as Lucie soon remarked, could be described as very impressive. There was the seductive and cynical Barras, a close-shaven, sinister-looking man both lazy and venal, whose dinner parties were the most sought after in Paris; the heavy-drinking Reubell, with legs too small for his body, reported by an English agent in Paris to his superiors in London to be of a 'rapaciousness without equal'; the feeble, doctrinaire La Revéllière, head of the new sect of theophilanthropists; Carnot, a military engineer known for his organisational skills; and a harmless naval officer called Le Tourneur. Of these, Barras was the man of the moment. 'He is playing the Prince,' reported an English agent to his masters in London. Lucie would later say that they had all been 'drawn from the dregs of the gutter'.

Mme de Montesson and Pulchérie de Valance greeted Lucie and Frédéric warmly. Lucie, noting that the French liked nothing better than someone new and different, found herself the centre of attraction. 'The strangeness of the life I had led in America and my wish to return there,' she noted wryly, 'made me quite the rage for at least a month.' One by one, the nobles who had survived the violence by living quietly on their estates, and the émigrés, returning from Holland, Spain, Italy, the German States and England, were making their way cautiously back into the city. Most were now very poor. As in Bordeaux, they found their former mansions boarded up, with signs that they were for sale; when they could, they moved back in, perching disconsolately in magnificent rooms, stripped of their mirrors, curtains and furniture. Six thousand private houses stood empty. In the Boulevard des Italiens, soon known as 'le petit Koblenz' on account of all the émigrés who gathered there, friends met again and cried in each others' arms. Because no one wanted to be thought wealthy, people professed total penury. 'The supreme *bon ton* was to have been ruined,' wrote the Baron de Frémilly, 'to have been a "suspect", persecuted, above all in prison . . .'

Many returning from exile spent their days searching for their ransacked possessions along the banks of the Seine, where stalls had been set up selling porcelain, glass, furniture, tapestries, all stolen or confiscated from the houses of the *çi-devant* nobility. The former convent of Les Petits Augustins had been renamed the Museum of French Monuments and was now a depository for church treasures, statues of the saints, sarcophagi, obelisks, bronze casts of kings and martyrs and stained-glass windows all piled on top of each other. As Marat had remarked early in the revolution, praising those who were setting fire to the fine mansions of the Faubourg Saint-Germain, liberty could only be built on equality, and how could mansions be shared other than by demolishing them? There were melancholy outings to the cemetery of the Madeleine, where the heads and bodies brought in baskets from the scaffold had been tossed into a 'ditch of the guillotined'.

Those who ventured as far as Versailles found more scenes of destruction, the avenues of trees cut down, the porphyry and alabaster busts cracked and chipped, grass growing in the courtyards, the magnificent furniture, pictures and tapestries sold off. In the park, the setting for so many elegant and magical receptions, the fountains stood silent, the statues mutilated, the lake a dark grey swamp. Only the Orangerie had survived, with its 1,200 orange trees, some dating to the 17th century. And the frescoed temple of Flora seemed strangely untouched, as if Marie Antoinette was about to arrive to dine beneath the painted garlands and arabesques, though the eight sphinxes crouching on the staircase had had their noses and ears chopped off. Frédéric Meyer, strolling through the Queen's apartments in Versailles, heard the sound of flutes and a harp, playing first an andante and then an adagio. He followed the music to what had been Marie Antoinette's bedroom, where a solitary clock was sounding the hour.

Lucie and Frédéric soon discovered that no quarter of Paris had been more profoundly altered by the revolution than the Faubourg Saint-Germain and the streets running from the rue de Varennes to the end of the rue de Grenelle. It was as if 'an army of Huns or Vandals' had passed that way. The Hôtel Biron, where Lucie had paid visits as a child, was now a dance hall; the Hôtel Vaudreuil, once renowned for its collection of paintings by David, Boucher, Greuze and Poussin, lay in ruins; the Hôtel de Conti was a horse market.Though Lucie's house in the rue du Bac was relatively unscathed, still owned by Mme de Staël's husband, the nearby convent of the Récollettes had been turned into a theatre, its saints defaced, its cloisters overgrown and full of rubbish. Of Lucie's former neighbours, the Maréchal de Mailly, at number 1 rue du Bac, had survived the attack on the Tuileries in August 1792, only to be arrested, to escape, to be caught again and finally beheaded, at the age of 86. The Duc de Clermont-Tonnerre and his son, who had once lived between Lucie's house and the missionary order in whose orchards and gardens she had walked as a child, had been among the last to die.

* * *

Soon after reaching Paris, Lucie went to call on Thérésia Cabarrus, married for the past three years to Tallien, by whom she had a child. She was living on the edge of the Champs-Elysées, in a charming small thatched house hidden by poplars and lime trees called La Chaumière, which looked like a country cottage from the outside but was extremely luxurious inside, with frescoes in the Roman and Etruscan style. The two women greeted each other affectionately. Recounting to Lucie the story of her ordeal in prison and her last-minute rescue by Tallien as she was about to appear before the revolutionary tribunal, Thérésia told of his subsequent furious jealousy, and how she sometimes now feared for her life. Tallien had much to be jealous about. Thérésia, whose tale of deliverance and whose exploits in Bordeaux had made her the toast of Paris, had become one of the most fashionable women in a city longing for extravagance and beauty. 'This woman,' observed Pitt on a visit to Paris, 'would be capable of closing the gates of hell.'

Thérésia was to be seen at public dances, dressed *à la sauvage*, in flesh-coloured clinging culottes, with diamond rings on her toes, or *à la Cleopatra*, or *à la Diane*, or *à la Psyche*, any dress with a Greek or Roman theme. Her high-waisted, white muslin dresses, simple straw hats, shawls, preferably from Kashmir, were admired and copied throughout Paris. When she dressed as Diana, said Barras, who was widely known to be her lover, she was the 'female dictator of beauty'. In 1795, Mlle Bertin had returned to reopen her shop in the rue de la Loi, but it was the new *Journal des Dames et des Modes*, started by a former abbé and professor of philosophy called La Mesangère, which was setting the tone not only for Paris but for much of Europe.

The *Journal de Lyon* of 21 February 1793 had been the first to use the word '*muscadin*' to denote the opponents of the revolution, the word taken from the apprentice grocers of Lyon who smelt strongly of nutmeg. Distinguished by their exotic appearance – long, powdered hair, cravats worn right up to their bottom lip, grey or brown redingotes borrowed from the English, tight breeches and white stockings – the *muscadins* were parading around Paris in the summer of 1797, staves in their hands, occasionally brawling with the remaining pockets of Jacobins. At their more fanciful, they were known as '*incroyables*' or rather '*incoyables*' because they lisped their 'r's. Mercier particularly disliked the fashion for enormous cravats, saying that the 'head reposes on a cravat as on a cushion in the form of a wash-basin; with others it serves as a grave for their chins'.

Their women companions, known as '*merveilleuses*', carried fans on which were painted portraits of Marie Antoinette; they left the feeding of their babies to goats, of which there were many, wandering around the streets, because wet-nurses were hard to find. Both men and women had fleurs de lys sewn on to their clothes. As in the years leading up to the revolution, people were once again wearing the signs of their political allegiances: royalists blond wigs and black collars, Jacobins red collars and pantaloons. In the *Journal des Dames et des Modes*, La Mesangère suggested that nuns, released from their convents, should dress like Roman vestal virgins.

Very little of all this touched Lucie, who was in any case too serious and too poor to indulge in fashions. She laughed when she was offered 200 francs for her long fair hair, a hairdresser telling her that fair wigs were much in demand. 'I naturally refused,' she wrote later, 'but from then on held my hair in great respect.' Like the other *ci-devant* nobles, meeting once again in salons to revive the art of conversation, she was witnessing a new Paris rising on the ashes of the old; power had shifted from the nobility and the Church to the new rich, the bankers, the suppliers of goods to the army and the speculators, many of them using their fortunes to buy up the châteaux and mansions lying empty. Many of these people, now, held receptions more sumptuous than those under Louis XVI. *Chez* Mme de Staël, so it was said, 'one sorts things out', *chez* Talleyrand 'one mocks', and *chez* Mme Tallien 'one negotiates'.

Speculation, born in the famine and misery of 1795, when the *assignats* lost their value as they multiplied to meet the expenses of the state, had made fortunes for those canny or ruthless enough to take risks. By 1797, Paris was in a fever of speculation, over food and soap, furniture and matches, even houses, sometimes

Fashion plate of 1790. The 'femme patriote' wears red, white and blue with a tricolour cockade on her hat. Her simple loose hair is in contrast to the powdered coiffures of the recent past.

bought and sold within days, without ever being seen, many of the deals concluded under the arcades of what was now known as Maison-Egalité, the old Palais-Royal. Here agents promised huge returns on fabulous scientific discoveries, like horses that did not need feeding, or mechanical carriages that ran on wheels. The Directoire was turning into one of the most corrupt periods in French history. Good living, complained Mercier, was making the Parisians 'insolent, lazy, amoral and greedy'. In the house of one of the richest speculators, the cook produced a dinner of exotic birds, pies and pâtés of Indian curlews, Java pheasants and ostriches.

It was at dinner with Pulchérie de Valance that Lucie met Talleyrand again, as enigmatic and wily as ever, who casually let slip that he had just been appointed Foreign Minister. Nothing about Talleyrand would surprise her, noted Lucie, 'unless, perhaps, it should be something lacking in taste'. Even serving

Fashion plate of 1798. The cropped hair – '*chevelure à la Titus*' – and ankle ribbons are in classical style.

an exceedingly corrupt government, she felt sure, Talleyrand would remain a very great gentleman. Talleyrand owed his appointment, so it was said, to Mme de Staël and Benjamin Constant, the pale-faced, carrot-haired romantic writer and ardent republican who was now her constant companion, and it was rumoured in Paris that Talleyrand had threatened to blow his brains out unless he was given a ministry. Talleyrand had needed help in securing power, Mme de Staël later observed, but once there 'he had no need of anyone to keep it'.

Talleyrand now provided Lucie with a most enjoyable event, one that reminded her pleasurably of the elegance of her earlier years. His new post had coincided with the arrival in Paris of Ali Effendi, an emissary from the Sublime Porte, with an entourage of 50 attendant Turks. Since London, Berlin, Vienna and St Petersburg all had Turkish ambassadors, Paris was keen to follow suit. Ali Effendi arrived in Marseilles from Constantinople after a stormy 55-day journey and was taken aback to discover that he was expected to spend a further month in quarantine; his hosts placated him with a constant supply of fresh olives, truffles, anchovies, newspapers and yoghurt. After a royal procession across France, he reached Paris in July where the mansion of the *ci-devant* Princesse de Monaco had been put at his disposal. He was an instant, enormous, public success.

Within days, Turcomania had replaced Anglomania. Parisians did what they had always done when faced by an exciting event: they turned it into a fashion. Soon, women were wearing turbans and dresses *à la Odalisque*. The *Journal de Paris*, describing Ali Effendi as somewhat above medium height and livelier than his appearance suggested, remarked on his superb two-tiered turban, the top half green, the bottom white muslin, with a gold button perched on top. The Turkish emissary, they noted, wore ermine, which was apparently a summer fur, and the sleeves of his outer garment fell to well below his hands.

Ali Effendi's visit lasted four weeks and began with a presentation of gifts to the Directors: a silk tent, ten magnificent Arab horses, precious stones, scents and essences. Night after night, balls and displays of fireworks were laid on for his benefit. Talleyrand invited Lucie to attend a lunch party in the ambassador's honour, having arranged the room with sofas and low tables, with a sumptuous buffet, rising in tiers halfway up the wall and laden with every kind of delicious and exotic food. Leading his august Turkish guest to a divan, Talleyrand asked, through an interpreter, which lady Ali Effendi would care to have seated next to him. The Turk did not hesitate, as Lucie recorded later, adding that she was not really surprised, 'for of all the ladies present, none could stand the brilliant light of a mid-August noon, whereas my own complexion and fair hair had nothing to fear from it'. Her gleaming skin never ceased to give her pleasure. Seated by his side, Lucie found the ambassador to be a handsome man in his 50s, who asked innumerable questions through his Greek interpreter, 'and paid me a thousand amiable compliments'. On discovering that she particularly liked aromatic pastilles, he filled a handkerchief with a selection from his own pockets. Next day arrived a flask of attar of roses and a valuable length of green and gold cloth. To add to Lucie's triumph – which created something of a stir – Thérésia Tallien had not been invited to the lunch.

It was very soon after her arrival in Paris that Lucie became uncomfortably aware that royalist sentiments, quite out of tune with the mood of the Directoire, were being expressed by people other than the foppish *muscadins*. In the salon of Mme de Montesson, deputies favourably disposed to the royalist cause talked openly about the prospects of a shift to the right and the eventual return to France of Louis XVIII, still in exile in England. Many of these people wore badges and ribbons by which to recognise each other, knotting their handkerchiefs in particular ways or wearing black velvet collars. At dinners given by their friend M. de Brouquens, who had survived the revolution in Bordeaux and returned to Paris, or in the drawing room of Mme de Staël, where Lucie spent part of every day, royalist deputies in the Assembly spoke freely of their hopes, despite the presence of the servants.

Lucie was treated as 'ridiculous and pedantic' when she pointed out that every word was making its way back to Talleyrand and to Fouché, the former Oratorian and arch republican recently appointed by Barras to run an unofficial secret police for the surveillance of the former nobility. The royalists had been greatly heartened by the elections in April, which had brought in a majority of avowed constitutional monarchists as new deputies, many of them men who had been imprisoned during the Terror. In May, after a plot by the left to overthrow the Directoire was uncovered and put down before it could do harm, the government swung further to the right.

* * *

At daybreak on 4 September 1797 – 18 fructidor, as it was known under the revolutionary calendar – Lucie was sitting on her bed feeding Charlotte when she heard loud noises coming from the street. Marguerite went to see what was happening and returned to say that soldiers and gun-carriages were pouring onto the nearby boulevard. Frédéric went off to find news. When he failed to return, Lucie and Pulchérie de Valance, modestly dressed so as not to attract attention, set out for Mme de Staël's house, making their way through streets crowded with anxious, silent people. Several of the roads had been barricaded. Pushing their way to the front, they were in time to see a number of heavily guarded carriages pass by, in which sat several of the royalist deputies Lucie had met with M. de Brouquens. Seeing Lucie, the men waved, immediately provoking shouts of 'Down with the royalists' from, as Lucie wrote later, 'a number of those horrible women who appear only during revolutions or disorders'.

They reached Mme de Staël's apartments to learn that the Director Barras, fearing a monarchist takeover, had turned for help to Napoleon Bonaparte, the young general of the army in Italy whose military exploits had become the talk of Paris. Napoleon had despatched one of his subordinates to Paris, and the monarchists had quickly been crushed. The leaders of the conspiracy were all under arrest. Carnot, one of the two Directors involved, had fled; the other, the moderate royalist Barthélemy, who had replaced Le Tourneur, was in prison.

Within hours, 60 right-wing deputies, including many of the men who had spoken out so freely in the salons in the Faubourg Saint-Germain, were on their way to the 'dry guillotine': this meant imprisonment in Guyana, the prisoners being sent across France in iron cages to the coast. Many would eventually die in Guyana. The results of the April elections were declared void and 177 of the new deputies banished. Members of the Bourbon royal family, such as the Prince de Conti and the Duchesse d'Orléans, had been rounded up and were to be deported to Spain. Over the next few weeks a military commission set up in the Hôtel de Ville passed death sentences on a number of plotters, several of them recently returned émigrés. In the name of suppressing a counter-revolutionary conspiracy, the remaining three members of the Directoire, Barras, Reubell and La Revéllière, with the support of the army, took over the administration of the state and outlawed all opposition. The first steps towards dictatorship had effectively been taken.

For the recently returned émigrés, just beginning to find their feet and retrieve what was left of their pre-revolutionary fortunes, the attempted right-wing coup spelt disaster. France's new leaders were not prepared to see a return to power of monarchists. Under threat of arrest and trial before a military tribunal, some 150,000 people were ordered to leave Paris within 24 hours and France within a week.

Lucie's first thought was how to warn the Princesse d'Hénin, who was staying just outside the city at Saint-Ouen with the

Princess de Poix. The barricades were up all around Paris. No one was being permitted to leave without a passport. Since she was still feeding Charlotte, she was forced to take the baby with her. Passing herself off as a nurse to a friend with a valid passport to travel, she reached Saint-Ouen on foot, exhausted by the long trudge with her plump daughter. The émigrés sheltering there together, some of them under false names – the Princesse d'Hénin had a false passport in the name of a Swiss dressmaker – were appalled by Lucie's news: many were in the middle of delicate negotiations for the return of their properties from the state.

For a while, Lucie and Frédéric, whose names had never been on any list of émigrés, hoped that they would not be affected by the new decree. They called on Talleyrand for advice, but found him too preoccupied with his own future to offer much reassurance. Even Tallien was unable to help, though he did provide them with passports. When it became clear that they, too, as *ci-devant* nobles at the court of Louis XVI, would have to leave France once again, they briefly considered going to Spain, and from there returning to America; but the Princesse d'Hénin, who intended to return to London, persuaded Frédéric that England would be better. The Faubourg Saint-Germain was in a state of turmoil, the streets full of unhappy and undecided people, wondering where to go and how to get there. No one could contemplate a return to the poverty and precariousness of exile without dread. Frédéric, who had been discussing the buying back of Hautefontaine, was forced to abandon all talks.

In a mood of profound gloom and uncertainty, with just two small trunks of clothes, everything else they possessed still at Le Bouilh, Lucie, Frédéric, the two children and Marguerite set out in a coach for Calais. This second emigration would prove more ruinous than the first.

CHAPTER ELEVEN

Hordes of Vagabond French

Their boat left Calais for Dover at eleven o'clock at night. Though the south-easterly wind was light and the sky cloudless, Frédéric immediately took to his bunk, overcome by his usual seasickness. Lucie went up on deck and found a hatch cover to perch on, holding Charlotte on her knee. Seated next to her was a young man who offered her a shoulder to lean against: he turned out to be the son of the editor of the *Edinburgh Review*, whom she had known in Boston. They spent the night talking about America; Lucie told him that if it proved impossible to return to France before too long, her plan – her wish – was to go back to the farm in Troy. With the pale dawn of an English September came her first sight of the cliffs of Dover.

By 1797, the English had become accustomed to their many French visitors, few of whom had risked, as Lucie and Frédéric had, too quick a return to France after the Terror. Fugitives from the revolution had been reaching the shores of the south coast since the summer of 1789, in waves that increased in response to each new incident of violence and each new repressive law against the French Church and the nobility. One of the first to come had been the Prince de Condé, who brought with him 28 servants; later had come soldiers, fleeing anarchy in the army, 'non-juror' clerics who refused to sever their vows to the Pope, aristocrats on the list of 'suspects', and all the wig-makers, chefs, valets and coachmen who had once looked after them. Many were unclear as to whether they were betraying the King in abandoning him to his fate, or had been driven into exile by his weakness. '*La patrie* [the fatherland] becomes a meaningless term,' noted the Comte d'Antraigues, 'when it has lost its laws, its customs, its habits . . . France for me is nothing but a corpse, and all one loves of the dead is the memory of them.'

Singly or in groups, among them entire congregations of priests and convents of nuns, these soldiers and clerics and servants had arrived, some bringing money in bags and accompanied by retainers, others destitute, bedraggled and disguised as women or sailors. When a party of 37 religious sisters from a convent at Montargin put ashore on Shoreham beach, curious spectators gathered to peer at the 'fugitive virgins'. For the most part, those arriving had been received kindly and with generosity. While they were unlikely to 'much contribute to our amusement', said Gibbon, these people were entitled to pity and esteem.

Though by 1797 France was again at war with England, for the third time in 40 years, the French and the English were inextricably entangled, importing each other's fashions and craftsmen, reading each other's Enlightenment philosophers, sharing notions of '*bon ton*' and good taste. When, before the revolution, the English aristocracy had visited Paris, they had mixed naturally with the inhabitants of the Faubourg Saint-Germain. Lucie's family was in no way exceptional either in its fluent command of English, or in its many close English relations, particularly among the Catholics. The sons of recusant families were still obliged to travel to the continent to get a Catholic education. The Duc de Lauzun had shared a string of racehorses at Newmarket with Fox; the Duchess of Devonshire brought French interior decorators to Chatsworth. For the Whig women in particular, the salons of Paris, with their scholarly and formidable hostesses, to whom the great men of the day paid homage, had considerable appeal.

When the Bastille fell, the initial reaction in England had there fore been to welcome the revolution, precisely because it was perceived to be supported by Condorcet, Lafayette, Frédéric de la Tour du Pin and his father, all people the English knew and admired. For the English religious dissenters, as for the opposition Whigs, the defeat of the despotic Bourbons was a blow against tyranny. Fox spoke of the fall of the Bastille as 'the greatest event that ever happened in the world'.

All this changed abruptly in 1792. With the attack on the Tuileries, the massacre of the priests in the prisons, and the execution of the King and Marie Antoinette, came the realisation that the bloodless revolution they had fondly imagined was an illusion. The stories of bloodshed carried across the Channel by the émigrés had shocked the English, bringing out the best in them, particularly as many of the French refugees displayed admirable stoicism in the face of their disasters. Those arriving throughout 1792 and 1793, among whom there were several thousand Catholic priests, were taken in and helped. The violence and confusion in France had lent itself naturally to caricature and provided Gillray with savage flights of fantasy, with his frolicking *sans-culottes* and sharp-toothed *poissardes*, perfect illustrations for Burke's vision of the madness and destruction sweeping Europe.

But the French Revolution had had its own effect on England. It had rekindled a radicalism largely dormant in British politics since the early 1780s, bringing to the fore questions about human rights and social justice, questions raised by Thomas Paine in *The Rights of Man*, and Mary Wollstonecraft in *A Vindication of the Rights of Women*. The ownership of land, popular education, the role of the press and the Church, had all rapidly become topics fit for debate among people who no longer saw themselves as trapped within prescribed social spheres, but free to learn, to advance themselves and to challenge the rights of others. At the same time, renewed war with France had brought in its wake inflation, press gangs and crippling taxes. With republican stirrings and whispers becoming more audible, Pitt had responded by introducing penalties for sedition, libel and treason. The passage of the Aliens Act, in January 1793 – which had caught Talleyrand in its net – had been designed to control republican spies and Jacobins; but for the French émigrés trying to reach England, it had meant more documents and passports and more daunting bureaucracy.

Burke had lived just long enough to see his warnings about a 'despotic democracy' in France, a 'strange, nameless, wild thing' liable to threaten all Europe, at first greeted as slightly absurd, at last taken seriously. In March 1797, not long before Lucie's arrival, *The Times* warned against those 'Foreigners . . . whose dangerous opinions, suspicious conduct and violent speeches call for the utmost vigilance and severity', and urged that steps be taken to curb the spread of Jacobin ideas. Reports of 'stout-made' men landing from fishing boats, uttering 'improper expressions relative to government', were despatched from customs officers along the coast to London.

The England reached by Frédéric and Lucie and their children, on a blowy September morning in 1797, was therefore not as welcoming as it had been earlier. Eight years of steady arrivals, bringing some 25,000 Frenchmen, Frenchwomen and their children to England, added to growing shortages and political repression, had not made people eager to face a fresh surge of émigrés, what a British politician described as 'hordes of vagabond French . . . pouring in upon us'. On the docks of Dover, as the de la Tour du Pins climbed ashore, they were treated roughly by the English customs officers, considerably worse, as Lucie observed, than they had been treated in Spain. It was not until she pointed out that she was an English subject, and gave them the names of her three English uncles – Lord Dillon, Lord Kenmare and Sir William Jerningham – that the brusque tone of the officers changed.

Lucie's first act was to send word to her aunt, Lady Jerningham, and by the time they reached London that night Edward, the young page at her wedding, was waiting to take them to the Jerningham house in Bolton Row off Piccadilly. There they found Sir William's brother, the Chevalier Jerningham, a frequent visitor during Lucie's childhood to the rue du Bac and Hautefontaine. Lady Jerningham was a forceful, good-hearted woman and much attached to Lucie, who noted with considerable pleasure how kind her aunt was to Frédéric and how genuinely taken she seemed to be with Humbert. 'We established ourselves,' wrote Lucie, 'as if we had been the children of the house.' Of all her relations, both French and English, Lucie felt fondest of her English aunt.

* * *

London, at the end of the 18th century, had seen vast changes. It was the age of the Georgian terrace house and the neo-classical architecture made fashionable by Robert Adam, James Gibbs and Sir John Soane. With just under a million people, London was the largest city in Europe, owning and building more ships than anywhere in the world. Like Paris, it was noisy, packed with wagons, carriages and markets, its streets full of itinerant vendors ringing their bells and singing out their wares: cigars and walking sticks, rat poison and shirt buttons, shellfish and sheep's trotters. As Paris did for France, London represented the heart of the country. Sydney Smith, obliged to live in his parish in Yorkshire, remarked that for him it meant being '12 miles from the nearest lemon'.

But while France had suffered greatly from its political revolution, England had been growing rich from its inventions in industry and agriculture, from its new machinery for spinning, smelting and pumping, an explosion of technology that had brought with it extraordinary wealth, but also brutal and dangerous conditions. England was the country where the basic inventions that would create modern industry were made, perfected and introduced, but children as young as 5 were going to work cleaning chimneys.

To house London's ever-growing population, the city had been moving west, past Mayfair and Piccadilly, and north, through Marylebone towards Islington and Highgate, where visitors were warned that the cold of the winter was so extreme that 'many constitutions cannot endure it'. Though Hampstead was still a heath, menaced by highwaymen, and Kensington and Chelsea were villages, surrounded by countryside, London was soon to engulf them. To amuse and simplify the lives of Londoners, there were new gadgets: toothbrushes, roller skates, ball-cocks, dumb-waiters and fountain pens. The sandwich had just made its appearance, as had toast, though the young Swedish scientist, Peter Kalm, maintained that toast had been invented because it was the only way to spread butter on to bread in an English winter.

Almost every French visitor, having dutifully praised the spaciousness and cleanliness of the fine new Georgian squares and the excellence of the street lighting, remarked on the tangled lanes that lay behind them, with their hidden courtyards and dark alleyways, home to the poor, of whom there were a growing number, hit hard by spiralling grain prices and the enclosures of common land. The year of Lucie's arrival, Irish rebels were sent off to Botany Bay, to join the English felons, men and women convicted for stealing sheep and poaching deer, in the harsh penal settlements of newly discovered Australia.

The French also remarked on the desolation of the London Sunday, the streets silent and deserted, the few passers-by 'like walking shadows'. As the Comte de Montloisier put it, the English were a 'semi-paralysed people'. And there were very few indeed who did not comment on the fog, the swirling greyish-white mists from the coal-burning stoves which obliterated the city for days at a stretch. It was the fog, remarked Montloisier, which was responsible for the lack of 'animal vitality' in Londoners, and for their permanent '*état de spleen*', an affliction that combined boredom and melancholy and often led to madness and suicide. In his seven years in London, he noted gloomily, he saw the grapes growing on a wall opposite his house change colour only once. Just occasionally, this moroseness was described as having something of the much loved French '*sensibilité*'; more often it was seen as seriousness so profound that some French visitors wondered whether the English, enveloped in fog and puffed up like balloons on butter and beer, were in fact capable of laughter at all.

The 'era of Jacobinism' that had preceded Lucie to London had brought with it more austere fashions. Gone were the buckles, ruffles and powdered wigs* for men, the 'plumpers' designed to fatten the cheeks of sallow women, the 4-feet-tall ostrich feathers, the hoops and the false buttocks; except at court, where George III and his German-speaking Queen Charlotte had been on the throne for more than 30 years, presiding over a staid Windsor, where hoops remained *de rigueur* on formal occasions. George III's first spell of madness had happened soon after the fall of the Bastille, in the summer of 1789. Compared to Versailles, the French courtiers found Windsor painfully cold and stupefyingly dull.

Shortly before Lucie's arrival, the Princess Royal had married the hereditary Prince of Württemberg, a man so fat that Napoleon would later remark that God had created him merely to see how far the human skin could be stretched without bursting. Whether at court or in the drawing rooms of London society, Englishwomen remained, to the surprise and annoyance of their French guests, firmly in their segregated and inferior places, expected to withdraw after dinner to allow the men to talk literature and politics. In England, a visitor smugly remarked, women were 'the momentary toy of passion', while in France they were companions 'in the hours of reason and conversation'. As Jane Austen put it, 'Imbecility in females is a great enhancement of their personal charms', something that Lucie, brought up to talk intelligently, would find extraordinary. The French were also disconcerted by the casual manners of

* Walter Savage Landor was reported to be the first Oxford undergraduate to give up powdered hair, in 1793.

their hosts, the way young people hummed, put their feet up and perched on tables.

* * *

With rich relations able to take the family in, Lucie's position in London was considerably easier than that of all but a handful of the other French émigrés. But it was also strewn with emotional traps. Her stepmother, Countess Dillon, had reached England some time before, bringing with her Betsy and Alexandre, her two children by her first husband, and 13-year-old Fanny, her daughter by Arthur. Two other daughters by Arthur, whom Lucie had never seen, had died in infancy.

Lucie was fond of the gentle, good-natured Betsy, whom she had known when at a convent school in Paris, and who had just made what was turning into an unhappy marriage with the well-connected but wayward Edward de Fitz-James. Betsy was pregnant and clearly miserable. 'She was a sweet girl,' Lucie wrote later, 'and deserved a better fate.' Edward's mother, the Duchesse de Fitz-James, ran what was described as the gayest of the French salons in London, the one where '*la haute émigration*', the most noble émigrés, gathered. Lucie felt rather less affection for Alexandre who, though carefree and charming, had, she said, 'little intelligence and even less learning', lacked all talent and was interested only in fashion, horses and 'small intrigues'. Like his mother, whom Lucie described grudgingly as a woman 'not without a certain natural wit', Alexandre had never been known to open a book.

Fearing to be greeted coldly by her stepmother, Lucie was relieved and pleased when Mme Dillon came to call, eager for an account of Arthur's last winter in Paris. Though they had much to talk about, Lucie felt surprised that her clever and well-educated father had chosen to marry such a woman.

More problematic was how Lucie would be received by her grandmother, the ill-tempered Mme de Rothe, who had fled from Koblenz to London with Archbishop Dillon after the defeat of the émigré army. To Lucie's enduring relief, they had not met since Lucie's departure for Holland nine years earlier, and there had been virtually no contact between them in all that time. Persuaded by Lady Jerningham that it was her duty to show her grandmother some mark of respect, Lucie, taking the bright and easy-going Humbert with her, but leaving Frédéric behind, went to call. What Frédéric did not tell her until later was that Mme de Rothe had been spreading malicious stories about Arthur and about himself around London, and that she had stipulated that on no account would she receive him.

Mme de Rothe and the weak but not unkind Archbishop were living in a modest house in Thayer Street on a pension of £1,000 a year provided by Arthur's eldest brother, Lord Dillon. One of the servants who had accompanied them into exile, Michel Esquerre, mistakenly believing it safe to return to France, had gone home and been guillotined in May 1794. The Archbishop was now 84; the authorities, who, through agents, kept a close eye on the émigrés settled in England, listed him as a 'rebel to his country and his Church'.

When his manservant, who remembered Lucie from Hautefontaine and burst into tears on seeing her at the door, announced her, the Archbishop greeted her most affectionately. He hugged Humbert 'again and again' and soon began questioning him in French and English, evidently much charmed by the little boy. He pressed Lucie to return next day to dine with him and the six elderly bishops from the Languedoc who shared his meals and whom Lucie remembered from her travels to Narbonne.

Mme de Rothe was icy. Lucie kissed her hand; her grandmother addressed her as 'Madame'. Learning that Lady Jerningham had invited the family to spend the winter at Cossey Hall, near Norwich in Norfolk, further displeased her and Lucie watched with mounting anxiety as Mme de Rothe began to mutter to herself under her breath, a sign Lucie remembered as heralding an outburst of ill-temper. After half an hour, trembling lest her grandmother embark on a long list of accusations against her father or against Frédéric, Lucie kissed her hand again and left.

Lucie never referred to her grandmother again, in anything she wrote, though she paid many visits to the Archbishop and cannot have failed to have met her. It was as if her profoundly unhappy childhood had simply never happened.

After this, came a visit to Lord Dillon, who greeted her with a 'cool courtesy', and offered her his box at the Opera; but nothing else. Then Lucie called on Lord Kenmare and his 18-year-old daughter Charlotte, who were both warm and affectionate.

There was one family visit left, and in some ways Lucie feared it the most. There was something in her relationship with Frédéric's forceful aunt that had always troubled and unnerved her. The Princesse d'Hénin was living in Richmond, where many of the aristocratic émigrés had settled, sharing a small house with her faithful and cowed companion, Lally-Tollendal. This second exile had not softened her forceful nature and she made no effort to conceal her envy at Lucie's invitation to Cossey, all the greater since Lally-Tollendal had spent many pleasant months there on his own as the guest of Lady Jerningham. But the Princesse d'Hénin was never cold and unjust in the manner of Mme de Rothe, and Lucie was grateful that she was now generous enough to see how important Lady Jerningham's support would be to the family. It was with considerable relief that Lucie, her round of visits completed, prepared to leave for the country. What little she had seen of the émigrés in London, gossiping and intriguing, while fluttering shamelessly around the 'pale constellation' of rich and fashionable hostesses, had depressed her exceedingly.

* * *

The party, consisting of Lucie, Frédéric, the children and Marguerite, as well as Lucie's stepmother and her children and various maids and grooms, travelled to Norfolk in a convoy of carriages, crawling slowly over terrible roads marked, for the first time since the Romans, by regular milestones. The mail coach service had recently been extended as far as Norwich, and couriers passed

Cossey Hall, the Norfolk home of Lucie's aunt, Lady Jerningham, with its chapel built some years later.

them on the road at the considerably faster speed of 9 miles an hour. Lucie, for whom new sights and new places never failed to raise her spirits and give her pleasure, greatly enjoyed a day they spent on the way at the races in Newmarket. They reached Cossey at the beginning of October; the weather was windy but mild.

Cossey Hall, not far from Great Yarmouth on the Norfolk coast, which had been in the Jerningham family since the middle of the 16th century, was a large, red-brick, partly medieval and partly Tudor house, with a single tower. It had projecting wings, angled buttresses and gables, topped by square pinnacles. More imposing and eccentric than beautiful, it was said to have been lived in for a while by Anne of Cleves, when banished from court by Henry VIII. But its setting, in the middle of a valley through which wound the river Wensum, surrounded by forests of oak, beech and chestnut, was delightful. The park was full of deer and, compared to the severe geometric parterres and gravelled paths of the great French gardens, seemed, with its grass, ponds and walks, very informal. The walled kitchen garden had glasshouses for tomatoes and peaches and a 60-foot run of cucumber pits. A hamlet of about 600 people, most of them living on an island of cottages contained in a loop of the river, provided staff for the estate. After the ruggedness of Albany, it was all very gentle. When the weather was good, Lucie, wearing a new riding habit given to her by her stepmother, went out on one of Edward de Fitz-James's horses, on a side-saddle he had thoughtfully provided.

As Catholics, Sir William Jerningham and his brother the Chevalier, a Knight of Malta, had been educated in France, Sir William staying on to serve at court and in the army. With memories of the Gordon riots of 1780, and their virulent attacks on Catholics by a burning, looting mob, fresh in people's minds, the Jerningham chapel at Cossey remained hidden away at the top of the house in the gables, though by the end of the 18th century restrictions against Catholics were beginning to ease. Sir William, who had returned from France to campaign on behalf of English Catholics, was planning a new chapel in the grounds, modelled on King's College Chapel in Cambridge. Since the beginning of the revolution, Lady Jerningham had been taking in as guests French refugee priests as well as the sisters from the Blue Nuns in the Faubourg Saint-Antoine, where she had studied as a girl, and her salon for the French had earned her the affectionate name of 'her Catholic Majesty'. Ever practical, she had written to her daughter that 'I like to have several People in the House, and a multitude cannot be had cheaper than with the unfortunate French: no Servants, no Horses, no Drinkings'.

About the inside of the house, with its mullioned windows, wood panelling and flagstones, Lucie would write only that it was 'old but comfortable'; she approved of the food, saying that it was 'plentiful and not too elaborate'. Coming from a Parisian childhood, where the nobility, though less wealthy than the English aristocracy, lived on a far more lavish scale, what she seems to have most liked was Cossey's unpretentiousness, though with £18,000 a year in income, the Jerninghams were one of England's small elite of privileged landed families. Accustomed to the acute cold of a North American winter, she does not appear to have noticed the chilliness and draughts for which English country houses were already famous.

They were a large group in the house, 19, including a Catholic chaplain, most of them in some way related. Charlotte, Lady Jerningham's only daughter, who had recently married Sir Richard Bedingfield, lived not far away in a moated 15th-century fortified manor house. Both Sir William and his brother spoke excellent French and Lady Jerningham took charge of Humbert's education, leading him off to her room every morning after breakfast to read and write in both French and English. Sir William, a warm-hearted, affable man, was writing a paper on mangel-wurzels. There was an excellent library for Frédéric to read in and among the pictures that hung on the panelled walls was a portrait of Queen Mary Tudor by Holbein.

Lucie had fled from France without warm clothes for either herself or the children, but when cold weather set in at the end of October, and it began to snow, Lady Jerningham sent out for winter wardrobes for them all. Knowing that Lucie could sew, and wanting to provide her with some occupation, she ordered lengths of different materials, tactfully pretending it would also encourage Fanny to sew. To distinguish her from another Dillon cousin called Fanny, Lucie's stepsister, who was turning into a pretty, somewhat forceful girl, with an oval face and determined brown eyes, was known as the 'little tall Fanny'.

A local parson, the Rev. P. Woodforde, coming to dine with

the family at Cossey, was shocked to find them eating pheasant, swan and ham on Fridays and fast days. Lady Jerningham, he noted, was a 'fine woman, thou' large and extremely sensible but much given to satire'. Part of her sensibleness was to insist that Humbert and Charlotte be vaccinated against smallpox, and she sent to Norwich for her own doctor to carry it out. When spring came and the Jerninghams prepared to return to London, they pressed Lucie and Frédéric to remain at Cossey, in a small cottage in the grounds. But Frédéric had business in London to attend to; and Lucie was once again pregnant – her seventh pregnancy in as many years – and feeling very ill. Fearing that she might again miscarry, she preferred to be near good doctors. The pleasant months in the country came to an end. They had suited Lucie well, just as her years in Troy had suited her. Among the many advantages of her equable nature was a genuine ability to make the most of wherever she found herself, and a refusal to spend time regretting or anticipating.

* * *

For all the generosity of their English hosts, the French émigrés were seldom very happy in their state of exile. Those who had not grown up speaking English – the great majority – found the language hard; they hated the cold; they considered the fruit 'bad, sour and half ripe' and the tarts watery under their crusts of 'half-cooked dough'. When the Marchese de Caracciolo wrote to the King of Naples that in England he had discovered a country of '22 religions and two sauces', these were not words of praise. Fretting about events at home, mourning those who had gone to the guillotine, many of the émigrés had become crabby and quarrelsome, feeding, as Lucie had remarked, on gossip and rumour. They were being devoured, observed Mallet du Pan, editor of one of London's three French language papers, *Le Mercure Britannique*, and briefly a spy for the British government, 'by an indomitable spirit of discord, malice and despotism'.

And many were extremely badly off, particularly the widows with young children, and the elderly clerics, living in damp dark lodgings in Southwark or Somerstown, mostly on a diet of potatoes, the staple food of the poor which the French much despised. Chateaubriand, who began his period of exile in a garret off Holborn and was capable of feeling aggrieved in almost any circumstance in which he found himself, wrote later that 'I was eaten up with hunger . . . I sucked linen rags dipped in water, I chewed grass and paper. When I went past bakeries I was horribly tormented.' Chateaubriand soon gravitated to smarter lodgings in Marylebone.

Some of the worst poverty was alleviated by a number of charitable ventures, both public and private. Jean-François de la Marche, Bishop of Saint-Pol de Léon, had fled to England from Brittany in the spring of 1791. Though he was over 60, his health was good and he set up a Committee for the Relief of the French Clergy and Laity in Golden Square, where he began distributing relief at the rate of 1 shilling a day for adult men, and half as much for women and servants. Reports to the committee, listing the worst cases, spoke of 'Mme de D . . . dead of hunger . . . she has left a paralysed husband and three sons, all three sick' and 'Mme de B . . . left without anything and five children who are completely naked'. As needs increased, the administration of the fund was taken over by the government. In its editorials *The Times* urged Londoners to respond generously: 'Should this country,' the paper asked, 'not afford some further protection to these unfortunate strangers, whither must they fly?'

But as the months and then the years passed, even those who had arrived with money began to run out of funds. Our fortunes, wrote the Marquis de Tremane to the Prince de Bouillon, a rich and philanthropic Frenchman who did a great deal to support those worst off, 'are simply not lasting as long as our persecution'. Some of the émigrés fell ill, others lost their jobs. M. de Rodire, who taught French, 'lost his scholars'. Mlle le Boucher went mad. The disastrous Quiberon expedition in 1795, after which many of those who were not killed in battle were guillotined under the émigré laws, had left many hundreds of widows and small children, reduced to selling off, one by one, all their possessions. For a single year, the committee estimated that it needed to raise at least £150,000 for the destitute émigrés, a quarter of it going to members of the nobility. Some of the money went on health, the Middlesex Hospital having agreed in 1793 to open two special wards for 'sick French clergy'. Most of them were reported to suffer from bad eyesight and '*grande faiblesse*', extreme feebleness, though there were official complaints about the number of leeches used in a single year – 36,100 – and the quantity of wine the French patients drank (49 dozen bottles). Could some of these French priests, asked *The Times*, not be persuaded to help get in the harvest?

But it was not all grim. Lucie was only one of many writers who later pointed out that wherever possible, even in the midst of misery, the French émigrés remained astonishingly cheerful and that when they got together, they laughed. The French, remarked Lucie, 'are by nature gay, so that although we were desolate, ruined and furious, we nonetheless succeeded in preserving our good humour'. Though few could afford a carriage – the Abbé Baston, who found London 'monstrously big', complained that his long walks were ruined by impudent women who buffeted him off the pavements – and very few owned the changes of dress necessary to go out in society, most nonetheless took pleasure in brightly coloured, noisy, smelly, bustling Georgian London.

They went, when they had the money, to the vast new Drury Lane theatre, or to the Vauxhall and the Ranelagh pleasure gardens, where they sat under arbours of honeysuckle and roses and listened to music or danced late into the night. They strolled along Charing Cross, through Leicester Square and Piccadilly, and watched street shows of freaks, midgets, women gladiators and even mathematical pigs, and if they were lucky they saw the 'amazing Learn'd Dog' which could answer questions on Ovid's *Metamorphoses*, knew the Greek alphabet and could tell the time. For the more robust, there was bear-baiting and cock-fighting.

The Abbé Tardy, who wrote a guide to help his compatriots navigate the hazards of this alien land, recommended excursions to Hackney, 'the richest village in Europe', and to the 'mountain of Sydenham' to admire the views over London. Those who wished to learn to swim he directed to the Peerless Pool, near Finsbury Square. Good coffee, he warned, was all but impossible to find; best to drink port wine, 'which the climate of England demands'. On the subject of boiled vegetables in white sauces, and cold boiled meat, which his readers should expect to encounter if they stayed in a *pension*, the Abbé was very gloomy.

When not weakened by illness, the French émigrés, whatever their class and background, were extremely resourceful. In workshops up and down Marylebone, women who had grown up surrounded by servants, never doing anything that resembled work, now embroidered chiffon dresses and made straw hats, sending the men to the Cornhill to buy the straw, and giving the youngest women the unenviable task of selling the hats to milliners. Two hundred priests made carpets for the Marquis of Buckingham, while monks entertained Lord Bridgewater's guests by strolling up and down the lawns of his country estate, reading their breviaries. Jean-Baptiste Cléry, Louis XVI's valet, gave readings from his account of the King's tearful parting from his family. Jean-Gabriel Peltier had a miniature guillotine built of walnut and, appealing to the macabre and ghoulish in Georgian England, announced: 'Today we guillotine a goose, tomorrow a duck.' Mlle Merelle gave harp lessons; M. de Gaumont bound books; the Marquis de Chavannes sold coal and the Comte de Belinaye wine. M. d'Albignac tossed salads at fashionable dinner parties. The Comtesse de Guery's ice creams became popular with the Prince of Wales; and when the Abbé Delille, the 'blind bard of emigration', reciting Milton in French in the Duchess of Devonshire's drawing room, reached the part about the King's execution, his listeners wept. Pierre Danloux, bitter foe in Paris of David, spent ten years in England painting portraits – among them that of Betsy de Fitz-James – charging 15 guineas for a bust and 50 for a full-length picture. Dispossessed, sad and fearful for their future, they complained, as Lucie noted, remarkably little.

Soho, first populated by the French Protestant Huguenots fleeing persecution in the 1680s, became a gathering place for émigré writers and artists. They congregated in a bookshop run by a former Benedictine monk, who used his own library to start his business, or at the émigré school opened by the Abbé Carron in the Tottenham Court Road; and they met after Mass on Sundays in St Patrick's Chapel in Sutton Street, consecrated in 1792 as the first Catholic chapel to be built that was not attached to an embassy. A few wrote novels, often heavily autobiographical in nature, full of English 'mylords' and examples of '*délicatesse*', which fed the new appetite for romantic fiction on both sides of the Channel. The Comtesse de Flahaut, Talleyrand's mistress and mother of his child, sick of making straw hats, wrote her immensely successful *Adèle de Senange*. Those who had money were expected to share it. When guests left the Duchesse de Fitz-James's soirées, they were supposed to slip some money under their plates.

And some young women, who could find no way to make money, contracted hasty marriages, like that of the Comtesse d'Osmond's daughter, Adèle, who at 17 accepted the hand of the 42-year-old General de Boigne, marrying him 12 days after their first meeting. The Comtesse d'Osmond was the Dillon relation who had come to Hautefontaine when Lucie was a child to beg the Archbishop's help at court, and who had much disapproved of the lax morals of the house. Both Adèle and her mother made very little effort to conceal their distaste for the General, who was both rich and generous; soon, their criticisms and mockery were being repeated around London and General Boigne's shortcomings became a topic for gossip throughout the European émigré world.

Lucie and Frédéric were beginning to experience money problems of their own. Forced to flee Paris at such short notice without returning to Le Bouilh, they had brought almost nothing out with them. They could have applied to the Bishop of Saint-Pol de Léon, but felt that this would embarrass their rich relations. Though Frédéric was considered a traitor by Mme de Rothe for his early enthusiasm for the revolution, the Archbishop offered to write to Lord Dillon on their behalf, pointing out that however badly her husband had behaved, this should not be used as a reason not to help Lucie. She was, after all, his niece. When he had been the head of the family, 'in my days of glory and opulence', he had felt an obligation to provide 'for all my relations in trouble'. The letter yielded nothing. The single invitation to his box at the opera was Lord Dillon's sole gesture towards his French niece.

On returning to London, however, Lucie and Frédéric learnt of a sad but very welcome legacy. A letter came from Martinique, informing them of the death of M. Combes, Lucie's much-loved tutor at Hautefontaine, whom Arthur had appointed Registrar of Martinique. Living in Arthur's house on the island, M. Combes had managed to save 60,000 francs. While the de la Tour du Pins had been in Albany, M. Combes had repeatedly tried to send them money, only to be thwarted at every turn by Lucie's stepmother, who had borrowed the capital – against interest – and found constant excuses for not paying it back. Not long before his death, M. Combes had written to say that the grief of knowing Lucie to be in a foreign country, without money, was slowly killing him. With interest accrued over the years, M. Combes's legacy amounted to slightly more than 71,000 francs.

The life of the French planters in exile was considerably more comfortable than that of the French mainland émigrés. Most had managed to bring out with them at least part of their considerable fortune and many continued to receive rents and revenues from their plantations in the West Indies. Since many had settled in Marylebone, in fashionable houses in and around Manchester Square, also home to the richer noble Parisians, the area was referred to as '*le Faubourg Saint-Germain*'. It was here that Lucie's stepmother had found a house, and where she entertained on a lavish scale. Until news of M. Combes's bequest reached London, she had treated Lucie with affection. Overnight, she became distant and increasingly hostile. Referring Frédéric to her Creole agent for the money, she pleaded that funds were not arriving and that the sugar harvest had been poor. Small sums were advanced, with extreme reluctance. 'We were given as alms,' noted Lucie bitterly, 'what was really our own property.'

Not willing to be a constant burden on Lady Jerningham, and in any case well aware of Mme d'Hénin's proprietary feelings towards Frédéric, they now accepted her invitation to share her house in Richmond. Lucie, worrying constantly about money, was also conscious of the need for clothes for the coming baby. She wished they had never left Cossey.

* * *

Richmond lay 9 miles west of central London, reached along the river, through countryside filled by market gardens and hamlets going down to the water's edge. The Abbé de Blanc, an early visitor, described it as an immense garden offering the eye 'a kind of image of earthly paradise'. A bridge had been built in 1777 to join Richmond to Twickenham, and Garrick's theatre, just off the Green, was popular with both residents and visitors. Richmond itself remained a charming small country town, with views across the fields to spires and the white gables of farmhouses, and in the distance the grey towers of Windsor. Up and down the Thames, watermen waited to ferry travellers across. Rich merchants, wanting summer houses close to the city, had bought land on which to build imposing new mansions and villas; cottages had gone up in the lanes and alleys behind. The mail arrived at 9 in the morning, and there was a regular coach service all through the day to London from the courtyard of the Old Ship Inn.

Since 1789, some 40 French émigrés had settled in Richmond,

most of them aristocrats and royalists. They found life cheaper and less crowded than in London. One of the first to arrive had been Amélie de Lauzun, wife of Arthur's faithless friend, though by 1797 she was dead, having returned too soon to Paris and gone to the guillotine for treason not long after her husband. Horace Walpole, living nearby at Strawberry Hill, befriended many of the émigrés, playing lotto with them in the evenings, and referring to Richmond as '*une véritable petty France*'. Though frequently impatient at the endless gossip about their 'absurd countrymen', he had been horrified by the news of the September massacres, and repelled by the fact that so many of the 'perpetrators or advocates for such universal devastation' had been philosophers, geometricians and astronomers.

When the Princesse d'Hénin had first fled to England in September 1792, she and Lally-Tollendal had shared a house in a damp valley in Surrey at the foot of Box Hill, 20 miles from London, with Mme de Staël, her lover the Comte de Narbonne and Talleyrand. Juniper Hall, a red-brick former coaching inn, had acquired a reputation for intrigue and scandal, both Mme de Staël and the Princesse d'Hénin living openly with their lovers in what were described as the 'elegantly disordered alcoves of *Les Liaisons Dangereuses*'. Though their view that despite the horrors and the excesses, the revolution had been both necessary and inevitable had made them suspect in the eyes of both Pitt's government and most of the exiled royalists, Mme de Staël's brilliance and fascination had drawn many visitors to Juniper Hall. But in the spring of 1793, Mme de Staël had tired of England, telling Gibbon that the Tories had crushed all serious argument and made London the most boring city on earth, and she returned to Switzerland and to her parents. The author Fanny Burney, who had recently left her position at court and lived nearby, met her future husband Alexandre d'Arblay at Juniper Hall. She described the weak but scholarly Lally-Tollendal as the Cicero of the French Revolution.

When Juniper Hall was given up, Princesse d'Hénin and Lally-Tollendal had moved to Richmond, to a small white gabled house in Osmond Row, and it was here that Lucie, Frédéric and the children were invited to occupy the ground floor. Though pretty, the house was extremely cramped. Lucie shared one small room with Charlotte, Frédéric a second with Humbert. It is not clear where Marguerite slept. There was nowhere for them to sit or to receive visitors, other than in the Princesse d'Hénin's drawing room. Lucie admired Frédéric's aunt; but she was not fond of her. She found her manners too autocratic and her tongue too sharp. The Princesse's bad temper and outspoken views, which had grown sharper with time and misfortune, had already alienated much of the émigré community. Every day, Lucie would write, she was forced to suffer 'showers of pinpricks'. Lally-Tollendal was scarcely better liked. In the royalist salons, he was referred to as 'the people's dregs'; Lucie, who appreciated his kindness, considered him the 'most timorous of gentlemen' and had little patience with the way that he never dared risk any amusing remark for fear that it offend the Princesse d'Hénin.

In the evenings in Osmond Row, they were joined by other disaffected émigrés, all of them at odds with the royalist Relief Committee. From time to time, Chateaubriand came to read aloud from his new novel, *Atala*. Lucie, at 28 considerably younger than the rest of the party, and describing herself later as 'laughter-loving', found these evenings very tedious. She was bored not only by the ceaseless political debate, but by Lally-Tollendal's insistence on discussing his recent *Défense des Emigrés*, a long tirade, not so much against the revolution itself as the arbitrary injustice of the émigré laws, which condemned them all, regardless of their views and positions, as traitors and cowards. Adèle de Boigne, in her acerbic memoirs, noted that his contemporaries referred to Lally-Tollendal as 'the fattest of sensitive men', whereas they might have done better to add that he was 'the flattest of humorous' ones, always weeping and snivelling over the past.

But Richmond was not without its pleasures. That summer, the Princesse de Bouillon, the extremely ugly companion of the Prince de Salm, of whom Lucie had grown fond in Paris just before the revolution, arrived in Richmond to collect a legacy left her by the unfortunate Amélie de Lauzun. In the eight years since they had last met, Mme de Bouillon had grown even uglier, her back more humped, her 'yellow, dried skin' clinging to her bones, her mouth so full of black and broken teeth that she was terrifying to look at. But the Princess had lost neither her charm nor her wit, and soon introduced Lucie to the Duchess of Devonshire, who gave a lunch party for the French émigrés in the newly done-up Chiswick House. The once famously beautiful Duchess was now 40 and worried about losing her sight; her looks were going, and she had become, as a friend maliciously wrote, 'corpulent . . . her complexion coarse, one eye gone and her neck immense'. The Devonshire House Circle, with its Whig aristocracy and its passion for the theatre and gambling, satirised by Sheridan in *The School for Scandal*, was winding down, and the Duchess spent more of her time at Chiswick House, calling it her 'earthly paradise'. She had planted lilac, honeysuckle and roses near the house, to scent the air. Later, the Duchess's sister, Lady Bessborough, invited Lucie and Frédéric to a dinner at her house in Roehampton. And there was a visit to Hampton Court, where Lucie again met Anne Wellesley, with whom she had played as a child. Lucie, though seven months pregnant, loved these outings.

The Princesse de Bouillon now offered to exchange her larger and more comfortable lodgings in Richmond for the two ground-floor rooms in Osmond Row, saying that she was lonely on her own. They had scarcely moved into their new home when Lucie gave birth to a son, a 'strong and beautiful child'. She called him Edward, after his godfather, Lady Jerningham's son and her former page at her wedding. There were now once again three children in the family: Humbert, who had turned 9, the plump Charlotte, almost 2, and the new baby. While Marguerite cooked, Lucie sewed and ironed; an English nurse was hired to help with Edward. Humbert's English was so good that on his way to school he called at the local shop to place the daily order.

One of her first visitors after the birth was the Chevalier Jerningham, who wrote to tell his sister-in-law that he had not found Lucie in good health. Lady Jerningham's response was to charge him with finding somewhere in Richmond for her niece's growing family, sending £45 to put down on a charming small early Jacobean terrace house at number 3 The Green. It belonged to a famous Drury Lane actress who never used it, but who had furnished its very small rooms with perfect taste. It was, wrote Lucie, 'beyond anything we could have wished'.

It is sometimes as if the lives of Frédéric and Lucie were marked with particular tragedy; they accept, they endure, they recover, only to be struck down again. They do not complain. Georgian London was a sickly, unhealthy place, even if vaccination was making huge strides against smallpox and the plague had disappeared. Fevers – typhus, dysentery, measles, influenza, tuberculosis – still struck out of the blue, rampaging unchecked in epidemic waves. At a time when far more people died young than old, parents could not expect to see all their children survive beyond infancy. But Lucie and Frédéric had already lost Séraphine, and Lucie had miscarried three babies and seen another die at birth. Even for the times, her losses were extreme.

The day that they were due to move into their new terrace house, Edward, just 3 months old, caught pleurisy. The autumn was damp and cold and Lucie blamed the English nurse for not taking better care of him. Within days, Edward was dead. Lucie, who had been breast-feeding him, fell 'very ill, and was' she would write, 'myself near death'. 'The grief', she wrote many years later, was such that 'it curdled my milk'. This time, she took many months to recover. In her memoirs, she made little of Edward's death, preferring to remain silent. But the gaps in her narrative, at moments of tragedy like this, were telling. It was as if she retired into herself, and simply concentrated on surviving; and, after a period, she picked up the thread again.

Getting rid of the English nursemaid, she now had only Marguerite to help her. Humbert went to school every afternoon, before going on to spend an hour or so at the home of a French émigré called M. de Thuisy and his four young sons. Lucie and Frédéric grew fond of this family and M. de Thuisy made a point of calling every week on the day Lucie did the ironing, when he would sit by the fire as they talked, handing her the hot irons, which he had first cleaned on brick and sandpaper. England was a country, Lucie was discovering, where those not possessed of a great fortune could still live comfortably. She delighted in the daily visits of the butcher to the house and in the fact that there was never any haggling over price or weight.

They were once again very short of money. Lucie's stepmother's repayments had trickled almost to a halt. The kind M. de Thuisy, observing their anxiety, found ways of bringing her sewing, particularly linen to be marked, at which she excelled. One morning, when they were down to their last 600 francs, a despairing letter arrived from M. de Chambeau, still in Spain waiting to be allowed to return home and himself left with nothing. His rich uncle had recently died in France, leaving him his entire fortune; but as a proscribed émigré, liable to be arrested if he returned to France, he was not permitted to inherit. Frédéric, taking most of their remaining money, hastened to a banker and took out a bill of exchange, payable to M. de Chambeau in Madrid. They now had a single £5 note left. As if by instinct, one morning soon after, Edward Jerningham rode to Richmond to call on them. Lucie, who was very fond of him, regarded him somewhat as the younger brother she had never had. Edward had just turned 21 and come into a sizeable inheritance. As he left the house, Lucie saw him slip something into her work-basket, but he appeared so embarrassed that she said nothing. After he had ridden away, she found an envelope with the words: 'Given to my dear cousin, by her friend Ned'. It contained a note for £100.

Bit by bit, Lucie started going out again. She was once again in touch with her childhood friend, Amédée de Duras, with whom she had often played music in the Faubourg Saint-Germain, and who had fled to London soon after the revolution. Amédée had recently married Claire de Kersaint, the rich only daughter of the naval hero in the American wars, who had gone to the guillotine at about the same time as Arthur, accused of spying for England. Kersaint had tried, but failed, to kill himself while in prison. He had sat briefly with Arthur in the Assembly.

The marriage, at which the Princesse d'Hénin stood in for Claire's mother, who was ill, brought together the cream of the Faubourg Saint-Germain in exile. Like Mme de Staël, Claire was too manly to be beautiful, but her somewhat wary expression suggested both intelligence and great determination. She was small, just over 5 feet, with brown eyes, black hair and a small mouth. Writing later about those who had been young in the Terror, she would say that they 'would carry to their graves the premature melancholy that filled their souls'. Lucie was devoted to Amédée but deplored his haughty manner. After the young couple moved nearby to Teddington, she spent many hours comforting Claire for his infidelities, calming her storms of fury and despair, while reprimanding Amédée for his waywardness. As Lucie noted, Claire, who was 22, wanted romance, while Amédée was the least romantic of men. Lucie counselled patience and forbearance: she told Claire to try to make their house more pleasant, so that Amédée would not wish so often to get away; but then Lucie's own marriage remained an exceptionally happy one, and her feelings for Claire would always be ambivalent.

When Claire gave birth to a daughter, Félicie, Lucie was asked to be her godmother. Though increasingly accepting of her new friend, she remarked, in the clear and critical tone that she frequently adopted when describing behaviour that was in some way wanting, that beneath the younger woman's apparent passion and intelligence, lay 'arrogance and tyranny'. In Teddington, Claire became increasingly distraught; she wept incessantly, while Amédée grew ever more bored. Lucie told her that Amédée hated scenes, and pointed out that love could not be commanded. 'Having lectured the husband,' she wrote later, 'I consoled the wife.' Her troubled friendship with Claire, which was to last for many years, would cause her much unhappiness.

Shortly after the birth of a second daughter, called Clara, Amédée, a First Gentleman of the Bedchamber, was summoned to do his tour of duty at Mitau, in the Duchy of Courland where Louis XVIII maintained his court in exile. The children were to be left with Mme de Thuisy. But on reaching Mitau, Amédée was informed that the King had let it be known that he would receive no one who had sat in the Assembly in Paris at the time of his brother's trial and execution. Admiral Kersaint had been present that day: Claire would not be welcome at court. In 1795, Louis XVIII had drawn up a list of names of those who had rallied to the new French government, according to the wickedness of their deeds and the punishment they would receive. Those who had voted for the death of the King had an 'e' by their name, for *écartelé*, drawn and quartered, then there were those with a 'p' for *pendu*, hanged, those with an 'r' for *roué*, to be put to the rack, and those he called 'heedless, pusillanimous and stupid', who had a 'g' for *galères*, the galleys. Lafayette and two of the de Lameth brothers had 'r's.

One of Lucie's neighbours in Richmond was a Miss Lydia White, a well-known local bluestocking, who, together with her unmarried sister, held musical evenings. Enjoying Lucie's stories of her life in Albany, they pressed her to borrow books from their well-stocked library. When the two sisters moved away from Richmond, Lucie, knowing that she could not afford one of the many costly circulating libraries, resigned herself to a bookless life. A little later, a box full of books arrived from Hookman's Library in London, with a letter telling her that she was entitled to request any number of French and English volumes from the stock of 20,000 books, and that if she put in her order by the seven o'clock morning coach, she would receive them the same evening. She never learnt who was responsible for the gift, but assumed it was the Misses White.

Other acts of kindness touched her. An alderman lived in the next door house on the Green. Having befriended the talkative Humbert, and learnt from him details of the family's misfortunes, he sent round a constant supply of fruit from his hothouses, with notes saying that they were for 'the young gentleman'. He also had pots of sweet-smelling plants placed all along their shared railings, so that their scent floated into Lucie's rooms. And there were occasional outings, to Teddington to practise music with Amédée, to London to visit Lady Jerningham or to hear Johann Baptist Cramer play at a private party in London, where Lucie, accustomed to the respect accorded music and musicians in France, was shocked that no one stopped talking to listen.

For a while, the émigré community was full of rumours of a possible invasion by the French, and it was said that Napoleon might land with his men on the Norfolk Broads. Lady Jerningham's reaction was to declare that she would raise an army of 'stout

Gillray's cartoon illustrates the fear of what might have happened if the French had invaded.

robust young female peasants, dairy maids, servants, field workers, the wives and daughters of rustics' and arm them with pikes. They were to wear neat plain uniforms and their task would be to drive the cattle and horses inland, away from the coast where they might be seized by the invaders.

One week, Lucie and Frédéric accompanied their old friend M. de Poix on an expedition to Windsor, Oxford, Blenheim and Stowe, pausing along the way to inspect the country houses open to visitors in the absence of their owners. With her not always well-concealed disdain for the English, Lucie observed that it was only in their country settings that English gentlemen 'really became "*grands seigneurs*"'. Enjoying herself, she also remarked on the improved weather: away from London it seemed to her no worse than that of Holland.

Though they had at last extracted some more money out of Lucie's stepmother – with the result that they were now cut by that entire side of the family – Frédéric was thinking of arranging for Lucie to return briefly to Paris to see what could be salvaged of their property. There was again news from France that in order to recover confiscated property, owners had to apply for it in person. Lucie would, they both thought, be able to travel in safety under her English name, though the English did not look kindly on the French who chose to cross backwards and forwards between the two countries. 'What loyalty can be expected from men, who on their arrival at Paris, proceed even to the length of taking the oath of hatred against Royalty,' asked *The Times*, 'while in this country they affect a strong attachment to Kingly government?'

Lucie herself was extremely reluctant to go; she feared not the dangers, but the parting from Frédéric and the children. It was on the day she was due to leave Richmond that news came of the fate of two émigrés, men who had partnered her at dances in Paris before the revolution, and who had recently returned secretly to Paris on the same mission. They had been caught, and shot. Lucie's trip was abandoned. The days passed very slowly. Her health was poor and she felt permanently listless. 'My life in Richmond,' she wrote later, 'was very monotonous.' She was once again pregnant. She was 29: this was her eighth pregnancy.

* * *

One of the many portraits of Marie Antoinette painted by Elisabeth Vigée Le Brun. The cradle to which the Dauphin points had contained eleven-month-old Sophie. When the baby died she was painted out.

Louis XVI in the Phrygian red bonnet he was forced to wear - a typical caricature of the French royal family drawn and widely circulated during the Revolution.

The march of the fishwives on Versailles, in early October 1789.

Camille Desmoulins, revolutionary friend of Lucie's father, Arthur. Desmoulins's defence of Arthur before the Tribunal saved neither his own life nor that of Arthur, who went to the guillotine the same day as Camille's wife, Lucile.

Thérésia Cabarrus, considered by many to be the most beautiful woman in France. Lover and later wife of Tallien, one of the harshest revolutionary leaders, she helped Lucie and her family escape to America.

An eighteenth-century view of the Hudson river in Upstate New York, where Lucie spent three years in exile.

The shrewd and scheming Charles-Maurice Talleyrand, friend to Lucie and Frédéric for most of their lives.

Lucie and six-year-old Humbert among the Indians on her farm near Albany, as imagined by a romantic French artist.

A Parisian tea party, at the height of the fashionable extravaganzas of the Directoire.

Josephine, kneeling to be crowned at the coronation of December 1804. Lucie declined to become her lady-in-waiting.

Napoleon, caught in a rare defeated mood.

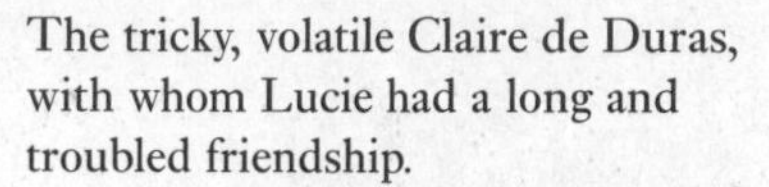

The tricky, volatile Claire de Duras, with whom Lucie had a long and troubled friendship.

Félicie de la Rochejacquelein, Claire's elder daughter and Lucie's much loved godchild, with whom she exchanged hundreds of letters during the last thirty-five years of her life.

The writer Germaine de Staël, painted as the heroine of her own novel, *Corinne*. A close friend of Frédéric, she kept a celebrated salon after the Revolution.

The Château de Vêves in Belgium, home to Lucie's son-in-law Auguste de Liederkerke Beaufort and his family.

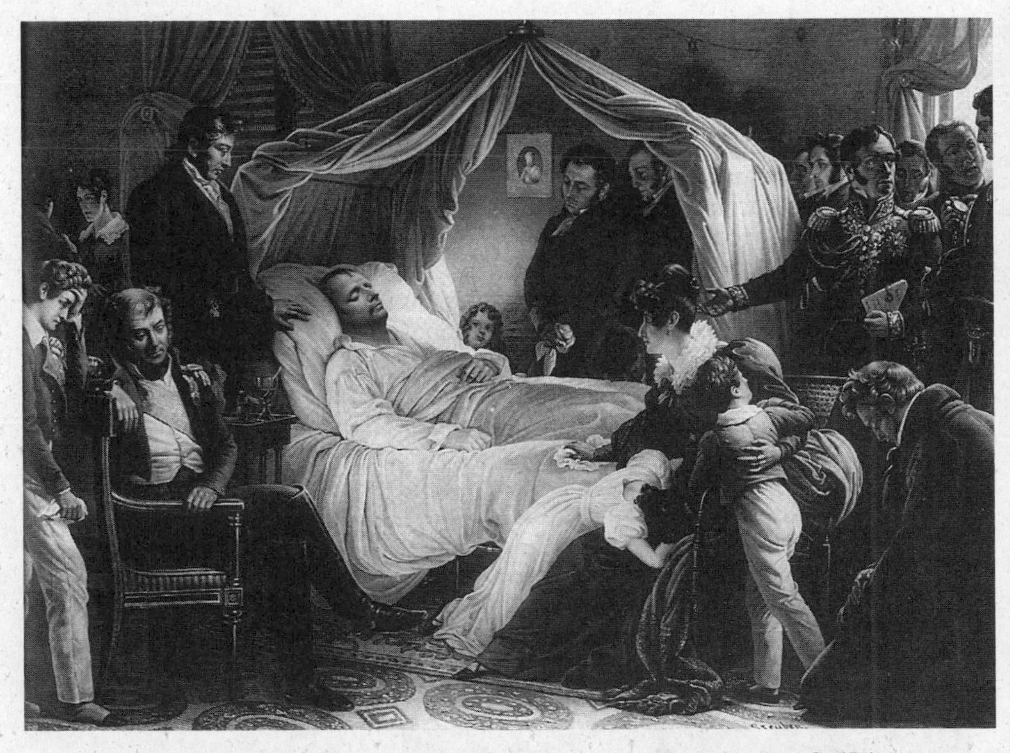

Lucie's half-sister Fanny and her children at Napoleon's deathbed on St Helena.

Lady Jerningham need not have feared a French invasion: Napoleon's eye was on Egypt, where the writ of the Ottoman Empire was weak and where the Beys who governed it were subjecting French merchants to constant vexations. Egypt also promised to be the route by which it would be possible to strike at England's coveted possessions in India. Having led his ragged, exhausted troops to victory after victory in Italy, Napoleon had returned to Paris a conquering hero, who had changed the map of Europe in France's favour and made the revolution seem respectable again. The Egyptian expedition, however, proved expensive in men, decimated by ferocious Mameluke warriors, eye diseases and the plague, while Admiral Nelson had laid waste to the French navy off Alexandria, effectively cutting off the French from their homeland. Napoleon, remarked General Kleber, was the kind of general who needed a monthly income of 10,000 men.

On 23 August 1799, leaving his army behind in Egypt and passing off his defeats as victories, Napoleon set sail for France. He landed in Fréjus on 9 October. The Austrians were driving the French out of Germany, the Russians harrying them in Italy; Paris was in a state of unease. Barras, the only one of the five original Directors still in power, was negotiating with the exiled Louis XVIII for a return of the Bourbon monarchy; the Abbé Sieyès, whose pamphlet about the Third Estate was widely regarded as one of the defining documents of the revolution, was plotting on the contrary for a stronger executive. (When asked where he had spent the revolution, Sieyès memorably replied: 'I survived.')

Napoleon, approached both by Barras and by Sieyès for his support, agreed to back Sieyès, on condition that a new Constitution be drafted. On 9 November – 18 brumaire – an emergency was declared. Napoleon was given the command of the troops in the Paris region, the Directoire was overthrown, and Barras was removed, though not without some resistance from the assembled deputies, which evaporated when confronted by the bayonets of Napoleon's grenadiers. The days of the corrupt, incompetent Directoire were over.

Ten years of constant war and political turmoil had made the French long for peace and order. Napoleon was a man untainted by the venality of the Directoire, someone who could put both the revolution and its chaotic aftermath to rest. In him, royalists chose to see someone capable of restoring the monarchy; the former Jacobins preferred to believe that he could prevent it. Napoleon deftly engineered that authority would be vested in three Consuls, elected for ten years. There would be a Council of State to draft bills, a Tribunate to discuss them, and a Legislative Assembly to vote on them. After a brief interim period, a moderate lawyer, Jean-Jacques de Cambacérès, and a disciple of Rousseau's called Charles-François Lebrun, joined Napoleon as Consuls. 'Bonaparte has his eye on a dictatorship,' warned an English agent, reporting to his superiors in London. 'He wants to play Cromwell . . .'

In London, the news of Napoleon's return and the fall of the Directoire was greeted with cautious optimism. The last two years had disillusioned many of the émigrés, France appearing to be as oppressive in peace as it was in war. The revolution, argued the émigré journalists, had turned out to be synonymous not with liberty but with destruction. But there was also a sense of excitement that it might at last be possible to go home in safety.

At Cossey, where Frédéric, Lucie and the children had joined Lady Jerningham to spend another winter in the country, letters arrived from de Brouquens and Augustin de Lameth urging them to return. Since England and France were still at war, they were advised to travel via Holland on German passports. Lucie was seven months pregnant and Lady Jerningham suggested that she let Frédéric go alone. She unnerved her niece still further by saying that she would be happy to bring up the new baby as her own. Lucie rejected both ideas: she had no intention of being parted either from her baby or from Frédéric. And she feared that if he were obliged to flee again, he would make his way to Le Bouilh, and then to Spain, where it might take her many months to join him.

Unwilling at first to believe that there could be any serious difficulties in obtaining French passports, Frédéric went to call on the Bishop of Arras, the sole accredited minister of the court in exile with the power to grant them. The Bishop was unhelpful. He was not interested in assisting anyone, he told Frédéric, who did not have the patience to wait until the counter-revolution and the restoration of the Bourbons. In the event, Dutch passports were secured, and passages booked for the family and Marguerite on a Royal Navy packet travelling from Great Yarmouth to Cuxhaven. Because of terrible weather, gale winds blowing from the north-west, they spent a miserable month waiting in dreary lodgings in Great Yarmouth, unable even to visit nearby Cossey lest the captain suddenly decide to weigh anchor. The day Lucie's baby was due was getting uncomfortably close, and she was terrified that she might give birth before reaching Paris. Eventually, at the very end of December, they woke to a fine morning and the captain summoned the passengers on board.

The sea was still very rough, and the 14 French, German and Russian passengers huddled in a single cabin. Lucie found a bunk, keeping Charlotte close to her, but the hatches had been battened down and there was no fresh air. Frédéric and Marguerite, overcome with sickness, 'lay like dead things'. Only Humbert remained on his feet. The boat heaved and tossed for 48 hours, and those passengers who were not actually sick got drunk on brandy and punch. Lucie would later remember these two days as among the most unpleasant of her life. For a while, it seemed as if they might have to put back or, if the ice was very thick in the estuary, land on a small island nearby.

But the weather cleared, the ice proved less thick than had been feared, and the boat anchored at Cuxhaven. A kind gesture on the part of the captain, by providing her with a small boat to take her closer to the jetty with Charlotte, nearly proved fatal. With the boat rocking from side to side, Lucie slipped and was only prevented from falling into the sea by two sailors who, while hauling her up on shore, at the same time yanked her arms in such a way that she was in the most terrible pain. Clutching Charlotte in aching arms, she was barely able to reach Frédéric, waiting close by with a cart.

Every inn in Cuxhaven was full of émigrés trying to reach France. Only after some hours did a landlord, taking pity on Lucie's condition, allow the family in, and provided straw mattresses and blankets for them to sleep on the floor. Lucie had by now developed a high fever. She became delirious, and Frédéric feared that she might miscarry. A doctor was found, and through an interpreter Lucie described her accident. He prescribed a strong sedative and applied a plaster made of barley boiled in red wine to her painful side. Twenty-four hours later, she woke, feeling perfectly well again. She was, as she frequently remarked, blessed with excellent health and a determined nature.

CHAPTER TWELVE

Toothless Dogs and Clawless Cats

While Lucie was recovering, Frédéric managed to buy a small barouche and a horse for 200 francs. On a fine morning, they set out for Paris, but the weather soon turned stormy, with squalls of driving rain; Frédéric and Humbert sat hunched at the front under a single umbrella, while Lucie, Charlotte and Marguerite cowered under the hood. At Bremen, they paused for two days in an inn to dry out by the warmth of one of Germany's immense wood stoves. When they took to the road again, deep snow lay on the open heath; the barouche turned over in the snowdrifts, leaving them shaken but unhurt.

Late one evening, they reached the Hanoverian garrison town of Wildeshausen, where a regimental ball was to be held that night. Every room was already taken. They were huddling around the fire in the town's only inn when an officer, saying that he expected to be dancing all night, offered them his room. Greatly relieved, they ordered supper to be sent up. It was now that Lucie realised that she was going into labour. Frédéric became distraught; Lucie tried to comfort him, saying that babies could, after all, be born anywhere; but, she wrote later, 'it is impossible to describe his despair'.

With difficulty, for none of them spoke German, a French barber, who had deserted during the Seven Years War, was found to interpret; he in turn fetched the doctor, an elegant young officer who came straight from the ball. Since Lucie lay swaddled in her cloak, it took some minutes to explain the nature of the problem. Once he had grasped the situation, the doctor, helped by the barber, Denis, efficiently arranged for the family to move into two unoccupied rooms belonging to a rich farmer on the edge of the town, Lucie as always supervising every step. 'As I was not yet in much pain,' she wrote later, 'I had time to attend to all our small arrangements.' Increasingly, it was Lucie rather than the anxious and good-natured Frédéric who tended to take charge.

It was here, on the morning of 13 February 1800, that a small, frail girl was born, 'so thin and delicate that I hardly dared hope she would live'. Lucie estimated that she was about six weeks premature. They called her Cécile, after Frédéric's sister. She was baptised in the Catholic church in Wildeshausen; Denis and his wife, who spoke not a word of French, acted as godparents.

Within hours of the birth, the local magistrate summoned Frédéric and informed him that they would have to leave the town within 48 hours, the Electorate of Hanover having strict rules about French émigrés travelling on what were clearly false Danish passports. But when he learnt their real name, he recalled how kindly his nephew had been treated by Frédéric when Minister at The Hague, after which the whole town opened its doors to the travellers and their delicate new baby. When, two weeks later, they left the farmer's house and set forth again, officers from the garrison, the elegant doctor among them, escorted them on the first lap of their journey towards Holland. Inside the barouche, Lucie clasped Cécile tightly to her, never once exposing her to the 'icy air of those northern plains'.

At Utrecht, reached after many slow, jolting days on the road, they were surprised to encounter the Princesse d'Hénin, who was on her own way back to Paris. She had paused in Utrecht to visit Lafayette, who, having been freed from five years' captivity, was living nearby at Vianen and now hoped to be allowed to return to France. During his years in prison he and the Princesse d'Hénin had kept up a secret correspondence, some of his letters written with a toothpick in vinegar and charcoal on scraps of paper.

With new passports from the French Ambassador to Holland, whom Frédéric knew from his time in the French Foreign Ministry – papers as false as all the rest, stating that the family had never been in England – they travelled on to Paris where Augustin de Lameth had secured lodgings for them in the rue de Miromesnil. The house belonged to a former mistress of the Duc de Bourbon; one of Lucie's first acts was to drape muslin over the full-length mirrors that covered the walls, saying that it irritated her to keep seeing her reflection at every turn.

It was not just a new century. Though the country was at a low ebb, its factories in ruins, its schools without teachers, its churches closed and its Treasury empty, and though thousands of destitute people wandered the streets of Paris, there was a real feeling that France had at last been delivered from a long nightmare. The revolution was truly over. And it was calm, without the frenzy that had marked the Directoire. Under the three new Consuls, genuine order was gradually being established. The Vendée, where civil war had simmered on, was being pacified. Several of the most unpopular and harsh laws had already been rescinded. Daily newspapers had been reduced from a chaotic 73 to 13, and these were under police surveillance. A new inspector-general of the gendarmerie had been named to fight brigandage, escort prisoners and 'ensure the safety of people and property'. It was becoming strict and orderly, but after ten years of uncertainty, most found it all extremely reassuring.

And the émigrés were finally coming home, even those who had not dared do so in 1797. Singly and in groups, many on false passports and under assumed names, some wearing the threadbare fashions of the *ancien régime*, others in disguise, they were making their way back to the outskirts of the city, venturing into the centre only when they perceived others doing so in safety. By the end of the Terror there had been 146,000 names of émigrés on the proscribed list, men and women prohibited from returning to France. But the Directoire had removed 13,000 of them and the new Consulat looked set to remove many more. The émigrés longed to be home. Whatever their politics and their parties, they longed to find out what was left of former possessions, to learn who had survived and who gone to the guillotine, to discover the shape and mood of the new France. Some were hoping to be reunited with children left behind with servants, fearing that they might have abandoned the old courtesies and gone the way of the *merveilleuses* and *muscadins*. 'It has become as fashionable to come home,' noted the Comte de Neuilly drily, 'as it once was to flee.'

It was not, however, as easy or as straightforward as it may have seemed. In positions of power were many people whose new fortunes rested on stolen émigré property, or who feared that they might lose their jobs to better candidates returning from abroad. Over a million people had bought property belonging to the Church or to the émigrés, who were said to have lost possessions valued at some 2 milliard francs. Fouché, the Minister for Police, who had helped himself to several large estates, intended to fight hard against all requests for '*radiation*', the removal of names from the hated émigré list. Fouché was using spies to gather incriminating information about anti-revolutionary activities during the years of exile.

Lucie and Frédéric were not on the list of the proscribed, but Frédéric's mother was, even though she had barely stepped outside her convent in the rue des Fossés Saint-Victor. Just the same, they needed certificates to prove that they had never left France. And though these documents were widely known to be fabrications, in due course Lucie was forced to present herself, with her 'cloud of witnesses', at her local municipal office. The mayor signed the papers, laughing as he whispered in her ear that he knew that every bit of clothing she was wearing came from England. Frédéric had written to Alexander Hamilton before leaving London asking him to sell their farm in Troy, and was hoping to be able to retrieve or buy back at least some of their former possessions with the money. Hautefontaine, condemned as a nest of aristocrats and 'non-juror' priests, had been sold in 1799 to a speculator, together with 1,709 separate pieces of furniture, glass, kitchenware and linen, and could not be got back. But its library of 3,000 books, including 112 volumes of Pancoucke's *Encyclopédie*, was intact, carted off to the Château de Compiègne with libraries from other châteaux; and this he was hoping to recover.

In the spring and summer of 1800, the Place Vendôme became Paris's most elegant social meeting place. It was here, in a building on the corner of the rue Saint-Honoré, that a *Commission des Émigrés* had been set up to hear cases and hand down verdicts, the favourable ones often the result of handsome bribes. Newly returned émigrés, from Koblenz, Lausanne, Madrid, St Petersburg or London, hastened to the Place Vendôme to greet friends, discuss strategies, talk over plans, enquire about jobs, and to seek, after many years of absence, lost friends and relations. Some had been away for ten years. Not all knew the fate of people they loved. By early autumn, 8,083 requests had been filed with the commission; Fouché managed to reject 1,747 of them. With his lank fair hair, long pale face, staccato voice and jerky movements, Fouché, who had sent some 2,000 Lyonnais to their deaths with the words 'no mitigation, no delay, no postponement of punishment', was turning into a man many Parisians feared.

* * *

When Chateaubriand returned to Paris in 1800, he expected to 'descend into Hell'. Instead, in the Champs-Elysées, he was greeted by music, the sound of violins, horns, clarinets and drums sounding out cheerfully to a people intent on normality and reassurance. Given his morose nature, Chateaubriand chose to dwell on the belfries stripped of their bells during the Terror, and the headless statues of saints, but most other émigrés, reaching the city from their years of anxiety and exile, were delighted by the liveliness and sense of order of Paris. It was tranquil, and stable, in a way which might actually last.

The revolution had done away with the salons of the Faubourg Saint-Germain, but now, searching for the familiar in a world torn apart by a hatred that many still found impossible to comprehend, bold aristocratic women began opening their doors once again to those coming from abroad or creeping back into the open after years of terrified concealment. What they longed for was the '*douceur de vivre*' that had once made their lives so pleasing. One of the first salons to reopen was that of Frédéric's friend, Mme de Montesson. Now in her 60s, the widow of the Duc d'Orléans had kept her youthful complexion and striking violet eyes, though she moved a little stiffly, telling friends that her bones ached from the 18 months she had spent in prison.

Announcing that she intended to give a dinner at two o'clock every Wednesday, at which the harp would be played and perhaps a book read aloud, Mme de Montesson soon gathered around her a dozen of the surviving luminaries of the *ancien régime*. It was said that it was in her salon that men wore silk stockings and buckles for the first time since the revolution. The food was excellent, the footmen wore livery and the china came from Sèvres. Elegant and understated, invariably wearing white or becoming shades of cream, Mme de Montesson let it be known that she would rise to greet no one, except for Napoleon's wife, Josephine, or in order to show to the door someone she did not wish to receive again, and that she would tolerate no politics in her salon. Guests quickly read into her 'open' or 'shut' expression whether they were in favour or not; just the barest lowering of her voice was enough to keep passions in check. According to her great-niece, Mme Bochsa, no one knew better how to freeze people out with 'nuanced and knowing politeness'.

Just the same, most people found Mme de Montesson more

intimate and welcoming that Mme de Genlis, who, though better educated, was prone to pedantry. In her new salon, Mme de Genlis was blaming Voltaire and the *Encyclopédistes* for having paved the way for murderers like Robespierre. In the rue de Cléry, Mme Vigée-Lebrun, allowed back from St Petersburg only after 254 artists signed a petition on her behalf, had opened her long gallery to a resident music society. Musical soirées were becoming fashionable once again, leading the Goncourt brothers later to observe that many in the audience, believing that it was *bon ton* to listen to music, were in fact 'heroically bored' at having to do so. It was in the rue de Cléry that *The Marriage of Figaro* and *Don Giovanni* were sung for the first time in the original Italian.

And in a house lent to Mme de Beauvau in the Faubourg Saint-Germain, the three surviving '*princesses combinées*' – d'Hénin, de Poix and de Beauvau – met again, receiving their visitors stretched out on chaises longues, drinking coffee from a gold coffee pot. Mme de Poix had gone completely blind. Mme de Beauvau was sharing her very small house with her sister-in-law, the former abbess of Saint-Antoine des Champs, as well as two former ladies-in-waiting to Marie Antoinette, and Ourika, the Senegalese girl given to her as a gift by the Chevalier de Boufflers. Lally-Tollendal, who was considered very greedy by the princesses, was often tearful when talking about the past.

In these salons, Lucie and Frédéric rediscovered a world and friends that they had almost forgotten. Mme de Staël, at work on a novel, was gathering around her political figures of every persuasion, artists, émigrés, old Jacobins and royalists, talking, arguing, reasoning far into the night as she had always done. Many of the old nobility, Lucie observed, were hastening to make their peace with Napoleon, though seldom taking the trouble to conceal their contempt for his rough ways when among friends.

It was not long before they encountered Talleyrand, enjoying his position as Minister for Foreign Affairs, who asked them whether there was anything they needed. Lucie, remarking wryly that Talleyrand had 'steered his fortunes with much skill', told him that their financial problems were such that they had no choice but to settle at Le Bouilh. 'So much the worse for you,' Talleyrand replied, 'but it's folly.' There was always money, he added, when you needed it. Talleyrand, noted Lucie, was as ever 'amiable, but not really helpful'. A young English visitor called

Catherine Wilmot, finding herself next to him at a reception, left a memorable if somewhat fanciful description of the man now rising to be the most important statesman of the age. He was wearing, she noted, an embroidered scarlet velvet coat and ruffles. 'At a distance, his Face is large, pale and soft, like a Cream Cheese, but on approaching nearer, cunning and rank hypocrisy supplant all other resemblances . . . Gobbling like a Duck in my ear about the vicissitudes of fortune and happiness . . . he presented me with his fat paw.' Talleyrand himself had given a magnificent ball in the rue du Bac shortly before Lucie's return. 'All these parties,' wrote a returning noblewoman, 'were lovely; above all, they possessed the charm of things that you thought you had lost and now discovered were still there.'

What soon became clear, however, was that a quite separate and very different kind of salon life was developing in Paris. It was more frivolous and considerably more luxurious and it was taking place, not in the faded surroundings of the *ancien régime*, with their chipped plates and empty rooms, but in the magnificent newly decorated houses of the *agioteurs* – the speculators – and the bankers. Foremost among these were the receptions held by Lucie's friend, Thérésia, who, though still married to the jealous Tallien, had spent the years of the Directoire as mistress to Barras, filling his apartments in the Luxembourg with works of art that had once belonged to Marie Antoinette. Thérésia, now 27, still charmed Paris with her yellow straw hats, crowns of flowers, white dresses *à la Mameluk* and scarves woven becomingly around her short curls. She had a rival in Mme Récamier, described as so beautiful that she had been made 'by the Creator for the happiness of man', a 'coquette' with all the 'charm, the virtue, the inconstancy and the weakness of the perfect woman', even if some considered her chest too flat, her arms rather thin and her hair, though silky, thin.

It was Mme Récamier who gave the first masked ball at the Opéra since the revolution. Her salon was described as being too small for the many émigrés, former Jacobins, liberals, artists and officers who flocked to see her perform her famous shawl dance, her face unpowdered, wearing white muslin and satin, her curly hair framed by a black ribbon placed low over her forehead. At Mme Récamier's, the dancing was so animated and lasted for so long that ladies changed their fans, bouquets and shoes several times during the evening. After dinner, favoured guests were given tours of her bedroom, with its white-and-gold bed, with bronze candelabras and violet damask falling in pleats, a white marble statue of Silence on one side and a golden lamp in the shape of a genii holding an urn on the other. In the bathing cabinet, the bath could be turned into a sofa edged with gold. Mme de Récamier, though considerably younger, had been befriended by Mme de Staël, and was said to be acquiring a distinct taste for intellectual conversation.

These new salons, observed Sophie Gay, who would later write two volumes about society in France after the revolution, were turning into places where the 'debris' of the *ancien régime* could meet the new political leaders of the day, each needing the other, the '*anciens*' because they longed to feel wanted, the parvenus because they liked to be admired. Republicans and royalists, attempting to reinvent themselves in this new Consular Paris, where everything changed every day and no one knew what would happen next, 'played together without liking each other, but without fearing each other either, rather as poor toothless dogs might play with cats which had had their claws removed'.

In February, Napoleon, the Corsican artillery officer whose valour at the siege of Toulon had won him glory, had used a plebiscite to have himself made First Consul, equivalent to the title of *Princeps* given by the Roman Senate to Augustus. He and Josephine had moved out of the Luxembourg and into the Tuileries, Napoleon taking the first-floor rooms overlooking the gardens. Josephine chose Marie Antoinette's apartments on the ground floor, where in just a few months the Queen's hair had turned completely white. Bronzes and tapestries were brought from Versailles, and the fashionable Percier and Fontaine decorated a drawing room in yellow and lilac silk. The Consul's apartments were widely praised as in surprisingly good taste. Napoleon was already talking of making Paris 'not only the most beautiful city that ever existed, but that could ever exist', and plans were being discussed to tackle the stench and murkiness of the city's labyrinth of dark alley-ways and its contaminated supplies of water.

Napoleon was not the first man to dream of a grander, cleaner, better lit Paris, but the public works he envisaged, for the immediate glory of his army and later for his own, were on a scale not contemplated before. The Louvre, reduced to a blackened, crumbling ruin, its outer walls sprouting a warren of ramshackle hovels, was to be cleaned up and joined to the Tuileries; the gardens were to be enlarged, and, where potatoes had grown during the revolution, lawns were planted with avenues of trees and trellises covered in flowering creepers. At the Palais-Royal, 496 lime trees had already been placed in long straight lines. Napoleon preferred engineers to architects, sculptors to painters; he liked colonnades and the regular contours of classical buildings. The Paris he planned would be symmetrical and monumental, full of triumphal arches, and it would have a necropolis, like Cairo's City of the Dead.

Lucie had not met either Napoleon or Josephine. But she had known and often danced with Josephine's first husband, M. de Beauharnais, just before the revolution, when he was reputed to be the finest dancer in Paris. Instinctively admiring of Napoleon's military victories, she felt ambivalent about Josephine's social pretensions, remarking, in her most lofty manner, that Josephine, the daughter of a plantation owner from Martinique, had not possessed the necessary degrees of nobility to make her a fully accepted guest at Marie Antoinette's court. However, Lucie and Josephine were distantly related through Mme Dillon, whose mother was Josephine's aunt, and soon after reaching Paris Lucie received an invitation to call. Mindful of her own more elevated social standing, and sensing that it was just a move to win over a former lady-in-waiting to Marie Antoinette, Lucie decided to delay. 'I determined,' she wrote later, with the strong sense of herself that was becoming more marked, 'to increase the value of my condescension by making her wait a little.'

When she did finally consent to call, she found herself captivated by Josephine's open and friendly manner, and by her evident desire to help as many of the émigrés return to Paris as possible. In the Tuileries, Josephine had three small black boys-in-waiting and a Mameluke in Turkish dress at the door. 'She bore herself like a Queen,' wrote Lucie, 'though not outstandingly intelligent, she well understood her husband's plan: he was counting on her to win the allegiance of the upper ranks of society.' It was not, she added, a difficult task: everyone was 'hastening to gather around the rising star'. In 1800, Josephine was 37; she was not exactly pretty, but she had delicate features and all who knew her remarked on the kindness of her voice and expression, and her 'perfect' figure, even if some could not help adding that her complexion was suspiciously dark and her teeth extremely poor.

Napoleon had begun confiscating art in 1794 while campaigning in the Netherlands and the Rhineland, when four commissioners – a botanist, an architect, a librarian and a geologist – had been despatched to the conquered territories to select and send back furniture, books, maps and pictures. In 1798, in the wake of his military victories in Italy, had come priceless books from Pope Pius VI's magnificent library. They had reached Paris in time to be paraded around the Champs de Mars on the anniversary of Robespierre's fall, escorted by a procession of curators and archivists, and by a number of ostriches, gazelles and camels, which happened to have arrived at the same time from North Africa. The arrival of carts, groaning under trophies, Veroneses from Padua, Leonardo da Vincis from Milan, pearls and precious stones from Bologna, had become a familiar sight on the streets of the capital.

Never had there been a greater appetite for the new, to be fed by new foods – salami from Bologna, sweetmeats from Egypt – by new meals – *le thé*, and a *déjeuner à la fourchette* of kidneys and pickled onions which could be taken at any time of the day – and even by new smells. Consular Paris smelt delicious, at least for the rich, the fashion for all things Greek including strongly scented baths: Thérésia took hers in strawberries and raspberries. Even language was new. The exaggerations that had marked the conversations in the pre-revolutionary salons of the Faubourg Saint-Germain with their flowery expressions of loathing and passion, had been replaced by shorter, sharper phrases. It made women, said Mme de Genlis, sound less affected, but at the same time colder and less welcoming. No longer did men lower their voices when addressing women or refrain from paying them direct compliments. Gone was the chaperone doing embroidery in one corner when a man came to call, and gone were the two lackeys, with flares, accompanying women out at night. Before the revolution, remarked Mme de Genlis wistfully, aristocratic ladies had required 'witnesses and light'.

In 1800 the very fabric of the city itself seemed to change day by day. Dresses worn white one day, were zigzagged in violet or blue the next; ribbons were first striped, then chequered. Carriages were higher off the ground, then lower. In the houses of the rich bankers and speculators in the Chaussée d'Antin – the Faubourg Saint-Germain had been abandoned as too shabby – Percier and Fontaine, drawing on the recent excavations at Pompeii, were producing interiors in violets and pale greens, pastoral scenes giving way to the geometrical, garlands to winged sphinxes, caryatids, Venuses, nymphs and Graces, painted, chiselled or sculpted, in marble, leather and bronze. Salons became atriums. The moribund silk industry in Lyons was undergoing a rapid transformation, as Napoleon insisted that Anglomania should be forsaken in order to revive French textiles. In the rue Mestrée, the Frères Jacob were charging exorbitant prices for furniture made from mahogany and maple, for ebony clocks with gold figures and Egyptian hieroglyphics, for imitation marble and porphyry.

Much of this new longing for luxury was being expressed in clothes, where wigs of every colour, from canary to orange, had suddenly come into fashion. Thérésia was said to own 30. The Chaussée d'Antin, said an English visitor disapprovingly, was full of men who looked like women and women who looked like prostitutes, even if, in the Tuileries, Napoleon let it be known that he was not in favour of so much nudity, and preferred to see his guests in decorous white silk and satin. Undergarments, abandoned during the Directoire, were reappearing. The men, on the other hand, he wished to see in uniform. One of the first decrees issued by the Consulate had been that Consuls, ministers, members of the Legislative and the Conseil d'Etat were all to have their own official uniforms, in varying shades and degrees of gold, lace, embroidery and plumes, outfits that would become gaudier and more theatrical as the years passed.

A special regiment of young volunteer hussars, in which the sons of the old noble families hastened to enrol, among them Frédéric's 18-year-old nephew Alfred de Lameth, were decked out in yellow, to be nicknamed 'canaries' by the people of Paris. August von Kotzebue, a German writer who visited Paris at around this time, remarked on the extraordinary amount of jewellery that women wore at night, seldom leaving their houses without seven or eight rows of gold chain around their necks, rings on every finger, medallions studded with diamonds and gold pins in their hair. Mme Récamier wore pearls, claiming that diamonds did not suit her as well. For those who wished to keep abreast of these changing styles, *Le Journal des Dames et des Modes* had separate supplements for furniture and decoration and was available in the *cabinets de lecture*, where issues could be rented by the hour. When the news was good, the *Journal* had its models smile. But whether good or bad, the models were always elegant, reclining on beds, watering flowers or feeding birds, for the main occupation of women now was to be elegant.

While Napoleon was in Egypt, Josephine had bought Malmaison, a three-storey, 18th-century stone house just outside Paris, with 300 acres of parkland, vineyards and fields running down to the Seine. She furnished it with sphinxes and a great deal of malachite, ebony, marble and bronze and a 'patriotic' bed in the shape of a campaign tent with drums for stools. When Lucie was invited to Malmaison, Josephine insisted on taking her on a complete tour of the house, stopping by every picture and sculpture to explain how each had been the gift of some grateful foreign court. Lucie was not just bored, but cross. 'The good woman,' she wrote later, 'was an inveterate liar. Even when the plain truth' – that they had all been looted at the point of a sword – 'would have been more striking than an invention, she preferred to invent.'

Formality and etiquette were already beginning to cast an oppressive hand over life in the Tuileries and at Malmaison, and Napoleon had turned to Mme de Montesson and Mme de Genlis for the *bon ton* he wished to see around him. At her new school in Saint-Germain, Mme Campan, Marie Antoinette's lady-in-waiting, was teaching the daughters of the speculators and bankers the manners of the *ancien régime*.

On the anniversary of 14 July, Frédéric and Lucie accompanied M. de Poix to watch a formation of veterans of the recent French victories at Marengo, many of them wounded and in tattered uniforms, carrying the standards and flags captured from the Austrians. They were surprised to find the crowds quiet and apathetic. A longing for peace and order, an exhaustion with turmoil and military valour, seemed to have settled over the city. Already people were beginning to talk of Robespierre and the 'reign of Terror' as if it had all happened somewhere else and a long time ago.

* * *

In September, taking with them as a tutor for Humbert M. de Calonne, a 'non-juror' priest who had spent the years of revolution in exile in Italy, they set out for Le Bouilh. They had neither the money nor the inclination for life in the new Paris. While Frédéric travelled on his own to see his mother's lands at Tesson on the way, Lucie hired a carriage large enough for herself, Marguerite, M. de Calonne, a maid and the three children. Humbert, who was 10, perched up beside the coachman. They made very slow progress, seldom covering more than 75 kilometres in 24 hours; the roads remained appalling, despite a new toll designed to raise money for repairs.

At Cholet, near Angers, they fell in with a woman on her way to Bordeaux to sell cloth. With brigands menacing many of the main highways, they were pleased to travel in convoy. It was from this woman, who had fought with the royalists, that they learnt the full story of the Vendée's long battle against revolutionary France, and the brutality with which it had been crushed, Cholet itself having been virtually destroyed in the fighting. For Lucie, the story of the Vendéen uprising came as an almost complete surprise. In her five months in Paris, she had heard little mention of it.

They found Le Bouilh, which they had left in the care of a reliable housekeeper, almost untouched; the land, however, had been neglected, and with the price of wine much reduced by continuing hostilities with England, the estate was worth almost nothing. The conscientious and hard-working Frédéric was neither a manager nor a businessman. The house in the rue du Bac had been sold for very little, Tesson and the estates at Saintes were either in ruins or had been sold, Hautefontaine had been taken by the state. To try to build up some kind of security, for Humbert was growing up and would soon be in need of a career, Frédéric bought a distillery, hoping that the higher profits from eau-de-vie might bring in a decent income.

The ungainly château began to fill up. A cousin of Frédéric's called Mme de Maurville arrived; she was the impoverished widow of an admiral and her only son had been a pupil at Burke's school for émigrés in London. Before long, they were joined by the Princesse d'Hénin, from whom Lucie never seemed able to escape for long, accepting with remarkable forbearance the volatile and domineering older woman. The Princesse would stay at Le Bouilh on and off for almost two years. Though the household was constantly thrown into turmoil by her capriciousness, and no one dared oppose her for fear of provoking a tantrum, Lucie accepted and appreciated her evident love for Frédéric, even if she doubted that the Princesse felt much affection for herself.

The Princesse d'Hénin brought with her Elisa, Lally-Tollendal's 14-year-old daughter, a docile, affectionate girl who had been learning the ways of the *ancien régime* at Mme Campan's academy. Lally-Tollendal wanted Elisa to live at Le Bouilh and agreed to pay the de la Tour du Pins the same fees as he had been paying Mme Camban, an arrangement Lucie found embarrassing but necessary. She was fond of the biddable Elisa, saying that her mind had been 'completely neglected', and she taught her English, leaving Frédéric to take history and geography and M. de Calonne Italian. With her usual clear and sometimes chilly eye, Lucie observed that Mme de Maurville and Elisa were 'both about equally lacking in intelligence', and she suspected that their feelings towards herself were probably closer to respect and awe than to affection. She added: 'Despite whatever may have been said of me, I am not a domineering woman.' Perhaps not; but she was certainly forceful, and becoming more so.

In the evenings, reverting to a habit started in the first years of their marriage, Frédéric read aloud to the assembled household. The vast château, with its high-ceilinged rooms and echoing halls, was filled with noise and children. There were occasional visits to Bordeaux, where the city's fortunes were gradually rising, as American ships were once again docking in the port, and ways were found to circumvent the British blockade. The wine harvest of 1798 had been one of the best of the entire 18th century. Under Napoleon's drive to revitalise the country, an envoy had arrived from Paris to run the police; the streets had been cleaned and lit by the new oil lighting; a fire service had been set up, a literary circle established, *lycées* were opening. The weeks passed, peaceful and contented. 'I was very happy,' noted Lucie, many years later, 'we were at last all united and in our own house.' Humbert, Charlotte and Cécile were all healthy.

Another visitor was Claire de Duras and her two small daughters, Clara and Lucie's goddaughter Félicie. Claire had returned alone to Paris, living with her mother-in-law while she tried to arrange for the names of Amédée and of her own mother to be removed from the list of proscribed émigrés. She was taking steps to reclaim the Kersaint lands and houses. Though it was some time before Amédée, First Gentleman of the Bedchamber to the exiled Louis XVIII, was allowed back, they were planning to buy the white, turreted medieval château of Ussé overlooking the Indre valley. Before leaving Paris, Claire had opened her own salon, gathering around her people who still dreamt of a Bourbon restoration and were irked by France's increasing subservience to Napoleon. Lucie found Claire agreeably changed; she observed crisply that Claire, having accepted that she would never win people with her looks, had fallen back on her intelligence and her wit, of which she had a great deal, and was clearly using them to good effect. Until now, Lucie, always moving, always alert for trouble, had had neither the time nor the place to enjoy close friendships with other women. With the difficult and self-obsessed Claire, she began to explore the possibilities of intimacy.

It was from her visitors that Lucie followed Consular life in Paris. To celebrate the first anniversary of 18 brumaire the famous ballooning couple, M. and Mme Garnerin, rose above the city before sending down a small dog suspended from one of Garnerin's fashionable new parachutes. Paris had lost none of its delight in spectacles. But soon after, as the First Consul was in his carriage on the way to hear the first performance of Haydn's *Creation*, an 'infernal machine' had exploded in a water wagon along his route, killing 20 people and destroying several houses, though Napoleon and Josephine were both unhurt. There was also news of Talleyrand, whom Napoleon, in keeping with the stricter moral tone of Paris, had obliged to marry his mistress of many years, Catherine Grand, the former wife of a British civil servant in Calcutta, who had been very beautiful but was reputed to be getting fat. It was said that she had once entertained Édouard Dillon, '*le beau Dillon*', naked, covered only by her immensely long hair. And the Consuls were finally allowing the return of all the émigrés, except those excluded on account of their open hostility to the new government, though those who accepted Napoleon's amnesty for their past misdeeds were kept under surveillance and warned that they might be expelled again at any time. Lucie and Frédéric's great friend M. de Chambeau was at last on his way home.

For some, it had all come too late. Of the 14 male members

of a family called de Jallays, all were dead, killed off by prison, the guillotine, penury or the war. And there were many thousands of families made permanently destitute by the forced sale of their lands. At a hospice in Paris could be seen elderly noblewomen, sitting on a row of chairs outside the door. Whenever they caught sight of someone they believed to have taken any of their former possessions, they withdrew silently into the chapel to pray.

Mme de Staël, too, was running into difficulties. She had published her novel *Delphine*, to considerable acclaim. But her glittering salon, at which she entertained ardent royalists with her disquisitions on platonic love, Protestantism and the Enlightenment, talking rapidly, never at a loss for words, her eyes sparkling with vivacity and wit, and looking, as one visitor remarked, 'more virile than feminine', had attracted the attention of Fouché's spies. After Napoleon abruptly dismissed 20 members of the Tribunat, among them her companion Benjamin Constant, Mme de Staël had taken to writing mordant epigrams about the 'sultan'. The day came when Napoleon decided to tolerate her no longer. Mme de Staël was ordered to withdraw to no closer than 40 leagues from Paris. It would be ten years before she was able to return and exile was extremely painful for her. It was only in Paris, she said mournfully, that 'French conversation could be found'.

Mme de Staël was not the only Frenchwoman who had turned to fiction. Claire de Duras and Mme de Genlis were both taking up their pens to produce novels about duty, devotion and unhappy love affairs, in a world suddenly devoted to luxury and ostentation, in which women were no longer the centre of their worlds but relegated to the edges. In many of these books – published, it was said, at the rate of four a day – it was sometimes as if only the classical myths could expiate a sense of guilt about the death of the King and the end of the monarchy.

Among the returning émigrés were many priests, 'non-jurors' who had escaped the guillotine and spent the revolution in the Catholic countries surrounding France. Over 24,000 priests had gone into exile, a far greater number than the nobility. Until the summer of 1801, the only religious cult tolerated in France remained that of the theophilanthropists, deists believing in God but not in a Church. But France had tired of civic and philosophical holidays, of the forced jollity of feast days celebrating the virtues of Marcus Aurelius or the heroism of William Tell.

When Chateaubriand, a man tuned to nostalgia, published his *Génie du Christianisme*, he found willing listeners not only in the Chaussée d'Antin and the Faubourg Saint-Germain but in the Tuileries. Like Mme de Genlis, Chateaubriand blamed Voltaire for inciting atheism. Dechristianisation, he said, had ruined the prestige and power of France; religion was all about the soul, conjugal love, filial devotion and maternal tenderness, and without God, men turned to crimes that no temporal laws could check.

Napoleon, seeking order and stability, was sensitive to the idea of a revived Church, seeing in it a certain way to win approval from ordinary people. In July 1801, he signed a concordat with the Pope. 'Go to Rome,' Napoleon instructed one of his generals. 'Tell the Holy Father that the First Consul wishes to make him a gift of 30 million French.' On Easter Day 1802, in the newly restored cathedral of Notre Dame in Paris, a *Te Deum* was sung for the first time in almost ten years. Few of the congregation could quite remember what they were expected to do. But it was not a total capitulation. The government would retain the right to nominate the higher clergy, and would be responsible for salaries. It was to be a submissive Church, despoiled of much of its great wealth.

In Bordeaux, as in all French parishes, Mass by 'non-juror' priests had until now been celebrated in secret, in private houses. When the local archbishop finally came back to the city, Lucie and Frédéric gathered together all their local parish priests to greet him. Later they accompanied him on a triumphal procession back to his see.

* * *

There was more to celebrate than the official return of Catholicism to France. For the first time in 10 years Europe was at peace. France had fought various battles with the Austrians, not all as victorious as Napoleon maintained, but victorious enough to sign a peace treaty at Lunéville in February 1801. Piedmont was annexed to France; Genoa became a French puppet state. Over the next few years, with total disregard for Italian aspirations for unity and independence, which Napoleon had warmly supported, various parts of Italy would be turned into mere tributaries of France. With the Peace of Amiens, signed with the British in March 1802, over a century of Anglo-French struggles were laid to rest. Fourteen thousand British prisoners were released from French jails. Martinique and Guadeloupe were returned to the French. Napoleon, eager to see the plantations prosper again, reintroduced slavery: the French slave traders had been dormant, but not ruined, having spent the fallow years as corsairs harrying British ships. Tom Paine and *The Rights of Man* were conveniently forgotten.

With the peace came a surge of English visitors across the Channel. They found Paris chilly, dilapidated and overwhelmingly fascinating. They came to find friends, to recover lost property, to hunt in the French forests and simply to look. The artists, Turner among them, visited the Louvre, where Vivant Denon, who had accompanied Napoleon to Egypt, had been appointed director of the Musée Central de la République. They wanted to see Europe's looted art and many spent their days at easels copying. Those with a taste for the macabre visited Mme Tussaud's waxworks to see Robespierre's death mask. Others went to inspect Victor, the Aveyron savage, a boy found in the woods near Toulouse, apparently deaf and dumb and living like an animal. In the college of surgery there was a wax head of the King of Poland's dwarf, Bébé, and wax representations of venereal diseases, the 'miserable objects of fatal libertinism'. Theatre lovers saw the beautiful Mlle Georges play in *Tancrède*, and Napoleon's favourite actor, the great Talma, in Molière's plays, though most complained that his style was too extravagant and declamatory, and said that his 'strutting bloated pomp, bombast gesticulation . . . trembling body and quivering hands' made them laugh.

For those with musical interests, there was Gossec, nearly 70 but still active in the new Conservatoire de Musique. The French ballet, visitors agreed, set to a mishmash of tunes taken from symphonies, sonatas and operas, was not what it had been in the days of Rameau. For the greedy, Paris had never offered more restaurants, clustered around the Palais-Royal, some serving food so elaborate and architectural that they felt repelled; there were truffles, they protested, in every dish. At Very's, you could choose from 25 different hors d'oeuvres, from herb salad to hog's pudding. And for the scientists, there was the chance to call on Mme Lavoisier, whose father had gone to the guillotine at the same time as her husband, the famous chemist, and who had filled her salon with his 'magnificent chemical and physical apparatus'.

The women of Paris, remarked the visitors, were all very pale; rouge had been given up in favour of white paint, *à la Psyche*, after a painting by Gérard. Those who called on Thérésia found that she had divorced Tallien and was living in the rue de Babylone with her new lover, the immensely rich army contractor Ouvrard. If fortunate enough to be invited into her bedroom, they discovered it to be more severe in style than that of her rival Mme Récamier, but even so its bed was shaped like a round tent, held up by the beak of a golden pelican, shrouded in white-and-crimson satin curtains with golden fringes falling in pleats to the floor.

Paris had become a vast bazaar, its quays along the Seine still full of bargains from sequestered houses, its fashions drawn from all over the world: waistcoats from England, linen from Holland, boots from Russia. The German dramatist August von Kotzebue, strolling past the Palais-Royal late at night, observed that the prostitutes were not as bold as before the revolution, but that there were several black girls, brought back by soldiers from Saint-Domingue and known as '*chats en poche*' – pocket cats – their 'black skins peeping from beneath their white dresses like flies in a milk pot'.

One day Kotzebue accompanied Mme Récamier to the ruined royal abbey of Saint-Denis, where the caretaker, ordered by Robespierre to burn the coffins of Louis XIV and Henri IV in the vaults along with other royal remains, had buried them instead, 42 kings, 32 queens and 63 princes and princesses piled one on top of the other. Most visitors at some point made a pilgrimage to the spot where Louis XVI and Marie Antoinette had died, and then went on to the Place du Carrousel, to see the bronze horses looted from San Marco in Venice. In the new botanical gardens, they watched keepers pouring buckets of cold water over 'the white bear from the North', in order to keep it in 'tolerable patience'.

But what the foreign visitors all wanted to do was to meet the French of whom they had heard so much, even if some were squeamish about taking up introductions to men they considered tainted by violent revolution. There was Joseph Fouché, who had appropriated the Hôtel Mazarin on the Quai Voltaire, and hung it with Gobelin tapestries, though his boots were said to be muddy and his linen seldom clean. And there was Cambacérès, the second Consul, who gave dinners on Mondays and Wednesdays to 35 men – never any women: meals of exceptional grandeur served by 50 footmen in blue-and-gold livery. Foreigners flocked to pay their respects to Talleyrand, the 'able and wily diplomat who dupes Europe'. Talleyrand's chef Carême, to whom he had set the test of creating an entire year of menus without a single repetition, using only seasonal food, was the most famous cook in Paris, the 'chef of kings and king of chefs'.

Above all, of course, the visitors wished to see Napoleon. There was much jostling for invitations to the Tuileries, from whose windows they could watch the First Consul – who had recently appointed himself Consul for life – review the troops, looking young and forbidding on his white horse, accompanied by his uniformed aides-de-camp and his guard of Mamelukes 'richly habited in the Oriental style'. The more fortunate, invited to an actual audience, found the rooms of the Tuileries packed, the men in court dress, the women in *décolleté*, the footmen in their new green livery edged with gold. As Napoleon walked slowly round the room, an equerry at his elbow murmuring the names of the foreign guests, Josephine followed a few paces behind, her hair dressed with a diadem of precious stones. After the audience came a 'sumptuous' dinner. It was known that Napoleon had an excellent sense of smell and that his head was so sensitive that his valet had his master's hats specially padded and broke them in himself.

Peace lasted just over 13 months. England, unwilling to accept France's supremacy over Europe, alarmed by its naval activities and wishing to keep French troops out of Holland, was the first to declare war, on 18 May 1803. In France, Napoleon ordered the arrest of some 7,000 English travellers, students, artists and merchants. Most were able to hasten away to the 'liberty, cleanliness and roast beef of old England'. For the next 11 years, all travel between France and England was suspended. Lucie was once again cut off from her much-loved aunt Lady Jerningham and the rest of her English family.

* * *

As the months and then the years passed, friends wrote from Paris to Le Bouilh urging Frédéric to consider applying for a post in the Consular administration. All the *ancien régime*, they said, was rallying to Napoleon. France was larger and more powerful than it had ever been. It had a new legal code, the Code Napoléon, that effectively put an end for ever to the feudal world and was bringing order to an administration that had varied chaotically from province to province; the roads were safer, the civil war in the Vendée was over, education was spreading, there were new programmes of public works and new measures to spur economic growth. Napoleon had even weathered – though with some difficulty – the brutal murder of the innocent young Duc d'Enghien, grandson of the Prince de Condé. Kidnapped from his house in Baden, he was executed by firing squad, in response to fears that royalists were again preparing a *coup d'état* to return the Bourbons to the throne. But Frédéric, always cautious, always unassuming, always weighing up all sides to every question, found the role of petitioner extremely distasteful. If the Emperor desired his services, he said, he knew where to find him. It was less a question of loyalty to the deposed Bourbons – for Frédéric, like Lucie, was impressed by what Napoleon had done for France – than of genuine lack of self-promotion.

France's new Constitution, amended in 1802, had given Napoleon every attribute of royal power, except the actual crown. Deciding that Malmaison was too small and modest, he and Josephine had restored the royal palace of Saint-Cloud. Here, at Sunday Mass, with the Bishop of Versailles officiating, Napoleon took the place in the Long Gallery once taken by Louis XVI. The two junior Consuls, Cambacérès and Lebrun, sat behind him. When Napoleon stood, Josephine, by his side, knelt. At night, they dined in state. Formality and manners were gradually consuming every facet of Consular life. At the Tuileries it was said that 'the château sweats etiquette'.

In the spring of 1804, a proclamation was posted on the walls of Saint-André-en-Cubzac, inviting people to say whether or not they would like to see their First Consul made Emperor. Frédéric, who had become President of the Canton, agonised over his vote, pacing up and down the garden at Le Bouilh, Lucie watching but saying nothing. For herself, she was clear: Napoleon had earned his crown. When Frédéric told her that he had signed with a 'yes', she was delighted. This referendum, like Napoleon's others, went resoundingly in his favour. At the end of November, Frédéric, as one of Bordeaux's leading citizens, was summoned to Paris for the coronation.

All through the autumn, by night as well as by day, 3,000 workmen had laboured to clean up and embellish the city. Around Notre Dame, the various buildings that had sprouted from its walls were torn down. Tiers of seats went up. Charlemagne's imperial insignia were brought from Aix-en-Chapelle. With extreme reluctance, Pope Pius VII agreed to travel from Rome to officiate.

On 2 December, the date set for the coronation, the day opened with a hailstorm. Courtiers wore costumes designed by David and Jean-Baptiste Isabey, a compromise between antiquity and the reign of Henri III, men who not so long before had sported red bonnets and *carmagnoles* decked out in laced boots, silk stockings, puffed white satin sleeves, short cloaks and plumed fur hats. Napoleon, his golden imperial coach groaning under carved garlands, medallions, allegorical figures and four great eagles, wings outspread, holding a golden crown designed by Fontaine, was so covered in diamonds and gold that spectators spoke of a 'walking mirror'.

It was reported that 500,000 people, almost the entire population of Paris, lined the streets. When Napoleon entered Notre Dame, he put on a crimson cape, sprinkled with gold stars and lined with ermine. Josephine's train was borne by Napoleon's sisters, who, ill-pleased by her preferment, wore somewhat sulky expressions in David's celebrated picture of the event. (Napoleon's mother, known as Madame Mère, who declined to attend, was painted in later.) That night, standing on the balcony in the Tuileries, the imperial family looked out across the snowy gardens at a brilliant display of fireworks. Napoleon had contemplated an elephant for his imperial symbol, but settled on a bee.

To oversee the splendour of court life, every detail of which Napoleon supervised, two prefects of the palace were appointed, chosen, it was said, because they were shorter than the new Emperor; but while they argued about what should be done, the *ci-devant* nobles tittered in the background. Gradually it was turning into Versailles, all over again. The royal army had never looked more splendid, in an array of bearskins and tigerskins, lace, towering plumes, helmets and breastplates. For courtiers, there was velvet in winter and taffeta in summer, the costumes designed by Isabey.

Ladies-in-waiting were being chosen for the Empress. A letter arrived at Le Bouilh inviting Lucie to come to Paris to take up the post. She did not hesitate: she declined, remarking later that of all the women approached, who, like her, had attended Marie Antoinette, she was the only one to refuse. No court could be more attractive to her than her life at Le Bouilh with Frédéric and the children.

While Frédéric was in Paris, he had talked to many of his old friends, now attached to the government and the court. Talleyrand, recently named Grand Chamberlain and Vice-Elector of the Empire, on learning that Frédéric was still reluctant to petition Napoleon for an appointment, said only, 'You will come to it' and shrugged his shoulders. Their friend from London, M. Malouet, had been appointed *Préfet Maritime* at Antwerp to administer the coastal area and set up a shipyard to build the great new ships of the line. When Frédéric returned to Le Bouilh, he brought with him an invitation from M. Malouet for Humbert to become his secretary once he reached 17. Lucie was both pleased and saddened. Strongly attached to her children, she could not envisage a day when they lived away from her.

Talleyrand, while urging restraint on further territorial acquisitions, had been forced to sit by and watch Napoleon absorb Piedmont into France, sacrifice Venice to Austria, and turn the rest of northern Italy into a republic. After Russia formed an alliance with Britain and Austria joined the coalition, Napoleon, waiting at Boulogne to cross the Channel and invade England, had turned to march east, taking the lame, protesting Talleyrand with him. Victory had followed victory. At the Battle of Austerlitz, the Austrian and Russian armies lost 26,000 men to Napoleon's 9,000. With the Peace of Pressburg, Napoleon became the master of western Europe, even if the annihilation of the Franco-Spanish fleet by Nelson at Trafalgar left Britain master of the sea. The coalition against France had been broken up, Austria was crushed, the Holy Roman Empire ceased to exist. Through battles and treaties, amnesties and deals, many brokered by the reluctant Talleyrand, Europe had been partitioned, then repartitioned.

Rewarded by the Principality of Benevento, a small enclave in the Kingdom of Naples, Prince Talleyrand was as efficient and canny as he had ever been, making millions of francs through bribes and backhanders. He continued to receive ambassadors, political colleagues, secretaries and friends during his daily public *levers*, when, emerging into his *grand cabinet de toilette* in several layers of flannel and nightcaps, he would sip camomile tea, while his long grey hair was combed, pomaded and powdered. 'For a credulous soul', remarked Aimée de Coigny, who had spent seven months of the revolution in prison and who knew him well, Talleyrand 'would be a satisfactory proof of the existence of the devil . . . a priest who denied his God, a prelate who ransacked his Church, a nobleman who betrayed his King . . . a diplomat and minister always intriguing and pushing back the boundaries of excess and treason'.

Early in 1806, Lucie realised that she was pregnant for the tenth time. She was now 35. She had miscarried the previous year – her sixth miscarriage – and she resolved to take more care of herself during this pregnancy. Women in the 18th and 19th centuries could expect to spend many months of their lives pregnant; but Lucie's constant pregnancies and miscarriages would have taxed a frailer woman.

In the spring of 1806, the family was faced with a very sad loss. Marguerite, whom Lucie had 'loved all her life as a daughter', fell ill. She grew weaker, but remained conscious. They bade each other the 'tenderest of farewells'. When she died, the entire household went into mourning. 'My grief,' wrote Lucie briefly, with the economy of phrase she reserved for her deepest emotions, 'was deep.'

There had been other deaths, less mourned. Mme de Montesson, having thrown a ball of great splendour for the marriage of Josephine's daughter Hortense and Napoleon's brother Louis, with ices shaped like fruit hanging in baskets from orange trees in flower, had died at the age of 68. According to a malicious friend, she had not aged well, her face 'crackled . . . like an old piece of pottery'. Napoleon, who had appreciated her advice on etiquette and her impeccable *bon ton*, ordered that her body lie in state for eight days in Saint-Roche.

From London came news of the death of Archbishop Dillon, at the age of 85, having been ill for two days with what Lady Jerningham described as 'gout of the stomach'. His funeral was held in the French chapel in Little George Street and was attended by the French nobility and clergy still in exile in England. Mme de Rothe had died not long before, behaving at the last with a dignity and courage she had seldom shown in life. Of all the relationships in her ill-tempered, embittered life, her 35-year-old liaison with the Archbishop had been the most harmonious. Not wanting the elderly Archbishop to be distressed, she had said nothing to him when she realised that she was seriously ill and had only a few days left to live; when she died, he was relieved to imagine that she had not suffered. The Archbishop spent the day of her funeral eating a hearty lunch and discussing Voltaire. Neither death caused Lucie much grief. She did not even remark on that of her grandmother.

The new baby came early. On 18 October, Lucie was dressing when she caught sight of Dr Dupouy, the doctor who had delivered Séraphine and Charlotte and who happened to be staying at Le Bouilh for a few days, walking past her window. Asking what he was doing up so early, she learnt that he had been called out to sign the death certificate of a neighbour who had died in the night. Lucie, who knew the woman well and had spoken to her the previous day, was shocked and upset. She went into labour. A boy was born; they called him Aymar. There were now five children in the house: Humbert, 16 and soon to leave, 10-year-old Charlotte who was turning out to be clever and full of curiosity, 6-year-old Cécile, and the new baby, Aymar. Elisa, Lally-Tollendal's daughter, was the fifth.

Lucie herself was ill again after Aymar's birth, with 'double tertian fever', presumably malaria, the fever recurring every day. She recovered slowly, but by Christmas felt well enough to take the baby to Bordeaux to be vaccinated against smallpox. The city's fortunes had revived, with a series of excellent years for wine, and a forest of masts stretched as far as the eye could see along the estuary. The whole family spent six weeks staying in M de Brouquens's house, attending balls and furthering Elisa's marriage prospects. Elisa was 18, small, with thick black hair, and, at a time when dancing was considered an art, was much admired as an exceptionally graceful dancer.

At Mass one day Elisa attracted the notice of one of Bordeaux's most eligible young men, a M. Henri d'Aux. M. de Brouquens helped Lucie arrange a number of meetings between the young couple; in due course, the d'Aux family came to ask for Elisa's hand. Lucie noted in her critical way that though indeed extremely rich and a gentleman of the old school, Henri's father was 'without a vestige of intelligence or learning'. Lally-Tollendal had promised Elisa a fine dowry if he were repaid a debt owed him by the state since the revolution. Napoleon, who wanted Lally-Tollendal in the government, secured the repayment and the bride's father arrived from Paris bearing a hundred sacks of money, each containing 1,000 francs. It was more money than Lucie had ever seen in her life.

The wedding was celebrated at Le Bouilh on 1 April 1807. For the dinner Lucie, Charlotte and Mme de Maurville made a centrepiece of red and white daisies, the flowers spelling out the names of Elisa and Henri against a background of green moss.

* * *

Soon after the grape harvest of 1807, 17-year-old Humbert left to take up his post as M. Malouet's secretary in Amiens. Lucie had been dreading this moment: when it came she was heartbroken. 'The gaiety of our house went with him,' she wrote. Charlotte, who was 11, and to whom he had been an affectionate brother and companion, felt his loss equally keenly.

Frédéric, who accompanied Humbert to Amiens, paused in Paris, where he was given a surprisingly warm welcome by Lucie's stepmother. Lucie's half-sister Fanny was now 23. Elegant rather than pretty, and somewhat capricious, Fanny had fallen in love with Prince Alphonse Pignatelli, and would have married him, had he not fallen ill. When Pignatelli realised that he did not have long to live, he had pressed Fanny to marry him, in order to leave her his considerable fortune. But Fanny had refused, and Pignatelli had died. Since then, General Henri Bertrand, one of Napoleon's aides-de-camp, had fallen in love with her and Napoleon was urging her to accept. But Fanny remained stubbornly opposed. When Frédéric called on Mme Dillon, she begged him to see the Empress Josephine on her behalf, in order to convey Fanny's final refusal.

In the Tuileries, Frédéric was received in Josephine's bedroom, where a deep alcove had been curtained off. He delivered Fanny's message, but the Empress detained him with questions about himself and Lucie, and about their future plans. Frédéric spoke out freely; it was not in his nature to be discreet. Later that evening, Frédéric called on Talleyrand who, he discovered, already knew all about the meeting from Napoleon. The Emperor had listened to every word, hidden behind the curtain in the alcove. 'Fortunately for you,' Talleyrand told Frédéric, 'your aristocratic airs have saved you. They happen to be the fashion in the Tuileries just now.'

That October Lucie learnt that a number of officers were to pass through Saint-André-en-Cubzac, on their way to the French border. Intending to bring the Iberian peninsula under direct French control in order to dominate the entire Mediterranean, Napoleon was pouring French troops into Spain, ostensibly under the guise of offering Portugal to Spain, which he accused of refusing to enforce the blockade against England. At Mass one morning, she caught sight of a particularly handsome young officer, splendidly turned out in the white cloak, baggy purple trousers and sabre of an aide-de-camp to General Murat. It turned out to be Frédéric's nephew, Alfred de Lameth, elated at the prospect of action in Spain. Lucie had always been fond of him. Before leaving, Alfred opened his writing case and presented her with a small knife with a mother-of-pearl handle, saying that he wished her to have something to remember him by.

While Frédéric was still away in Amiens, the Princesse d'Hénin wrote from Paris to say that the Emperor would be passing through Saint-André-en-Cubzac on his way to Bayonne. After the French had entered the Iberian peninsula, the Portuguese royal family had fled to Brazil. Napoleon, playing off various factions against each other, had sent 100,000 French troops to Spain and had taken the citadels of Pamplona and Barcelona. In March 1808, King Charles IV abdicated in favour of his son Ferdinand, but then revoked the abdication, and Napoleon had summoned a meeting of all concerned to Bayonne.

Lucie, Mme de Maurville and Charlotte joined the crowds thronging the banks of the river, where a brigantine waited to carry the Emperor across the Dordogne. The scene, wrote Lucie, was one of extraordinary excitement. 'A madness, a kind of delirium' seemed to descend on people for whom Napoleon had acquired an almost godlike heroic status. They waited all day, Lucie cursing that Frédéric would miss what might have been a chance to attract the Emperor's notice. She and Frédéric were increasingly worried about money: the eau-de-vie venture had not proved profitable and the turnover barely covered their daily expenses. When Napoleon finally arrived, followed by a detachment of *chasseurs*, it was all rather disappointing. The Emperor appeared sullen and bored while the mayor delivered his speech of welcome, then hastened away as soon as he could. 'We returned to Le Bouilh,' wrote Lucie, 'tired and very out of temper.'

It was therefore with some surprise that Lucie received a summons next day to attend on Josephine, who was following Napoleon to Bayonne. The court was in mourning for the King of Denmark, and she had no time to do more than add black ribbons to a grey satin dress before setting off for Bordeaux. 'Hesitation,' she wrote, 'was unthinkable.' Lucie had not lost all pleasure in her own appearance. She thought her improvised costume 'admirably becoming to a woman of 38 who, I say it without vanity, looked less than 30'.

That night, Napoleon, no longer as sallow and emaciated as he had been at the time of her stay in Paris, his angular features softened and his appearance considerably neater, entered the great dining room of the palace in Bordeaux where crowds gathered to greet him. While waiting, Lucie had been one of those marshalled into groups in a 'military manoeuvre' that had all the ladies standing in perfect straight lines. Catching sight of her, Napoleon asked her, laughing, why she mourned the King of Denmark so little that she was not in black. Replying that she had no black gown, she said that she had decided to come anyway as she was not prepared to renounce the pleasure of meeting her Emperor. Josephine, who had Lucie called out of the receiving line into a small salon, greeted her with great friendliness.

Next day, Lucie was informed that Napoleon had ordered that she spend each evening of Josephine's stay in Bordeaux in attendance on her. What struck Lucie was the fact that even while conducting a war with half of Europe, Napoleon found time to oversee every detail of his administration and his court, down to the clothes that Josephine should wear and the conversations she should hold; and that Josephine would not have dreamt of disobeying him. Lucie was 'cruelly disappointed' when she learnt that Napoleon himself had already left for Bayonne. Something about him had always charmed her. Each time she saw him, she wrote, 'my heart beat fast with excitement'.

Frédéric had returned and was present at the evening receptions where he played backgammon with Josephine. While in Paris he had heard from Talleyrand that Napoleon, wishing for an heir to his throne, was beginning to talk of a divorce from Josephine, who had not borne him a child and was now 43. That he could father one himself was clear from the fact that one of his mistresses had given him a son in 1806. Frédéric felt very uneasy when Josephine, pleased to find someone not at court with whom to talk, wanted to confide in him her terror of being discarded. She felt particularly bitter towards Talleyrand, whom she suspected – rightly – of encouraging the divorce. Napoleon had gone on ahead to Bayonne and Josephine was waiting and longing to be summoned to join him. One evening, she asked Lucie to help her decipher a scribbled note from the Emperor, saying that he was having trouble in Bayonne, where both King Charles IV and his son Ferdinand were assembled.

When Josephine was finally called to Bayonne, Lucie was not sorry to return to Le Bouilh. Ten days of imperial life, in full court dress, missing her children, were enough. She was only sorry that Frédéric seemed no nearer to preferment, and that, living with no fortune so far from Paris, it was extremely unlikely that anything would ever come his way.

* * *

In 1807, the royal families of Europe, jockeying for power and prestige and territorially ambitious, had between them descended to levels of intrigue, profligacy and lack of dignity that made the excesses of Louis XVI and Marie Antoinette seem insignificant. Charles IV, who had been on the throne of Spain for almost 20 years, was a credulous, intellectually feeble man, who preferred hunting and his own poor playing of the violin to government, and who allowed himself to be led by his coarse and vicious wife, Queen Maria Luisa, by whom he had 14 children, several of them subjects of Goya's finest portraits. For some years, real power had lain with her greedy and dissolute lover, the present Chief Minister, Manuel Godoy. Like some of the other enlightened despots of Europe, the Spanish royal family believed that reform, should any be necessary, must come from above, and not from the people. By the spring of 1808, however, Charles IV had signed away his claims to the Spanish throne in favour of Napoleon; in exchange he was to receive a pension and a comfortable exile.

In May word came from M. de Brouquens that the deposed royal family would be spending a few days in Bordeaux on their way to Fontainebleau, and that Napoleon had ordered that Lucie was to serve as lady-in-waiting to Queen Maria Luisa during her stay. Elisa had also been appointed to serve her. When a chamberlain accompanying the royal party showed Lucie into the Queen's apartments, he whispered that she should try her best not to laugh. The doors opened: Lucie felt more inclined to cry. The Queen was dressing. Apart from a very short cambric skirt, all she was wearing was a muslin fichu across her bosom, 'the driest, leanest and darkest bosom you can imagine'. She was short, stout, with a peeved expression and a loud, harsh voice, and her grey hair was piled high with red and yellow roses. Napoleon found her very ugly, as he told Talleyrand, and 'with her yellow skin, just like a mummy'. Her expression, he went on, was 'nasty and treacherous', adding that what he particularly disliked was her deep *décolleté* and her bare arms. Having brought very little with her from Madrid, the Queen was putting on a totally unsuitable dress in yellow crêpe and satin lent to her by Josephine. Lucie found her pitiful.

Over the next few days, Lucie was constantly at the little court in exile. She presented Bordeaux's dignitaries to the royal couple, took meals with them, and listened to Maria Luisa railing against her sons, denouncing them as monsters and the cause of all their misfortunes, and saying that no punishment would be too bad for them. 'It made me shudder,' she wrote. The two young men had been sent as prisoners, under house arrest, to Talleyrand's château at Valençay. Lucie also took strongly against the dissolute Godoy, who had been made Prince de la Paix, and who was very rude to her. The court consisted of a number of attendants, both Spanish and French, who found Lucie's agreeable nature extremely helpful, and kept repeating that they would like Napoleon to appoint her a permanent member of the Spanish court in exile. Lucie's one fear was that the Emperor might be persuaded to agree. One evening, knowing that the unregal Charles IV regarded himself as a fine violinist, she arranged a musical soirée, discreetly employing a proper musician to take over his part in the quartet whenever he lost his place. She was very relieved when she finally 'bade farewell to these unhappy foreigners' on the banks of the Dordogne, where a brigantine was to carry them across the river on the next leg of their journey into exile.

The King had left turmoil behind him in Spain. In May 1808, the people of Madrid rose up against the occupying forces. Though Joseph Bonaparte, Napoleon's brother, proclaimed King of Spain in June, had ruled decently in Naples and promised to do so again in Spain, the Spanish felt furious that a vast French army was in occupation. The profoundly conservative Spanish clergy and the great feudal landowners were united in their dislike of the Napoleonic Code, in their opposition to liberal reform and in their refusal to contemplate the loss of their feudal dues. Small local uprisings were fast turning into a full-scale popular insurrection, all sides acting with extreme brutality. With each ambush of French troops, brutal reprisals were visited on the Spanish people. News reached Le Bouilh one day that Alfred de Lameth, Frédéric's handsome nephew, had been attacked and killed while crossing a village on his way to lunch at Maréchal Soult's headquarters. The French immediately set fire to the village, killing as many of its inhabitants as they could find. Until her death, over 40 years later, Lucie kept Alfred's little knife, with the mother-of-pearl handle.

Late one evening, when Frédéric was in Bordeaux on business, and Lucie alone with the children at Le Bouilh, a messenger arrived bearing a note from him. It contained a single sentence. 'I am Prefect of Brussels . . . Brussels lies just ten leagues from Antwerp.' Preferment had finally come.

They learnt later from Napoleon's secretary, M. Malet, that the Emperor, in keeping with his practice of monitoring every appointment and decision even when on campaign, had been presented with a name for a possible new incumbent for the vacant post of *Préfet de la Dyle*, crossed it out, and written Frédéric's name in its place. 'Like Cincinnatus,' Lucie would write, Frédéric 'had been summoned from his plough and given the finest prefecture in France.'

Lucie's first thought was not of the salary and prestige of the job, but that she would soon be reunited with Humbert. Then she set about preparing for the move. 'In the important moments of life,' wrote Lucie revealingly, in words that said much about her decisive and determined nature, 'any thought which has not occurred in the first 24 hours is either pointless, or idle repetition.'

By the time Frédéric reappeared next day, Lucie had already mapped out their future. 'I was ready to tell him the plans and arrangements which seemed to me necessary.' Eleven-year-old Charlotte, evidently taking after her mother in energy and determination, had studied all available maps and gazetteers of the Dyle and was waiting to discuss them with her father. The eight years of peaceful domestic inactivity had come to an end. Lucie had been a happy and loving mother; but in her pleasure and eagerness, one senses a longing for something new.

CHAPTER THIRTEEN

Not a Court, but a Power

'My prefects, my priests, my rectors,' replied Napoleon, when asked how he planned to administer his Empire. In 1790, as part of the reorganisation of the country, France had been split up into 83 departments, their names chosen from local rivers and mountains; as conquered countries were brought into the fold, so the numbers of the departments grew. When he let it be known that he would appoint prefects from all political sides, ambassadors, soldiers and ministers hastened to offer themselves. Few shared Frédéric's reticence about seeking preferment. However, from his earliest days as Consul, Napoleon had also been much taken with the idea of nobility. He was appreciative of those of the *ancien régime* who rallied to him, and constantly sought ways to bring more back into the fold. Increasingly, he wished to see fewer former Jacobins and more aristocrats among his prefects. Frédéric, '*ci-devant Comte, Colonel, Ministre Plénipotentiare du roi en Hollande*', was a natural choice. Hiding behind the curtain in the alcove in Josephine's room, Napoleon had been impressed by Frédéric's 'aristocratic airs', his evident honesty and his obvious attachment to his family. Lucie, with her training at the court of Marie Antoinette, her excellent English and her pleasing ways, would serve France no less well than her husband.

Formally appointed on 12 May 1808, by Napoleon, *Conquéreur des Français, Roi d'Italie et Protecteur de la Confédération du Rhin*, Frédéric listed his fortune as 'none'. He explained that his revenues before the revolution had indeed come to 44,000 francs a year, but said that all he had left were a few worthless vines in Bordeaux. 'I have often been told,' he added, 'that one should not allow oneself to become poor. But I do not believe in such things. I offer myself simply as I am.' Frédéric's poverty and his rectitude both promised that the new prefect would prove hard-working and biddable. One of his first acts as prefect was in fact to send out his signature, so that it would be recognised as authentic on official documents; he signed it Latourdupin, as a single word, in a rounded, firm, legible hand. With the post came pomp, splendour and considerable power, for, in keeping with his ideas about authoritarian rule, Napoleon placed local administration in the hands of a single man, assisted by two councillors. He called his prefects '*des empereurs au petit pied*', emperors in miniature, and gave them uniforms in blue, red and white, their blue coats embroidered with silver, their red sashes trimmed with a silver fringe. The wheel had turned full circle: the prefects were almost indistinguishable from the *intendants* of the *ancien régime*.

Frédéric's job, however, would not be easy. Over his prefects as over all other aspects of his Empire, Napoleon liked to retain control. He personally vetted and selected every prefect, as well as many *sous-préfets* and mayors, and paid them according to the size and strategic importance of their departments; he sent inspectors to check up on them, scrutinised their reports, criticised their personal lives and sacked them at will. From his prefects he expected absolute submission and blind zeal. If they were seen to assume excessive power, they were called to order; if they abused it, they were removed. When the Prefect of Montenotte was reported to have taken his mistress on a trip rather than his wife, he was informed that the Emperor insisted on '*la plus grande décence*', total respectability. Their job, Napoleon told them, was to rally hearts to a common love of the motherland, no easy task in an empire that stretched from the Baltic sea to European Turkey and the Kingdom of the Two Sicilies, and which included, among its vassals, 32 German kings, princes and dukes. And Frédéric was not quite the loyal, unquestioning servant Napoleon appeared to take him for. There was a streak of independence in his character, and more than a little yearning for the lost monarchy; and he could, when pressed, be very stubborn.

* * *

Frédéric had gone on ahead. Lucie, Aymar, the two girls and Mme de Maurville reached Brussels in the summer of 1808. They found waiting for them a palatial mansion with extensive gardens, filled with servants, and a handsome, cultured, prosperous city, the streets paved and well lit, the buildings in good repair, the theatres playing to full houses. In the upper town, the *ville haute*, behind its 14th-century walls, were the houses of the aristocracy, the royal palaces and the guild houses, some identified not by numbers but by the names of flowers and animals, others painted pale green or yellow, with Ionic columns. In the *ville haute*, French was spoken.

Below, in the *ville basse* in the valley, along the banks of the canal and the river, life centred around the markets for wheat, flax, tobacco, linen, lace and beer, conveyed to customers throughout Europe by canal or road. In the new greenhouses grew grapes and pineapples. In the streets of Brussels could be seen dogs harnessed to wheelbarrows, and on fine evenings, people strolled along L'Allée Verte by the Palais Royale. Brussels was a city of pigeon fanciers, of public fountains with excellent water and of horse-racing; Lucie's uncle Lord Dillon was still remembered as one of three English lords who had brought their horses to race in Belgium long before the revolution. When Voltaire had visited the city in 1742, he observed that while Brussels had very few interesting people, it had exceptionally interesting books.

Belgium had been an early victim of the passion for conquest that gripped France during the revolution. A part of the Spanish Netherlands at the end of the 17th century, it had been French, Austrian, briefly independent, then Austrian again. With Dumouriez's victory at Jemappes in 1792, it became French; a year later, it fell again to the Austrians. But in 1794, General Pichegru, advancing over frozen water in an exceptionally cold winter, had brought Jacobism to Brussels, along with the cockade, a Tree of Liberty planted in the fine Gothic and baroque Grand' Place, and the loss of cart-loads of art plundered from the city's richly endowed churches and monasteries.

On 1 October 1795, the Austrian Netherlands, Flanders and Liège were all annexed to France and split into nine departments. Brussels, the seat of the prefecture of the department of the Dyle, was the most important. Napoleon had since done much to restore Belgian prosperity, ordering dozens of Brussels's famously comfortable carriages, with their glass windows, and instructing Josephine and her ladies-in-waiting to buy Belgian lace.

By 1808, Brussels had been French for 13 years. The first prefect, Louis-Gustave de Pontécoulant, was a Norman aristocrat and a cartographer who had worked hard to provide an efficient economy, based on Paris. He had welcomed as visitors to their native home the two composers Gossec and Grétry, restored churches and encouraged the production of lace. After five years, Napoleon had recalled him to Paris and made him a senator, sending in his place M. de Chabon, an even better administrator, who had set about restoring to their rightful owners property that had been confiscated. De Chabon earned a reputation for justice. He was, Lucie conceded, 'worthy, enlightened and firm', all virtues to which she was much attached.

She felt, however, that neither of these two men had done much to woo the affronted Belgian nobility, a tight-knit circle of some 300 families, most of them related in some way to one another, living in haughty and faded seclusion in and around the Grand' Place. Lucie, who immediately identified her role as ambassadress to these people, reflected that de Pontécoulant had failed on account of his wife, a former mistress of Mirabeau's, whose origins, she concluded, 'held little attraction' for the Bruxellois, and de Chabon on account of his own, an ailing, obscure woman 'from a very modest level of society'. Lucie herself, she noted with pleasure, was already known to the *beau monde* of Brussels through the Abbé Delille's popular poem, *La Pitié*, published in 1802, which talked of a young lady-in-waiting to Marie Antoinette, who had milked cows in the forests of the New World. In the *haute ville*, Lucie was received 'with the greatest kindness'. It was made clear to her that the Belgians were delighted that their salons were again to be 'presided over by an aristocrat'. Only Lucie's extreme directness excuses a tone that sometimes sounded extremely snobbish.

Along with her impeccable breeding, Lucie held another important card. Before the revolution, in the salons of the Faubourg Saint-Germain, Lucie had often met the Duchesse d'Arenberg, the last descendant of a famous Belgian 15th-century warrior baron. Mme d'Arenberg was known in Brussels as the Dowager, and regarded as the Queen Mother. She had perfect 18th-century manners, rode out in all weathers in an open carriage, was devoted to her grown-up children and particularly to her blind son, and received every evening between seven and nine. The Dowager was the arbiter of Brussels society. Letters of introduction had been sent ahead by the Princesse de Poix and Mme de Beauvau; the day after she reached Brussels, Lucie went to call. She was invited to dinner, whereupon 'the whole town signed our book'.

Lucie was an assiduous worker. Whether milking cows, sewing linen or making butter, she had always risen at dawn. Her job in Brussels, as she saw it, was to understand the nobility and charm them. 'To overcome their aloofness', she decided, needed influence, and influence would come from her salon. As people came to call on her, she began to put together a list of their names, their relations, their occupations and their interests. She borrowed books on the peerage from the Bibliothèque de Bourgogne, and she called on the services of a former Commander of the Knights of Malta, who came to the house every evening to answer a list of questions she had drawn up during the day. By the end of the month, she felt that she knew Brussels well, down to its 'different liaisons', its 'animosities' and its 'bickerings'.

Refusing to pay or receive visits in the mornings, which she spent with Cécile and Charlotte, she gave up the rest of the day to making friends with the *ville haute*. For the first time since the ministry in The Hague, she had a large staff to help her: a butler-valet, two liveried footmen, an office boy, a porter, an office footman, two stable boys and any number of cooks, maids and kitchen servants. Unimpressed by grandeur, largely indifferent to comfort, she took great pleasure in her own competence at organisation.

Just the same, it was not always smooth. Brussels was a city of intrigues, in an age of uncertain power and precarious fortunes. Though she and Frédéric made friends readily, there were people who were envious of a position apparently won without effort. In Brussels were two Frenchwomen, senior to Lucie in that their husbands held higher-ranking positions. One was Mme de Chambarlhac, the wife of the General commanding the division garrisoned in Brussels. She was a handsome woman in her 40s from a noble family, but she was rumoured to have been plucked out of a convent by her husband during the Italian campaign and to have followed him around, dressed in a hussar's uniform. Living rough among soldiers, Lucie decided, had ruined Mme de Chambarlhac's manners beyond repair. The second was Mme Betz, the wife of the First President. Mme Betz was in her 50s and wore, Lucie thought, unsuitably deep *décolleté*. Neither of these two ladies was ever invited to dine with the Dowager. Lucie, who was, did not care to receive them either. 'Naturally,' she wrote, 'I could not consider any connection' with such people, while the General she considered 'a fool'. The jealousy and animosity of these two women, along with that of their husbands', simmered.

Soon after reaching Brussels, Frédéric received a visit from the Comte de Liederkerke, who had been a junior officer in his regiment many years before. The count, who came from a very wealthy old Belgian family, asked whether his only son, Auguste, a probationer in the Council of State, might work for him as a secretary. Frédéric agreed. Humbert had left Antwerp and now arrived to join them, in order to study for the Council of State himself, and the prefecture, to Lucie's intense pleasure, was soon full of children and young people. Brussels, with its good food, its wines from Flanders and the Rhine, its games of whist and lotto at the Dowager's suppers, was turning out to be a pleasant posting.

* * *

General Henri Bertrand, Napoleon's favourite aide-de-camp, had not stopped yearning for Lucie's half-sister Fanny. He was a military engineer by training, a good-looking man with curly dark hair and regular features, sturdy, reserved and somewhat stubborn. Fanny held out against him for two years, displeasing Napoleon greatly one day by exclaiming, when he yet again asked her to consider the match, 'What, sire! . . . Why not the Pope's monkey?' But one day the Emperor sent her 30,000 francs with which to buy a wardrobe so that she could accompany Josephine to Fontainebleau. Josephine was very fond of Fanny, though Lucie herself found her 'extremely frivolous'. It was while at Fontainebleau that Fanny relented and agreed to marry Bertrand. Napoleon, remarked Lucie, was someone to whom one could refuse nothing, 'so gracefully and so winningly did he set about obtaining what he wanted'. He was not as graceful with everyone.

Fanny's elder half-sister, the unhappy Betsy de Fitz-James, had died recently, leaving four young children. But Napoleon would not hear of Fanny's marriage being delayed; he allowed her eight days, as he wished to be present, after which he was leaving to rejoin his army. Early in September, Lucie was suddenly ordered to Paris: the Emperor had announced that she alone would be able to arrange

Henri Bertrand.

the ceremony in time. She hastened to obey. 'Although the letter was a very friendly one,' she wrote, 'it was clearly a command.'

Leaving Brussels at once and travelling all night – the new four-horse *vélocitères* travelled faster than the old *turgotines* – Lucie reached the Princesse d'Hénin's house in the rue de Miromesnil at dawn. She paused only long enough to change her dress and send for a carriage to take her to her stepmother's house in the rue Joubert. There she found that Mme Dillon and Fanny had left for Beauregard, Mme de Boigne's house not far from Saint-Cloud. Lucie had not met her cousins, Adèle de Boigne, wife of the unfortunate General, or her mother, Mme d'Osmond, since her childhood at Hautefontaine. Hurrying on, she was at Beauregard by 11.30. 'Ah,' exclaimed Fanny, on seeing her, 'we're saved. Here is my sister!' Mme Dillon was still in bed, and very reluctant to be rushed. Lucie urged her to dress quickly and to accompany her back to Paris, so that they could begin making plans. While waiting, she strolled around the park with Bertrand, whom she found initially shy and awkward, but soon won over. By lunchtime they had come 'to understand one another well'. Lunch was taken with Mme de Boigne and Mme d'Osmond, neither of whom, remarked Lucie briskly, 'could abide me'. She ate heartily, having had nothing all day.

There was much to arrange. A judge had to be found to concoct a new birth certificate for Fanny, her own having disappeared somewhere in Flanders. Napoleon himself signed the marriage contracts of his favourite courtiers, and as her stepmother kept neither paper nor pens in her house, confirming Lucie's opinion of her as an uneducated and foolish woman, she had to send out to d'Expilly, Paris's finest stationer, for supplies. Napoleon had said that he would attend the wedding, so every detail had to be perfect. Lucie was informed that she personally was to be presented to the Emperor at Saint-Cloud next morning, and that she was to wear 'court dress and plumes'.

Lucie's intrepid life in the wastes of North America was well known to Napoleon, who clearly admired her boldness and was appreciative of her aristocratic origins. For her part, Lucie was surprised at how little frightened she was of a man of whom most of his courtiers were terrified. At Saint-Cloud, he greeted her, she reported later, 'graciously', and asked her about Brussels and its 'high society' as if to suggest that it was the only kind that might interest her. He teased both her and Mme de Bassano – wife of Napoleon's recently ennobled secretary M. Maret – about having made them rise so early; Mme de Bassano pouted, later telling Lucie that Napoleon had become 'most attentive to her'. Lucie was not the only intelligent woman to be dazzled when singled out by the Emperor.

The Emperor had given orders that Fanny and General Bertrand were to be married in the chapel of the Château de Saint-Leu, which belonged to his stepdaughter Hortense, Queen of Holland. At 3.30 in the afternoon of 17 September, Lucie followed a long procession of plumed, beribboned, decorated and gaudily dressed grandees from the château to the chapel, where the nuptial blessing was given by the Bishop of Nancy. Talleyrand and Maret, respectively Prince de Bénévent and Duc de Bassano, stood as witnesses for Fanny; Maréchal Berthier and Grand Maréchal Duroc for Bertrand. 'I am happy for you,' wrote Josephine in a little book for Fanny, which was also signed by Hortense and Napoleon's sister Pauline. Afterwards came a dinner and dancing; Hortense, who normally loved dancing, sulked because Napoleon had ordered her to give up her own set of emeralds and diamonds as an additional wedding present to Fanny. Though the Emperor himself did not attend, his presents were lavish: three properties, one in Poland, a second in Westphalia, a third in Hanover. Lucie found the whole event somewhat 'insipid'. Without the lustre of Napoleon's presence, the artifice of the imperial court held little attraction for her.

* * *

The year 1808 was the one in which Napoleon's rule reached its most glorious phase. The continent of Europe lay at his feet, with Austria and Prussia his allies, Russia, who was at war with the Ottoman Empire, neutral, and only Britain, Sicily and Sweden holding out against him. Three of his brothers were on thrones: Joseph in Spain, Louis in Holland, Jerome in Westphalia. Since French was now the language of Europe's courts, there was no need for Frenchmen to learn another language. Though France had indeed lost most of its colonies to the British, and the French merchant fleet was either paralysed or destroyed, the economy was healthy, with manufacturers finding outlets throughout the French-dominated continent, while agrarian reforms and the abolition of internal trade barriers, both legacies of the revolution, were beginning to show results.

At home, the Senate and the Corps Législatif were subdued and impotent; the Tribunat had been closed down. Almost all traces of republican liberty had vanished. It was no longer possible to gather, to talk or to write freely. The press was censored, and newspapers had been reduced to theatre reviews and paeans of praise for Napoleon. Though *Le Journal des Dames et des Modes* had illustrations showing women driving coaches and flying balloons, in practice the Code Napoléon had returned them to their hearths, forbidding them to travel, inherit or hold a bank account without their husbands' permission. Art in all its forms, from literature and poetry to painting and the theatre, colluded in glorifying the Emperor, artists having learnt that only lofty moral subjects would keep them out of prison. Secret police infiltrated every corner of private and public life. Troublemakers like Mme de Staël had been banished to internal exile; in eight state prisons languished those suspected of conspiracy. Even from his entourage, Napoleon would tolerate no contradiction. Many of the worst repressions of the *ancien régime* were being surpassed.

But the city of Paris itself had never looked more imposing. Like the Pharaohs, Napoleon intended to leave great monuments – paid for out of the heavy contributions levied on conquered enemies – by which his deeds would be remembered. Three thousand metres of new quays stretched along the banks of the Seine. There were new bridges – du Louvre, des Arts and d'Austerlitz – new museums and new schools. On top of the Colonne d'Austerlitz, cast after 1,200 cannons captured at the battle were melted down, the bronze statue of the Emperor was seen leaning on a sword, holding in one hand a globe, surmounted by the figure of Victory. The old palaces of the *ancien régime* were being restored, to house ministers and the imperial nobility. The Canal de l'Ourcq was at last bringing fresh water to Paris from three separate rivers, while the Louvre overflowed with most of the masterpieces of European art.

It was after the Battle of Austerlitz in 1805 that Napoleon's court had grown truly sumptuous. Having seduced the nobles of the *ancien régime* with honours and positions, bestowed on them his new *Légion d'Honneur*, and elevated his family and generals with imperial titles and fortunes pillaged across conquered Europe, the Emperor had imposed a military dictatorship on social life at court. Courtiers wore so many decorations that these looked not like orders and medals but 'a new style of dressing'. 'What I want is grandeur,' he announced. 'Whatever is grand is always beautiful.' It was indeed grand; but it was also stultifyingly formal and often extremely boring, a pale and exquisitely orchestrated shadow of pre-revolutionary spontaneity and gaiety. 'Fear assumed the mask of respect,' wrote the Vicomtesse de Noailles. 'It was not a court, but a power: but what a power!'

With the Comte de Ségur as Grand Master of Ceremonies – later described by Stendhal as a 'dwarf . . . one of the Emperor's weaknesses' – presiding over a court of several thousand people, many of whom moved with the Emperor from the Tuileries to Saint-Cloud, Fontainebleau and Compiègne, had come other *ci-devant* nobles to coach his clumsy generals and their ignorant young wives in the delicate subtleties and nuances of Marie Antoinette's court. The women learned faster. 'Adopt neither the posture of a fawning slave, not that of an insolent tribune,' the Comtesse de Bradi instructed them. 'Both roles make one look silly and ridiculous.' In what Claire de Rémusat, who became lady-in-waiting to Josephine when she was 22, called 'a daily growing despotism', the parvenus were taught how to walk, stand, sit and bow by Despréaux, former dancing master to Marie Antoinette. They were told how to address each other – princes rated an '*altesse royale*', grand marshals an '*altesse sérénissime*' – and how many plumes to wear. Rules and etiquette became a form of protection, against malicious remarks and embarrassing blunders. 'These debutantes in the career of manners,' observed Mme de Staël, still in exile and pining for Paris, 'would make a mistake if they confused the outer appearance of things with good taste.' Court balls had become cold and stiff. Napoleon, who did not care for dancing, had tried, and failed, to learn to waltz. 'I am sorry for you,' Talleyrand said to M. de Rémusat, *Superintendant des Spectacles*, 'for you are in charge of amusing the unamusable.'

There were, however, compensations, as Fanny Dillon had discovered. Fortunes were lavished on favourite generals to be spent on furs, carriages, silver, liveries, all the trappings of splendour. Josephine's increasing acquisitiveness – for diamonds, dresses, jewellery – found a ready following in her ladies-in-waiting, and in the separate establishments run by Napoleon's sisters, Elisa, Pauline and Caroline, and his stepdaughter, Hortense. Josephine herself changed her costume three times a day, never wore a pair of stockings twice and owned some 400 cashmere shawls, having them made into cushions for her dogs as she tired of them. A dress covered in thousands of fresh rose petals was designed to be worn only once. 'No language,' wrote the Duchesse d'Abrantès, whose husband General Junot had been made a duke, 'can convey a clear idea of the magnificence, the magical luxury . . . of this plumed and glittering circle', even if imperial receptions sometimes seemed more like 'reviews at which there happened to be women'. Napoleon's review of his guards in front of the Tuileries, held every Sunday when he was not away at the wars, was widely regarded as the finest military spectacle in Europe.

Mme de Rémusat, who like many of the old nobility – and like Lucie herself – continued to regard her own culture, manners and education as greatly superior to that of the newcomers, left a description of Napoleon at this time. He was a man, she wrote, 'ill-made, the upper part of his body . . . too long in proportion to his legs', with thin chestnut hair, greyish blue eyes, very white skin, and a thin-lipped mouth, who in repose looked melancholy and meditative and when angry menacing. Napoleon was awkward, 'deficient in education and manners', but his 'intellectual capacity' was 'vast', and his long monologues, which took the place of ordinary conversation, were a pleasure to listen to. Napoleon, said Mme de Rémusat, was selfish, but possessed 'fleeting tenderness'; and, just occasionally, he would appear in Josephine's rooms in the evening and tell ghost stories.

Mme de Rémusat was as clear, and as chilling, when contemplating Josephine, a woman 'easy to move and easy to appease, incapable of lasting feeling, of sustained attention, of serious reflection'. It was Napoleon, she maintained, who had taught Josephine to despise morality and also 'the art of lying, which they both practised with skill and effect'. By 1808, as Lucie quickly discovered, there were few details of the imperial couple's intimate lives that were not observed, picked over, criticised and recorded, from Napoleon's taste for fricasseed chicken, called Marengo, after the battle, to Josephine's careful smile, deliberately restrained in order to avoid revealing her bad teeth. At Saint-Cloud, as in the Tuileries, all life hung on Napoleon's pleasure.

More celebrations followed Fanny's wedding, each of the four witnesses holding a dinner in honour of the new couple in their mansions in Paris, where the mood these days was all for action rather than reflection, mirroring Napoleon's military exploits. Lucie stayed on in Paris for some time. At the Salon des Expositions that summer, David, Gérard, Girodet, Gros and Guérin all exhibited pictures, most with heroic martial themes, cast in antiquity, the parallels with Napoleon's feats clear to see. In the absence of new thoughts, and anxious not to dwell too heavily on the pleasures of the *ancien régime*, people were returning to the classics. Italian opera buffa, full of 'improbabilities and follies', had become very popular. Children were being christened Ossian, Porphrye and Zaphlora. Grimod de la Reynière was at work on his *Manuel des Amphitryons*, daringly taking menus from pre-revolutionary chefs. To sample the dishes which he invited readers to send in, Grimod had set up a *Jury Dégustateur*, which was attended by his large white angora cat; the jury, which spent five hours at table, sampled the dishes *à la Russe*, one by one, rather than having them all on the table at once, which was still the custom in France. Tasting a ragout of thrush, he declared that 'one would eat one's own father in this sauce'. The feet of a turkey, he added, were good for insomnia.

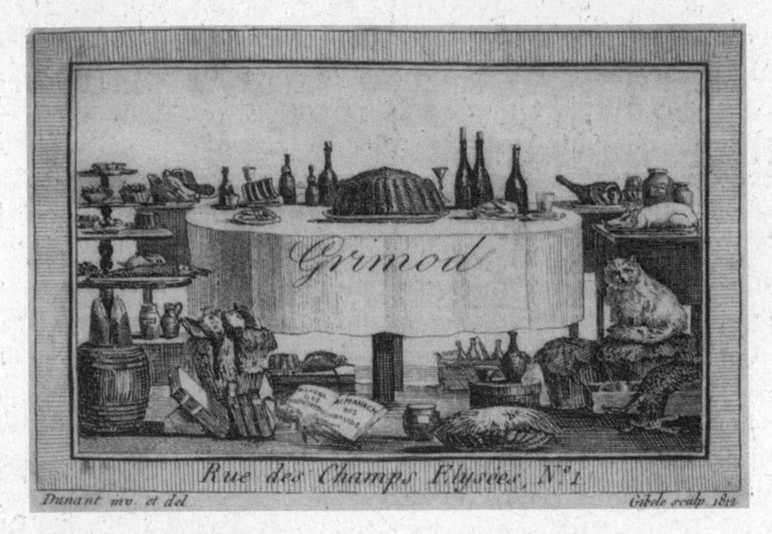

Visiting card of Grimod de la Reynière.

Mme de Staël, whose romantic novel *Corinne*, published the previous year, had made Italy fashionable in France, was at Coppet, at work on her study of German culture, *De l'Allemagne*. She continued to long to return to Paris, telling Mme Récamier, herself out of favour with Napoleon on account of their friendship, that 'one is dead when one is exiled. It is merely a tomb where the post arrives.' But she remained extremely indiscreet about her opinion of the Emperor, referring to him as 'Robespierre on horseback'.

Before returning to Brussels Lucie again met Claire de Duras, who wrote to a cousin that her friendship with the 'beautiful and affectionate' Lucie was one of the most 'delightful things' in her life. After Mme de Montesson's death, Napoleon had turned to Claire to advise him over matters of etiquette at court. The two women now became close again and when Lucie returned home, relieved to have escaped the 'tedium' of Paris, they wrote to each other frequently, Lucie sending presents of lace from Brussels, Claire red and blue bonnets from Paris. Amédée's name had remained on the proscribed list until 1807, and he still preferred to spend most of his time at Ussé. Claire remained Lucie's closest friend, perhaps the nearest she had ever come to intimacy with another woman. But their friendship was about to be tested in ways that revealed more of Lucie's nature than of Claire's frailties.

Not long after Lucie returned to Brussels, she heard from Claire that she had at last been introduced to Chateaubriand, whom she had been angling to meet for some time. Appointed Minister to the Valais in Switzerland by Napoleon in recognition of his support for the restored Catholic Church, Chateaubriand had resigned his post after the murder of the Duc d'Enghien. He had since given serious offence to the Emperor with an article comparing him to Nero. Like Mme de Staël and Mme Récamier, he spent much of his time banished from Paris.

An '*amitié profonde*', a deep friendship, quickly sprang up between him and Claire; within days, as she wrote to Lucie, they were addressing each other as 'dear brother, dear sister'. The trouble was that Claire's feelings for Chateaubriand were far from sisterly, and she was soon complaining bitterly that it was clear that 'he cares only for qualities of the mind'. With Amédée at Ussé, and the self-effacing Mme de Chateaubriand seldom in Paris, they went for long walks around the city. In the Plaine de Grenelle, where Chateaubriand's cousin had been executed by revolutionary guards, Claire knelt on the edge of the ditch where the 'blood of martyrs' had flowed, prayed, and picked a primrose. To Lucie,

Chateaubriand.

she wrote that had it not been for 'other commitments', she would simply have devoted her life to 'trying to please him'. She sent Lucie Chateaubriand's letters to read.

Lucie was appalled. 'Good God!' she wrote quickly and anxiously back. 'My dear Claire, how long must go by before you learn to be reasonable!' What shocked her was her friend's tone of self-abasement. Chateaubriand was not Socrates; if he sought adulation, that was because of his pride, and Claire would think more of herself if she kept company with good and serious people who recognised her true worth. 'I am devoted to you, my dear,' she wrote, 'devoted enough to be able to tell you the truth, even when you doubt my motives . . .' Such frankness was perilous.

When, soon afterwards, she learnt that after their third meeting, Claire had written a love letter to Chateaubriand, Lucie, who like many of their mutual friends considered Claire excitable and impetuous, returned to the attack. Letters went back and forth between Paris and Brussels, Lucie adopting an ever more disapproving tone. But Claire, by now fully embarked on a romantic passion, was unwilling to be deflected. It was not simply that the moral, faithful, sometimes prudish Lucie found her friend's excesses distasteful: it was about the responsibilities of friendship, the necessary compromises in love, as in life, the importance of holding steady in a world buffeted by vanity and fashion, of keeping faith with some inner certainty. Even the women who ran the 18th-century salons and believed in the ultimate power of language had seldom sounded so blunt. The effect was to give Lucie a vulnerability that she otherwise seldom permitted herself. So straight herself, she found moral confusion in others both disturbing and distasteful.

In reply to Claire's exaggerated outpourings about her devotion to Chateaubriand, Lucie urged restraint, caution, seemliness. 'Why do you say I detest him? I hate him only in that he is dangerous for you. Until you tell me that I am wrong or reassure me, I shall go on disliking him.' Lucie did not, in fact, much care for the brooding Chateaubriand, from whose vanity and self-indulgence she instinctively recoiled. She warned her friend that she was 'making a spectacle of herself before the whole of Paris'. She told her that it was a fantasy, that she should 'flee to Ussé . . . Calm your heart . . . Stop dwelling on this man who torments you . . . You must not think, my dear, that love at 40 is the same as that at 20 . . . Stop reading this eternal *Génie du Christianisme*, which you know by heart . . .' Urging Claire to take exercise, to study, but not poetry or metaphysics, 'a dangerous topic for women', she wrote: 'To look inside your heart, see what needs destroying and then not have the strength to do it: that is more dangerous than useful; one grows accustomed to one's enemy, and by making it familiar one loses the desire to get rid of it . . . I want to persuade you that there are a number of things in life that one must pass by without looking at.'

There were several occasions every day, admitted Lucie, when even she had to stop and say sharply to herself: 'No, don't think this way.' She begged her friend not to agonise over a man for whom she was not the most important person. 'M. de Chateaubriand seems to me like a flirtatious woman who wants many different men to pay attention to her; he has a little harem in which he seeks to bestow his favours equally in order to reign as Emperor.' Recommending patience and greater efforts not to neglect her young daughters, she wrote, as she had once written in Richmond, many years before, urging the then just married Claire to make fewer scenes about the faithless Amédée: 'You must not brood, nor look dissatisfied.' These were manners of the heart that for Lucie could never be neglected.

Though she made little progress with Chateaubriand, who remained firmly committed to their '*amitié profonde*' and nothing more, and was in any case pursuing other women, Claire eventually tired of Lucie's scolding. Her replies became distant, formal. Lucie's final words were bleak.

> Your last letter was so dry, so glacial that it crushed my heart. I saw in it something that has been threatening me for a long time: it is that M. de Chateaubriand will get what he wants, and that you will no longer love me. I am convinced that this is exactly what he wants; and he is right. He doesn't like me and he knows why. It doesn't matter to him that you are in fact fond of me: we are fighting over your heart. He is more skilled than I am . . . and you have begun to listen to him; if once you give way to him, then I am lost. When he tries to persuade you that you no longer love me, dear friend, do not believe him . . . But what is certain is that if I lose your love (your real love, for I want nothing else), then I do not know what will be left for me.

Lucie was right to fear Chateaubriand's influence: Claire did not reply. The correspondence ceased. It had, while it lasted, been intimate and loving, and it had exposed Lucie's own feelings in ways not apparent before. The woman glimpsed from behind the daily round of efficiency and self-abnegation was both softer and more self-reflective than the picture she would later draw of herself.

* * *

In Brussels, even if she complained about the rain and the grey skies, and the ignorance of the Bruxellois who were perfectly capable of 'thinking that Racine's Hector was slain in the Seven Years War', Lucie was in fact content. In the mornings, she taught Cécile history and Charlotte physics. She was devoted to her daughters. At 12, Charlotte was a modest, pious girl, 'level-headed, mature and confident'; there was nothing she could not talk to her about. Cécile, nearly 9, had a good singing voice. 'I take care to please my daughters,' Lucie wrote to a friend. 'I want them to think that theirs is the company I prefer to all other, and that I am never as happy as I am when I am with them.' Humbert was a studious 18-year-old, reading nothing but Italian and Latin. Aymar, not yet 3, was seldom mentioned. While the girls studied, Lucie painted miniature scenes of castles and landscapes on to plates and sugar bowls, or did her tapestry.

The side of her life in Brussels she least enjoyed were the receptions and balls she was obliged, as the prefect's wife, to give. The people who came to them, she wrote to a friend, 'bore me to death, but I try not to let it show . . . the human beings who pass before my eyes are like a magic lantern which I have forgotten as soon as I see it'. Meeting so many different people made her feel 'disgusted' with the human race. 'Every day I draw a little more into myself and I shall finish my days in a Trappist convent.' Sometimes, her own detachment from all but her immediate family worried her. When her friend Pauline de Bérenger gave birth to a first child at the age of 36, and Lucie found herself unexpectedly anxious about her, she wrote: 'I never realised that I was so interested in her, but it goes to show that I really do care. This makes me relieved: I worry that, like Mme du Deffant, I am not capable of loving anyone.' Since her love for Frédéric and the children was so deep, it is hard to believe that she meant the words in any but the lightest sense; but it revealed a concern that the incessant public life was taking a toll on her nature that she did not much care for. Even so, it did not stop her adding that she found Mme de Bérenger's refusal to read any novel she deemed 'inferior' lest it corrupt her own perfect taste 'totally absurd'. Even when meek, Lucie remained sharp.

* * *

In the spring of 1809, Austria, defeated three times in 12 years by Napoleon, rose up against the French; it took Napoleon a month to fight his way to Vienna, with heavy casualties on both sides. Austria was obliged to pay vast indemnities; it also lost 3 million subjects in lands ceded to the French. But Napoleon needed an heir, to carry on his empire, and the boy he had designated, his stepdaughter Hortense's son Louis, had died in 1807, at the age of 4. After months of rumours and uncertainty, the Emperor finally announced that he would divorce Josephine and marry the blue-eyed, blonde-haired, somewhat stolid 19-year-old daughter of Emperor Francis of Austria, Marie-Louise, whose aunt happened to have been Marie Antoinette.

The French court was fond of Josephine, who, while frivolous and spendthrift was also good-natured and very generous; she had put on weight and her face was more round than oval, but her voice remained warm and pleasing. With regret, they saw her retire to Malmaison, to cultivate the peonies, roses and dahlias that grew in profusion in the hothouses that had been built behind the house, to enjoy her collection of mostly looted Titians and Raphaels, and to sleep in her red tent bed, under the squatting gold eagle. As Louis-Antoine Bourrienne, Napoleon's secretary, remarked, for Josephine it was acquisition and not possession that gave her the most pleasure. She was sad, but not lonely. Her agreeable nature continued to attract many visitors, and the handsome settlement after the divorce enabled her to acquire ever more shrubs and flowers. Josephine, as Claire de Rémusat had noted, was not much of a reader; but she was passionate about horticulture, and about the scents and colours of her plants. Many years later at St Helena, Napoleon would say that Josephine had been '*une vraie femme*', a real woman.

The proxy marriage ceremony with Marie-Louise was held in Vienna. Napoleon awaited his bride at Compiègne, where 10 pages of instructions for the ceremonial of her reception had been circulated. It was from Fanny and Bertrand, who, together with the entire court, had been ordered to Compiègne, that Lucie heard the details of Marie-Louise's arrival. In the castle courtyard, a barouche stood ready, the horses already harnessed, waiting for word of the approaching convoy of carriages from Vienna. When it came, Napoleon set out to meet his bride. Not giving her time to alight, he jumped into Marie-Louise's coach, pushed his sister Caroline, Queen of Naples, who had been sent to greet her new sister-in-law, into the front and rode back by Marie-Louise's side to Compiègne. Flares and an orchestra of wind instruments awaited them. By the time the many formal presentations had been made, it was dark. Napoleon led his bride to her apartments.

The first impressions had been good: Marie-Louise, though not exactly beautiful, and standing half a head taller than Napoleon, was judged to have a 'fresh face'. The courtiers waited, longing for dinner, expecting at any minute to see her reappear in a new gown. Finally came a message to say that Their Majesties had retired for the night. Later, Bertrand discovered that the Archbishop of Vienna had given Marie-Louise a document stating that the proxy marriage was valid in the eyes of the Church: Napoleon had taken his new Empress straight to bed.

For the civil marriage at Saint-Cloud on 1 April, the early heavy rain had cleared and the spectacle of so many kings and queens, in full robes and uniforms and jewellery, was dazzling, even if Napoleon's sisters and Hortense made a fuss about having to carry Marie-Louise's train and dropped it to show their displeasure. But the start of the marriage was marred by a terrible fire in the Austrian Embassy, where Prince Schwartzenberg was holding a ball in honour of the Emperor and his new Empress, attended by the whole court. Two temporary tented rooms had been put up to house the 1,200 guests; a fire broke out and tore like lightning across the canvas; struggling to escape through one door, many people were trampled to death. The women's ball dresses went up in flames, which spread from one dress to the next. The ambassador's sister, who was 22 and pregnant, was killed when a chandelier fell on her head; the Russian Ambassador, the frail and corpulent Prince Kourakin, was severely injured. The finest jewels in Paris had been worn for the ball. Some of the women who managed to reach the streets were attacked by thieves. Prince Kourakin, who was wearing diamonds to the value of many millions of francs, lost them in the fray. 'I never saw such a blaze,' wrote an American visitor to Paris who happened to be nearby. 'It seemed as if half the city was on fire', the flames casting light on an incredible scene of half-dressed frantic people running around looking for friends and relations, while passers-by, pretending to help, snatched diamonds from their fingers, ears and hair.

* * *

In Brussels, the old nobility was not proving altogether easy to court, in spite of the best aristocratic efforts of Lucie and Frédéric; but they were growing slightly warmer towards the French. And when Napoleon, less than a month after his marriage, decided to bring his new bride to what had been the capital of her father's Belgian possessions, many of the remaining reservations disappeared. Brussels went into a frenzy of preparations: houses were painted and redecorated, a new staircase was built up the side of the Hôtel de Ville, with a grand new entrance, and, along the roads the imperial party would take, sand was spread, and even carpets were laid. The walls of Marie-Louise's apartments in the Château de Laeken were hung with pleated pale pink satin, looped with garlands of roses in carved silver. For days before the great event, musicians could be heard rehearsing in different parts of the city.

On 28 April, Frédéric and General Chambarlhac went to receive the Emperor and Empress at Tubize. On a fine spring day, Napoleon and Marie-Louise rode into Brussels in an open carriage, accompanied by a special guard of honour in green and maroon, raised by Frédéric from Brussels's leading families, Humbert and Auguste de Liederkerke among them. With them came the kings and queens of Westphalia and Naples, Prince Metternich and Prince Schwartzenberg of Austria, and all their suites of ministers, officers and attendants. The cannons roared, the muskets fired salvoes, and bells were rung in every church. Behind came the musicians, playing and singing in harmony. There were shouts of

'*Vive l'Impératrice*', who, it was generally agreed, had a dignified carriage and a sweet expression.

Next day, Lucie was ordered to present the ladies of Brussels to the Empress. Marie-Louise was pleasant, but said nothing. She had, wrote Lucie later, no trace of haughtiness but was, on the contrary, 'gentle, good, obliging' and extremely shy, 'which is proper for her age'. She was also 'remarkably naive'. At court, Marie-Louise had already earned a reputation for prudishness, making clear that she would tolerate no *doubles entendres* or risqué remarks.

The city of Brussels had organised a ball in honour of the Emperor and Empress and, to her immense pleasure, Lucie was invited first to a small private dinner at the Château de Laeken. It would mean a rapid change of dress, but the prospect of spending two hours with Napoleon charmed her, and she liked this kind of challenge. She positioned a maid with her ball dress at the Hôtel de Ville and pressed the mayor, the Duc d'Ursel, into service to act as her escort. At dinner there were just eight people; Lucie was on Napoleon's left. He spent most of the meal questioning her in great detail about the people of Brussels, about the lace-makers and the convents of the Béguines, women living in communities but not bound to perpetual vows.

It was all over in 45 minutes, as Napoleon never liked to waste much time on food; but in the drawing room afterwards he continued to talk to Lucie, while teasing his brother Jerome about all the 'rascally' men he had taken into his service in Westphalia. To Lucie, seeing that she was well versed in history, he talked about the French kings and the style of their reigns, saying that he thought that Louis XIV had only proved 'truly great' towards the end of his life. When he reached Louis XVI, he paused, then said, 'sadly and with respect, "an unfortunate Prince!"'.

Then came the dash to the Hôtel de Ville with the Duc d'Ursel. Lucie was standing in her allotted place, in full ball dress and jewellery, when Napoleon and Marie-Louise, holding a bouquet of flowers made of precious stones, arrived. The Emperor, complimenting Lucie on the speed with which she had changed, asked her whether she would be dancing. He laughed when she replied that, being a woman of 40, she no longer danced. The fact that the Empress did not dance either was seen as a hopeful sign that she might already be pregnant. Next day, the imperial suite left for Antwerp, Frédéric accompanying them by boat on the canal. News did indeed soon come that the Empress was expecting a child.

The visit had passed off very well. At least in part due to Frédéric's diplomatic skills and instincts, the right people had been noticed, the right honours bestowed. When, not long afterwards, Humbert went to Paris for his examinations for the Council of State, Napoleon, who was present – no detail of his administration being too small – took over the questioning. Discovering that Humbert was fluent in Italian as well as English, he gave him the sub-prefecture of Florence.

Life in Brussels continued, in its pleasant, social way, broken only by a painful operation to remove a tumour from below Frédéric's ankle, for which the Princesse d'Hénin sent a trusted Paris surgeon. It was just as well, for the wound had become inflamed after a Belgian doctor applied caustic in error, and there had been talk of amputation.

Frédéric was not yet able to walk when news arrived that a British fleet of 40 ships of the line and 30 frigates had entered the mouth of the river Escaut, and laid siege to Flushing. As a first step to challenge Napoleon's domination of continental Europe, the British plan was to move on to take Antwerp. Napoleon was away from Paris. Arch-Chancellor Cambacérès, alerted by telegraph, began marshalling troops from nearby garrisons and turned out the *Garde Nationale*. In Brussels, Frédéric gathered together the Guardsmen from the Dyle. Every day, leaving Brussels at five in the morning, he and Lucie went by carriage to Antwerp, changing horses along the way, to inspect the defences. After a late breakfast with Malouet, they returned to Brussels, a round trip of some six hours on the road. It was taken for granted that Lucie would accompany Frédéric: their marriage had become a partnership in every aspect of their lives.

The British campaign started well, but soon faltered; the plan to take Antwerp was abandoned. By now the French, who had been caught off guard, had regrouped, though it was widely agreed in Brussels that had the British adopted a different strategy they could indeed have entered Belgium unopposed. As it was, both armies were struck down with Flushing fever, the surrounding dikes being full of mosquitoes, which, arriving on top of dysentery and typhoid, put four in every ten of the 40,000 British soldiers out of action. The French *gardes*, bivouacked on the island of Walcheren, came down with the same fevers, and all the surrounding hospitals were soon overflowing with sick young men and boys. The prefecture in Brussels became a depot, where families handed in spare mattresses and linen. Lucie was working harder than ever.

But trouble was brewing for Frédéric. Napoleon was looking for culprits for France's slow response to the British invasion. M. Malouet, recently appointed *conseiller d'état*, warned that people were beginning to intrigue against him. The very qualities that made Frédéric so remarkable – his directness and incorruptibility – were ill-suited to the slippery world of Empire politics. In Brussels, people came forward to accuse him of having failed to summon the Guardsmen with sufficient haste. Only the courageous intervention of a young officer, who named the true culprits, deflected Napoleon's wrath. Frédéric had also made a dangerous enemy of Anne-Jean Savary, newly elevated Duc de Rovigo and appointed Minister of Police, by refusing to further the ambitions of one of his relations.

Early in 1811, Lucie and Frédéric went to Paris to see Humbert off to Florence. 'This departure,' wrote Lucie later, 'was the beginning of a long separation and was exceedingly painful to me.' Fanny had been staying in Brussels and now, like everyone else connected to the court, was hastening home to be present at the birth of Marie-Louise's baby. Hers was turning out to be a very happy marriage and Lucie felt closer to her. Fanny was already the mother of two children, Napoleon and Hortense, born without mishap despite her insistence on accompanying Bertrand all over Germany, when their carriage turned over several times on the bad roads.

In Paris, Lucie was cautiously reunited with Claire, who, to her alarm, was still pursuing Chateaubriand, with as little success as before. She had two serious rivals and Lucie found the petty jealousies and adulation of these three 'priestesses' 'ludicrous', each plotting behind the others' backs to ingratiate herself further in the eyes of their complacent god. Subterfuge and malice remained foreign to Lucie's nature. One evening, she was delighted to discover that she had been invited to attend a play at the Tuileries, where luxury shone ever more brightly, with much silver and lapis lazuli, and all lines straight and regular, everything that was 'capricious and irregular' out of fashion.

There was a small and select audience of just 50 lavishly dressed courtiers, who expressed surprise when Napoleon and Marie-Louise greeted the more simply dressed Lucie so warmly. To boost the French lace and silk industries, men without an office at court were required to wear the old *habit habillé*, the heavily embroidered silk coat of the *ancien régime*: rugged old soldiers, veterans of the republican wars, were to be seen in shimmering pinks, greens and pale greys.

Once Humbert had left for Florence, Lucie longed to return to Aymar and her daughters. Frédéric, evidently uneasy about his enemies, decided that they should remain in Paris to await the delivery of the Empress's baby. On the evening of 19 March, news came that Marie-Louise had gone into labour. The ladies-in-waiting hovered, were sent home, recalled, then sent home again. In the late morning of the 20th, a first cannon shot, fired from the Invalides, was heard. Paris froze: carriages stopped, people hurried to their windows, the streets filled. There were to be 20 shots for a girl, 25 for a boy. By the 19th, an expectant hush silenced the whole city. When the 21st boomed out, there was an explosion of cheering.

That evening, at nine o'clock, the court was summoned to the Tuileries to attend the provisional baptism of François-Charles-Joseph-Napoleon, King of Rome. Lucie, Fanny and Mme Dillon joined the throng jostling for positions close to the aisle where Napoleon and the baby would pass. They found an excellent spot at the top of the staircase, just by the row of the much decorated veterans of the *Vieille Garde*. When the baby appeared, carried by the royal governess on a satin and lace white cushion, Lucie was able to observe him very closely. Lucie had seen many newborn babies, not least her own six. This baby, she was certain, was not newborn. But whose it was, and whether he had been switched with another child, was a mystery she discussed only with Frédéric.

Towards the end of summer, though Marie-Louise had still not recovered her health after the birth of the baby, Napoleon returned with her to Belgium. While he went on a tour of inspection of his northern defences, the Empress stayed at the Château de Laeken. Lucie was expected to spend every evening with her, playing lotto. This time, she found Marie-Louise excruciatingly dull. Every day the Empress, holding out her wrist to have her pulse taken, asked whether Lucie thought she had a temperature;

every day, Lucie replied that not being a doctor, she could not tell. When the Duc d'Ursel tried to interest her in an outing to the Fôret des Soignes, she said that she did not care for forests; offered a portrait of her illustrious grandmother, Maria-Theresa, to hang on the walls of her apartments, she pronounced the frame too old-fashioned. 'In short,' concluded Lucie briskly, 'this insignificant woman' was utterly unworthy of 'the man whose destiny she shared'. Lucie did not meet Marie-Louise again until after Napoleon had fallen: and then 'she was still just as stupid'.

* * *

On St Helena Napoleon would say that he had wanted to 'make Paris the true capital of Europe'. By 1811, Rome had been annexed to the French Empire as a free imperial city; the lands lying on the north and east coasts of the Adriatic had become the Illyrian provinces; the Hanse cities of the North Sea had been absorbed; and when Louis abdicated, Holland had been incorporated into France by imperial decree. All that stood in the way of French hegemony were Britain and Russia, and victory in Spain, where French forces were stalled. Napoleon needed soldiers, to replace the tens of thousands who had died, and those now struggling wounded and crippled back from Spain. For yet more recruits he turned to his prefects to carry out levies in their departments.

Some of the attacks on Frédéric at the time of the British invasion had been justified: though he had indeed tried to raise men for the *Garde Nationale*, he had noticeably failed to enthuse the Belgians for Napoleon's wars. Nor had he done much since his arrival to chase up the growing number of deserters. Ordered by Paris to seize the parents who could not pay the fines of young men who had deserted or were in hiding, he refused, replying that it was not only illegal but pointless to send parents to jail for their sons. While mobile units scoured the countryside for fugitive soldiers, Frédéric kept repeating that fines, threats, forfeits and the taking of liens on entire communes seemed to him morally and legally wrong. Together with his transparent honesty, Frédéric was a stickler for doing things according to the law, and he was never afraid of quarrelling with anyone over matters of principle.

None of this made him popular with the Ministry of the Interior. Nor did his extreme reluctance to force Belgian families to send their sons to *lycées* and military academies in France, a suggestion greeted by the *ville haute* with horror. Frédéric had also been less than helpful when instructed by Rovigo to draw up a list of well-born, unmarried, rich Belgian girls over the age of 14, together with descriptions of their appearance, any deformities, talents and religious views, as possible wives for Napoleon's senior officers. As Frédéric saw it, these were simply forms of hostage-taking.

M. Réal was the member of the French Council of State in charge of the 'surveillance' of the nine Belgian departments; Lucie referred to him as a 'superior kind of spy'. In 1811, Réal arrived in Brussels on a tour of inspection, telling Lucie as soon as he set foot in the prefecture that to his mind Belgium was nothing but a 'country of fanatics and fools'. A number of Belgian priests had taken the lead in opposing Napoleon's stand against the Pope and Belgium was full of priests forced into hiding by their refusal to recognise the concordat. The Archbishop of Malines, M. de Pradt, however, far from protecting his recalcitrant priests, was in league with the Commissioner of Police, and when asked to give them up had been reported as saying: 'You want eight? Well, I will give you 45.' Neither the Commissioner nor the Archbishop approved of Frédéric, whom they suspected – rightly – of doing nothing when asked to track down the leader of the rebels. Nor did the Archbishop care for Lucie, whom he considered responsible for reporting back to Paris some injudicious remarks he had made to her. She had indeed repeated them to the Princesse d'Hénin, hoping to amuse her, but the letter had been intercepted and its contents shown to Talleyrand who used them in jest with the Archbishop. Lucie herself was not always tactful, and it had been foolish to take such risks.

While M. Réal was in Brussels, he stayed at the prefecture. One night Lucie gave a dinner in his honour, to which she invited all the people known to be particularly well disposed towards Frédéric. After dinner, there was a glittering reception for the *ville haute*. When the guests left, M. Réal expressed his disgust at so many useless elegant and illustrious people. Frédéric replied that he was sorry that he felt that way, but that he was certain that Napoleon saw things differently.

Réal returned to Paris to file a highly critical report. Not only was the Prefect weak and ineffective, but he was under the sway of his domineering and wilful wife, who had created a 'court' for herself in Brussels. 'If called upon to act in difficult circumstances, he would probably lack the necessary resourcefulness and severity . . . Married to a woman of considerable spirit, who has the greatest influence over him . . .' Frédéric was, the report went on, too liberal, too lacking in administrative firmness and too directionless.

There was some reluctance now among Belgian guests to attend Lucie's receptions in the prefecture, lest they be perceived as too closely associated with Frédéric's stubbornness in his dealings with Paris. The Comte de Mérode, whom Frédéric had helped avoid being sucked into Napoleon's net by overlooking false details in his residency papers, described Lucie's soirées as 'splendid' but added that 'one went to them with apprehension, so frightened were we of being seen and conscripted'. When Talleyrand came to Brussels to preside over the election of a senator and two deputies, Lucie found him refreshingly charming and entertaining, and noted that in comparison M. de Pradt looked like Scapin, Molière's comic buffoon, noted for his cowardice, in a purple cassock.Talleyrand was himself increasingly at odds with Napoleon over their very different views on the future of Europe, and by now Napoleon had already made his much repeated remark about his Foreign Minister being nothing but 'shit in yellow silk stockings'.

Most of the prefects had their protectors at court. Whether Talleyrand was a protector to Frédéric is not clear, but he was noticeably present at most of the crucial moments of Frédéric's life. He would, however, have been able to do nothing to shelter Frédéric from the spies, agents and secret police who were active in every corner of the Empire, often spying not only on their neighbours but on each other. 'Spy on everyone except for me,' Napoleon had ordered his police; and gendarmerie, guards, aides-de-camp, prefects, generals and mayors all duly sent constant reports back to him from wherever they were.

What they had to say in the winter of 1811 was not encouraging, either about events in Belgium or in the rest of the Empire. Despite constant rises in taxation, the deficit in the Treasury was such that Napoleon had cancelled the arrears of pay owed to soldiers killed in action. Led by the Banque de France, banks had been drawn into a cycle of competitive discounting: businesses were failing on all sides. There were several bankruptcies in the

Dyle. A terrible summer, with storms in some places and scorching heat in others, had combined with the British blockade to produce acute shortages of flour, and the potato so hated by the French was again back in evidence. From the provinces came reports of increased brigandage and ever more beggars. Posters were seen with the words 'bread, work or death'.

In Paris, censorship was so extreme that the *Gazette de France* was reduced to talking about religion and the *Journal de Paris* about entertainments. The press, Réal informed the Conseil d'Etat with satisfaction, 'is in a state of absolute servitude'. Though Mme de Staël agreed to remove 11 passages considered anti-French from *De l'Allemagne,* her long awaited book on German culture, Rovigo commanded that the entire edition of 10,000 copies be seized and burnt, and ordered her to leave France once again. Mme de Récamier, threatened with exile herself should she pay her friend a visit, said to Rovigo: 'One can forgive a great man the weakness of loving women, but not that of fearing them.'

* * *

In June 1812, troops of French soldiers began passing through Brussels on their way east across the river Niemen. It was the start of the Russian campaign. Soon groups of officers, with just a few hours in which to rest and eat, appeared at the prefecture, where Lucie gave them dinner before they disappeared into the night. Orders arrived to requisition farm wagons and horses, with forage, sometimes as many as 80 or 100 carts a day.

The *Grande Armée* – some 600,000 men and 200,000 horses, the largest army ever assembled in Europe, with Italian, Austrian, Polish and German soldiers fighting alongside the French – was soon advancing against a retreating Russian enemy, whose scorched earth policy drew them ever deeper into Russia. At the Battle of Borodino, at the end of August, the French lost 30,000 men, dead or wounded; among them were 43 generals. Napoleon continued to advance on Moscow. How the fire that destroyed four fifths of the largely wooden city started no one discovered. By the time the French were forced to retreat, the Russian winter was on them. Less than a sixth of the *Grande Armée* came home.

It was some time before the full scale of the disaster became known. In Brussels the autumn passed uneventfully, broken only by Auguste de Liederkerke's attentions to Charlotte, now 16, a tall, accomplished girl, not pretty, as Lucie noted in her straightforward way, but quick, cheerful and always helpful. 'Her qualities of heart,' wrote Lucie, 'surpassed even those of her mind.' Such was Charlotte's love of books that her mother removed her light each night before going to bed, to prevent her daughter from reading and writing until dawn. 'You will laugh at me,' Lucie wrote to a friend, 'but at the moment I am in love with a young man of 22 . . . whom I hope to make my son-in-law.'

On New Year's Day 1813, Auguste's mother called at the prefecture to ask formally for Charlotte's hand for her son. Cécile was away in a convent, preparing for her First Communion, but Lucie promised that she could return in time for the wedding. Better still, Humbert, whose 'aptitude for business matters, zeal and love for work' had greatly endeared him to his superiors in Florence, was to be transferred as Sub-Prefect to nearby Sens, but not before Lucie had irritated Fanny and Bertrand by her blatant attempts to advance her son's career. Where Frédéric and the children were concerned, Lucie had very little shame. To add further to their happiness, their dear friend M. de Chambeau was not only back in France but had been posted to Antwerp, and so was able to visit them in Brussels.

It was now that a blow fell. Frédéric's enemies had proved too powerful. One morning, when Lucie was in the prefecture and Frédéric away trying to raise more soldiers for Napoleon, a courier arrived from Paris with the news that Frédéric had been dismissed. He had, in the euphemism of the Ministry of the Interior, been 'called to other duties'; Rovigo was thought to be behind the sacking. Lucie's first thought was that the Liederkerke family might break off the engagement, but as soon as he heard the news Auguste insisted that nothing would cause him to change his mind. Lucie's next step was, given her nature, predictable. M. Réal had been right to mistrust her forcefulness. Without even waiting for Frédéric to return, she left for Paris, travelling by coach through the night. 'I decided,' she wrote later, 'not to give in without a struggle.' That Paris might be consumed by the news from Russia does not seem to have occurred to her. Nor, apparently, what Frédéric would feel about her journey. It was almost always Lucie, now, who was the more forceful and resolute of the two.

Reaching the Faubourg Saint-Germain at ten o'clock next

evening she went straight to Claire de Duras's house, to find 14-year-old Félicie and 13-year-old Clara in bed and their mother out. A servant was sent to fetch her. Soon Lucie, who had developed a fever during her journey, was put to bed in a nearby apartment; a doctor arrived to give her a 'calming potion'.

She slept all next day, rose at five, dressed in her most elegant clothes and ordered a livery carriage. Taking Claire with her, she set out for Versailles where she believed she would find Napoleon, who had hastened home ahead of his straggling small army of survivors when he learnt of a *coup d'état* being planned against him. The Emperor was in fact in the Trianon, recently restored for his use. Lucie took rooms in an inn and wrote out, in her careful, neat hand, a copy of a draft letter that she had brought with her, and addressed it to Napoleon. She and Claire got back into their carriage and had themselves driven to the Trianon, where the chamberlain on duty happened to be an old friend, Adrien de Mun. He promised to deliver her letter in person. That evening, a palace servant in lace and gold brought word that the Emperor would see her next day. Lucie slept soundly, rose, fortified herself with a large cup of coffee, and set out for the Trianon.

The interview with Napoleon lasted, she calculated later, exactly 59 minutes. From Napoleon's opening words – 'Madame, I fear that you are very displeased with me' – it went extremely well. Later, Lucie would describe this encounter as one of the most important occasions of her life. Having listened carefully to her account of Frédéric's difficulties in Brussels, including his brushes with M. Réal and General Chambarlhac and his wife, the former nun, he said: 'I was wrong. But what can be done about it?' As they talked, Napoleon walked briskly up and down the room, Lucie trying to keep pace with him. On his desk were papers relating to vacant prefectures. He paused, looked down the list, then asked: 'There is Amiens. Would that suit you?' It suited Lucie perfectly: Humbert, at Sens, would be very near. Before she left, bearing with her instructions to tell the various officials of his decision, Napoleon asked Lucie whether she had forgiven him; she, in turn, asked him to forgive her for her outspokenness. 'You were right to do so,' he replied. She curtsied and he walked to the door to see her out. This, too, was an honour.

A few days later, at a drawing room reception at court, she met the Emperor again. Catching sight of her standing behind a row of ladies, he came over and smiled: 'Are you pleased with me, Madame?' The ladies parted to let Lucie come forward. It was the last time, she wrote, 'I was to see the great man'. Later, she would describe the charm and sweetness of his smile, all the more remarkable because of the contrast with his normally severe expression.

Frédéric's appointment to Amiens was posted in *Le Moniteur*. Lucie was warmly congratulated by friends who had heard of his disgrace and feared that nothing could be done to save him. Napoleon was not known to change his mind, and colleagues were amazed by Lucie's success. She returned to Brussels to pack up the prefecture and restore the rooms to precisely the way they had been before her arrival five years earlier. Charlotte and Auguste's wedding had been set for the end of May, and to Lucie's delight Auguste had been appointed Sub-Prefect of Amiens, so she would not lose her daughter. Not wishing to offend the incoming Prefect by too great a show of popularity, Frédéric asked the mayor, their old friend the Duc d'Ursel, to conduct the civil ceremony at ten o'clock at night, when the streets were usually empty. Just the same, the family emerged to crowds of well-wishers, and when they returned to the house that had been lent to them by friends, they found that all the musicians of Brussels had gathered to serenade them.

CHAPTER FOURTEEN

Oh Unhappy France!

Lucie, Frédéric, Cécile and Aymar, taking with them the newly married Charlotte and Auguste, arrived in Amiens towards the end of March 1813. A prosperous wool-manufacturing town, Amiens was set in the flat, somewhat marshy, malarial plains of Picardy, 150 kilometres north of Paris. Frédéric's brother-in-law, Augustin, whose home at Hénencourt lay not far away, was waiting to introduce them to the town's leading families. The prefecture was a charming two-storey, grey-stone, 18th-century mansion, set back from the road and reached through an imposing stone gateway; behind were pretty, landscaped gardens stretching away to fields and forests.

Lucie was delighted with the house and its grounds, saying that they made her feel that she was living in the country; but she felt rather less warmly towards the officials who governed Amiens. For the most part, she noted, the men were 'utterly vulgar', while their wives were remarkable chiefly for their 'grotesque appearance and ridiculous behaviour'. They also had an absurd habit of addressing their husbands as '*ma poule*' – my chicken – or '*mon rat*'. The years of poverty and turmoil had done nothing to diminish Lucie's taste for *bon ton*. Nor had they altered her belief in the innate superiority of the aristocracy, views that sometimes sat oddly with her instinct for fairness or her genuinely kind heart; or, indeed, her very real support for the liberties proclaimed by the revolution. She and Charlotte briskly resolved to mix with none but Amiens's older families and long-established merchants.

Angélique de Maussion, whose husband had been in prison with Augustin de Lameth during the revolution, met Lucie and Frédéric soon after their arrival. Frédéric, she wrote later, was likeable, clever and warm-hearted, while Lucie was '*belle*, like the whole of her family, but also blessed with a strong character and a good mind'. Having been through so much, 'she had become what by birth she was destined to be: a great and noble woman'. What most people remarked on now, when they were introduced to Lucie, was how formidable she could be.

One of Frédéric's first and most unpleasant jobs was to recruit yet more soldiers for Napoleon's struggling armies, no easy task in a country openly longing for peace. Commissioners, sent to all parts of France by the Senate not long before to help with recruitment, had returned with reports of apathy, exhaustion and resentment. Over 2 million men had been called up since 1805, and nearly a million were dead, in Spain, in Russia, in Italy, killed in campaigns fought backwards and forwards across Europe. It would later be said that half of all French boys born between 1790 and 1795 had either been killed or wounded in Napoleon's wars. As Prefect of the Somme, Frédéric had orders to call up all remaining healthy, well-built men under the age of 35, both single and married, regardless of whether they had already served in the army, providing that they were of good moral character and stood at least 1 metre 65 centimetres in their bare feet. So acute was the shortage of young men that nearly all the guests at balls were women. The 'sun of Austerlitz', remarked Aimée de Coigny, had turned into a 'ball of fire devouring France'.

When not away reluctantly seeking men for the army, Frédéric spent much of his time finding pallets, blankets and linen for the troops, and also money, through the compulsory sale of lands. The former nobility, he noted, often seemed more willing to yield up their sons than their money. As in Brussels, Frédéric did not hurry to send off ever younger boys to the front, which was once again earning him a reputation for royalist, anti-Napoleonic leanings.

But Frédéric, for all his patrician airs, remained, like Lucie, convinced that the roots of the revolution had lain in the errors and profligacy of the court and the *ancien régime*. Both of them regarded all talk about a possible return of the exiled Bourbons with profound misgivings. 'All the errors and vices which had been at the root of the first revolution were still too vivid,' wrote Lucie. 'Their weakness would bring in its train abuses of every kind.' Even if many of their friends – Amédée de Duras, Mathieu de Montmorency – were known to be organising a secret royalist network in central France, meeting at Ussé, Frédéric would have no part in it. The most he was prepared to do to defy Napoleon was to hold back, as he had in Brussels, on the levies of young men for Napoleon's armies.

Amiens, surrounded by pleasant countryside and home to several of her aristocratic friends, with Humbert not far away as Sub-Prefect in Würtemberg, suited Lucie well. With false news of victories despatched back to France by Napoleon, it seemed as if the Empire remained safe. Soon after their arrival, the theatre season opened with a production of Molière's *Le Misanthrope*. There was a lecture on botany, given at one of the excellent local scientific academies, and Frédéric was asked to address the chamber of commerce on the subject of a new machine for slicing beets.

Frédéric was often away, travelling around Picardy on tours of inspection, worrying Lucie by the amount of time he spent in low-lying, mosquito-infested areas. Lucie, left alone with Cécile and Aymar – Charlotte, already pregnant, was frequently absent with Auguste – wrote to Claire de Duras, hoping to mend their friendship. She described her immense pride and pleasure in Cécile, now 14 and her closest friend and companion since Charlotte's marriage. Cécile, wrote Lucie, was clever, thoughtful and loving, and she sang extremely well; like Charlotte, she spoke and read fluently in both English and Italian. 'Her one fault,' wrote Lucie, 'is that she is too good.'

For her own part, she was occasionally overcome by melancholy and despair, for which she blamed the weather of northern France and her own weakness, though her age – 43 – may also have been to blame. 'This illness,' she wrote, describing what can only have been depression, 'for it is indeed an illness, to which I would infinitely prefer a fever or gout, makes me the unhappiest person in the world. Among other things, I think that I am going mad; I feel terrifyingly oppressed, my heart beats and my head feels heavy, and my thoughts are scrambled . . .' It was the first time – indeed, the only time – that Lucie admitted openly to such feelings. For a woman whose face was so determinedly set against self-pity or self-indulgence, these frailties were both disturbing and very puzzling. She did not plan to indulge them.

Lally-Tollendal and the Princesse d'Hénin were spending the summer with the Princesse de Poix at the Château de Mouchy, not far from Beauvais. In the summer of 1813 Lucie went to join them, planning to return by way of Paris. There she hoped to see Talleyrand in order to make sense of a mysterious message that he had sent Frédéric via a colleague neither of them quite trusted. Claire was also at Mouchy with her two daughters and her new son-in-law, Léopold de Talmont. In August, Félicie, Lucie's goddaughter, though just 15, had agreed to marry the good-looking and charming only son of the Prince de Talmont, whose father, one of the royalist heroes of the Vendée, had been killed fighting revolutionary troops. Claire, who had always loved Félicie more than Clara, had strongly opposed the match. But Félicie, who was as wilful and excitable as her mother, and also very much in love, had gone ahead regardless, and was now openly more affectionate towards Léopold's mother than towards her own.

At Mouchy, Lucie found Claire in a terrible mood, having already fallen out with Léopold, spending her days writing long angry letters to her new son-in-law, sending them by footman from one end of the château to the other. Once again resorting to what she somewhat disingenuously referred to as 'the frankness of affectionate and sincere friendship', Lucie argued Léopold's case. But she was unable either to deflect Clare's wrath or sweeten her temper, and both she and her hostess wearied of Claire's continual scenes. Lucie was much attached to the capricious Félicie, whose liveliness and good heart she valued.

In Paris, pausing at Lally-Tollendal's apartment only to change her dress, Lucie went straight to see Talleyrand in the rue Saint-Florentin, off the rue de Rivoli. He received her, she wrote later, 'as always, with the amiable courtesy of long friendship'. For all his slyness and treachery, Talleyrand possessed, she wrote, 'greater charm than I have ever known in any other man. Attempts to arm oneself against his immorality, his conduct, his way of life, against all the faults attributed to him were vain. His charm always penetrated the armour and left one like a bird fascinated by a serpent's gaze.' Aimée de Coigny would also write of serpents in the context of Talleyrand, saying that, like a snake with its skin, he changed his moods, but with a swiftness that could be terrifying. Talleyrand, as enigmatic as ever, asked Lucie to delay her departure for Amiens by 24 hours; but he refused to say why.

That night, she heard salvoes announcing Napoleon's return into the city. Next evening, Talleyrand failed to appear. Lucie had spent the day visiting her half-sister Fanny, living in splendour in the Tuileries since Bertrand had been appointed Grand Maréchal du Palais. When Talleyrand finally appeared, at eleven o'clock at night, he said nothing but wandered idly around Lally-Tollendal's rooms, admiring the portraits of English kings that hung on the walls. Lucie asked him about Napoleon. 'Don't talk to me about your Emperor,' Talleyrand replied. 'He's finished.' Pressed further, he added: 'He has lost all his stores and equipment . . . It's all over.' It was, as Lucie knew, a long time since Talleyrand's complicated relationship with Napoleon, which had started out with mutual attraction, had turned to hatred and contempt. Plotting with Austria and Russia in 1808, and Britain in 1810, Talleyrand had remained committed to the goal of a peaceful and equally balanced Europe rather than a supreme and conquering France. His treachery, a historian would later write, was 'sustained and remorseless, occupying all his waking hours'.

Then Talleyrand handed Lucie a cutting from an English newspaper, giving an account of a dinner in London held by the Prince Regent for Louis XVI's daughter, the Duchesse d'Angoulême. Lucie, puzzled, asked him to explain. 'Ah, how stupid you are,' exclaimed Talleyrand, putting on his coat and disappearing into the night. 'Give Gouvernet my good wishes. I am sending him this news for lunch.' When, next morning, Lucie repeated the conversation to Frédéric, he found it just as perplexing, saying only that if it was by such means that the Bourbons hoped to recover their throne, then 'they would not remain on it for long'. For the moment, Lucie and Frédéric remained loyal to the Emperor, and in any case both were extremely nervous of Talleyrand's intrigues, Lucie noting that they felt that he 'would stop at nothing and had no scruple whatsoever in leading people into danger and then abandoning them in order to save himself'.

It was towards the middle of the autumn of 1813, wrote Lucie, that 'the first real rumbles of the approaching storm were heard'. News began to reach Amiens of French military defeats. Since the summer, when the Emperor Francis had declared war on Napoleon – his son-in-law – over half a million Allied troops had been pouring into central Europe, pledged to impose peace through a coalition masterminded by the British Foreign Minister, Viscount Castlereagh, and using military tactics successfully learnt from Napoleon. In Spain, Wellington was routing the French troops. Bavaria had joined the coalition; Germany was now largely liberated. With the Battle of Leipzig on 16 October, against a combined force of Russians, Austrians, Prussians and Swedes, in which the French forces were outnumbered by two to one, much of what remained of the *Grande Armée* was lost, though the news sent back to France continued to cast defeats as victories. On 22 December, the Allied armies crossed the Rhine; offers of peace – and even the throne – in exchange for a return to the borders of 1791, were rejected by Napoleon. By January 1814, Allied soldiers, fighting for the first time on French soil, occupied a long line stretching from Langres to Namur.

On 17 February, Amiens woke to learn that there was fighting not far away, at Montdidier. A week later, Allied troops reached nearby Troyes. Then came news that a party of Cossacks had broken away from the main cavalry and were looting the surrounding countryside, carrying away pigs, cows, wine, wheat and sugar and raping any women they could find. In the town people began to bury and hide their most valuable possessions; wine merchants and jewellers removed the signs from their doors. Charlotte was nearing the end of her pregnancy. Lucie and Frédéric moved her, Cécile and Aymar, together with many of their own belongings, into a safer apartment in the centre of the town, but remained themselves in the prefecture to await developments. Men started to fortify the ramparts and chop down trees for a palisade, while a squadron of Chasseurs was despatched to confront the marauding Cossacks, who eventually galloped away, terrifying villagers by their strange uniforms and wild appearance.

Würtemberg had already been taken by the Allies, and Humbert, ill with pleurisy, had only just managed to escape in time to Paris. Having recovered, he made his way to Amiens, where Frédéric soon sent him back to Paris to find Talleyrand and try to discover what was happening. By now Paris was full of royalists, watching and scheming, men who had never abandoned their preference for a restored Bourbon monarchy and who had hated Napoleon and now wished to place Louis XVIII on the throne of France. In the rue Saint-Florentin, Talleyrand told Humbert to wait in an anteroom. At six o'clock next morning, while Humbert was still asleep on a bench, Talleyrand emerged, fully dressed and in his wig, tapped him on his shoulder and said: 'Go now. Wear a white cockade and shout: Long live the King!'

For Frédéric and Lucie, constitutional monarchists at heart and loyal servants to Louis XVI and Marie Antoinette, but who had agreed to serve Napoleon, the return of the Bourbons was at once a solution to France's current ills, and personally unsettling. They were not alone in their fears that retribution might follow against those who had supported Napoleon, nor in their deep misgivings about what kind of a King Louis XVIII would turn out to be.

* * *

As a young man, long before the revolution, Louis XVIII had been very fond of his food. Descriptions of him at 20 invariably included the word 'stout'. By 1814, when he was 59, he was enormous, a short, waddling, cold, calculating man who believed, with total certainty, in the divine right of kings. When Louis moved, observed Grenville, it was like watching 'the heavings of a ship'. But he was also witty, clever and extremely well read, with a particular liking for the worldly Horace; unlike his brother, Louis XVI, he preferred literature and the arts to machinery and the sciences. He had, with a great deal of patience, endured 22 years of exile, moved on from state to state by rulers embarrassed by his presence, kept afloat by the fluctuating generosity of foreign governments. And all the time he had held on to his conviction that the throne of France belonged to the descendants of St Louis. The long years of tedious and impoverished exile had somewhat softened his earlier inflexibility. While continuing to maintain that the revolution had been an aberration, perpetrated by murderous usurpers, he had become more moderate, accepting that there could be no total return to the *ancien régime*, but only to some kind of constitutional monarchy. Pacifist by conviction, willing to recognise the errors of the past and retain whatever sensible gains had emerged with the Consulate and the Empire, he would say about himself: 'I may not have much strength of character, but I believe myself to be more timid and easy-going than weak.'

For the last years of his exile, Louis XVIII had lived in England, at Hartwell House in Buckinghamshire, where he was wheeled around the gardens in a rolling chair, surrounded by elderly aristocrats, setting policies for his government in exile. On 1 February 1813, he had issued a proclamation: if he were returned to France, there would be no vengeance and no conscription. France would again become a country of peace, happiness and unity, and a country of manners, for Louis knew about manners and the way that kings should behave.

In 1799, his niece Madame Royale – the only surviving child of Marie Antoinette and Louis XVI, who had lived in exile in Vienna after her cousin, Emperor Francis of Austria, exchanged her against French prisoners – had married another cousin, the Duc d'Angoulême, elder son of the Comte d'Artois. During her years of seclusion, locked up alone in the Temple from the age of 14, and told of the death of her mother, aunt and brother only later, the Duchess had learnt to appear at all times impassive. She had a long, somewhat large nose and her eyes, which tended to look red as if she had been crying, had an enquiring, wary look. Her failure to have children was said to have 'dried her heart' and made her gloomy and pious. Louis XVIII's wife, Marie-Josephine of Savoy, had died in 1810, at which time the Duchess had taken her aunt's place as royal hostess. Like her uncle, she had been in England, waiting for the day when she would return to the court of France.

This campaign's last battle between Napoleon's troops and the Allied forces was fought on 30 March 1814, just below the heights of Montmartre. Next day, in bright sunshine, the King of Prussia, together with Tsar Alexander of Russia, wearing a coat trimmed with fur and gold epaulettes, and Prince Schwartzenberg, representing the Emperor of Austria, rode down the Champs-Elysées. It was the first time since the Hundred Years War that a foreign army had entered Paris. Watching the long lines of soldiers in their different coloured uniforms, the Parisians who lined the streets were at first silent, filled, noted a diplomat called André Delrieu, with 'passive and lugubrious consternation'; but then, bit by bit, the crowds came alive, women climbing out of their carriages to walk with the soldiers.

The Allies had somewhat conflicting views as to what should become of France and its fallen Emperor. The Tsar wanted to see General Bernadotte, Crown Prince of Sweden and former French Maréchal, on the French throne; the English favoured Louis XVIII; while the Emperor Francis would have liked to see his grandson, Napoleon and Marie-Louise's son, eventually become King. And there was still the question of a possible settlement with Napoleon. In the event, a bandwagon began to roll for the Bourbons, the Allied leaders declared themselves for Louis XVIII, and on 1 April a provisional government for France was appointed, with Talleyrand at its head. The return of a Bourbon king to the throne of France had become, as Tsar Alexander remarked, a 'necessary consequence imposed by the weight of circumstances'.

Next day, the Senate voted to depose Napoleon. In return for abdicating not only his own rights but those of his family, the former Emperor was offered Elba as a sovereign principality; for the Empress Marie-Louise, who soon left Paris for Austria with her son in a convoy of carriages, weighed down with treasure, there would be the duchies of Parma, Piacenza and Guastalla. Just before signing, Napoleon swallowed a dose of poison he had carried around with him for years. But he recovered, and on 13 April signed an Act of Abdication, before leaving for the south, in 14 carriages. He was accompanied by four Allied commissioners, 400 soldiers, a number of courtiers and servants, and a large library of books, including a complete edition of *Le Moniteur Universel* and the works of Rousseau, Voltaire, Plutarch, Cervantes and Fénelon.

At Saint-Raphael, having been jeered and threatened along the way, he boarded a British warship, the *Undaunted*, which took him to Elba. The faithful Bertrand had opted to stay by Napoleon's side. Fanny, to her great despair, joined them soon after, taking with her to Elba the three children, aged 6, 5 and 4. She was again pregnant. A boy, Alexandre, born at the end of August, lived only 3 months. Fanny hated Elba, spending most of her time with Madame Mère. Her extreme lack of punctuality was soon reported to be exasperating Napoleon.

Because of Louis XVIII's gout, which he spoke of as 'an enemy with whom I must live and die', it was his elegant, ambitious younger brother, the Comte d'Artois, who rode into Paris first, to install himself in the Tuileries. Over the next few days, Talleyrand, Maréchal Ney and the Duchesse de Courlande all gave balls, at which victors and vanquished mixed. At the Opéra, ladies wore the white lilies of the Bourbons in their hair and carried them in bouquets and as garlands. Napoleon's bees and eagles were scrubbed and scratched away, his portraits, which hung on innumerable walls all over France, were taken down; the white cockade and the fleurs de lys were back in evidence. 'Once again,' wrote the Vicomtesse de Noailles, granddaughter of the Princesse de Poix, 'we climbed back on to the throne with the Bourbons.' The Vicomtesse was one of the many aristocratic ladies who had longed for the fall of Napoleon, saying that his ambitions had hung over her 'like a guillotine permanently in place'. The Cossacks, with their small horses and high saddles, which they mounted even to cross a square, had set up their bivouacs in the Champs-Elysées, and here, as night fell, they could be heard singing mournful airs round their camp fires.

Four days before the *Undaunted* pulled out of Saint-Raphael, Louis XVIII had been helped on to an English ship to bear him across the Channel, accompanied by the Duchesse d'Angoulême in her prim, old-fashioned clothes, and the 80-year-old Prince de Condé. With them went the Duc de Blacas, Louis's taciturn confidant, a man with very short legs and a long body, bald under his blond wig. As the boat neared Calais, cannons were fired to salute their return; from afar the beach looked black with people, who cheered and wept. The royal party proceeded slowly towards Paris, every step of the way marked by triumphal arches, speeches and choirs singing *Te Deums*.

In Amiens, Humbert's words from Talleyrand had been greeted with uncertainty, no one sure as to their exact meaning. But once it became known that Napoleon had abdicated, and that he had accepted a large settlement, there was widespread anger that so many young men had died apparently in vain. People came to the prefecture, shouting '*Vive le roi*', and Lucie handed out white cockades, brought back from Paris by Humbert in his barouche. By the light of flares, and to the sound of ringing bells, Amiens declared for Talleyrand's provisional government. In the theatre, players hastily staged a production of *The Hunting Party of Henri IV*, which had been censored by Napoleon, and when the actor reached the line '*Vive Henri IV*' the entire audience rose to its feet and cheered. On 14 April, a Prussian troop of 2,000 men clattered through the town to much applause. The ten years of Napoleon's rule were shed with surprising ease.

Having been informed that the King intended to spend a night at the prefecture in Amiens, Frédéric set off for Nampont-Saint-Martin, where the royal party would first enter the Department of the Somme. Lucie was visited in the prefecture by people offering pictures, flowers, shrubs and ornaments to decorate her rooms for the great occasion. Next came a number of elderly courtiers, anxious to find favour with Louis XVIII and put behind them any service rendered to Napoleon. Though fond of these old friends, whom Lucie had known since childhood, she was exasperated by their display of 'prejudice, hatred, pettiness and bitterness' towards the fallen Emperor. Her first glimpses of the Restoration did not impress her.

On 28 April, the carriage drawn by eight magnificent white horses and bearing Louis XVIII, the Duchesse d'Angoulême and the Prince de Condé rumbled into Amiens, where the streets had been hung with white sheets and scattered with flowers. As it entered the town, the millers of Amiens, according to an ancient tradition, asked to be allowed to unhitch the horses and draw the coach themselves, in their new grey suits and white felt hats, bought specifically for the task. Twelve young girls, led by Cécile, presented bouquets to the Duchess. In the cathedral, where the *Te Deum* was sung, people cried with joy. The doors were kept shut until the King was seated: when they were swung open, a great roar was heard 'as of a flood breaking its banks' as people surged inside.

That evening, Lucie was seated next to the King, who pleased her with the courtesy and charm he showed the ordinary people of Amiens. They were 25 at table. Louis XVIII declared that, like his brother, he would eat in public, those waiting to be presented standing around the room at a little distance from the table. Angélique de Maussion, who was an accomplished painter, was allowed to sit nearby and sketch the King. Lucie was considerably less taken with the Duchesse d'Angoulême, who chose to ignore Angélique, even after learning that Angélique had offered to help Marie Antoinette escape when they were prisoners together during the Terror.

Lucie's much-loved cousin, Edward Jerningham and his new wife Emily had accompanied the royal party to France, Edward having pleased the King by articles he had written in English newspapers praising the Bourbons. She had not seen him since she had left England 15 years earlier. Together, they discussed how the Duchess might be helped to take more trouble with her appearance, abandon the heavy style of an earlier age in order to create a more elegant first impression on Paris, a city for which elegance remained essential. 'The obstinacy of the Princess,' noted Lucie, 'was immovable.' In the cathedral, observing Louis XVIII and the Duchess, brother-in-law and daughter to Marie Antoinette,

Lucie had been filled with emotion. Now she felt only foreboding. 'Alas,' she wrote, 'my illusions were to last less than twenty-four hours.'

Next morning, leaving 600 men wounded in fighting in the north to recover in Amiens's hospital, the royal cortège set off for Compiègne, where Claire de Duras, Chateaubriand, Talleyrand, Maréchal Ney and many other of Napoleon's generals and courtiers were waiting to greet them. Most had quickly and conveniently forgotten their years of service to Napoleon. Compiègne itself was in some chaos, its fine library of books partly destroyed by cannon fire. After discussions with senators come out from Paris to greet him, Louis XVIII issued a first proclamation, promising to adopt a liberal Constitution, to preserve liberty and to honour existing pensions, titles and decorations.

The château was seething with anxious people, come to press their suit with the restored Bourbons, scrambling over each other for preferment, for rewards, for places at court. Those who had held positions under Napoleon were desperate to hold on to them, those who had lost out by remaining monarchist now hoped for recognition and restitution. When, on 4 May, Louis at last entered Paris, to a *Te Deum* in Notre Dame and the obligatory balloon ascension by Mme Garnerin, Mme de Boigne remarked that it was so 'like a party that it is a pity it is a conquest'. That evening, such were the fireworks that the Seine looked like a river of fire. 'These excellent princes delight us all,' Claire wrote to her daughter Clara. 'We are very happy. It all seems like a dream.' For Claire and Amédée, the return of the Bourbons was all they desired; for Lucie and Frédéric, it was all rather more complicated. Leaving aside their own personal feelings about any restoration of the Bourbons, there was still the question of how the new King would view those who had held prominent posts under Napoleon.

* * *

The question, too, was just what kind of a monarchy it would turn out to be. By education, instinct and experience, Louis XVIII belonged to the 18th century, to a world in which an elite of aristocrats and churchmen ruled from *de haut en bas*. In exile he had often spoken of the necessity of not taking revenge, and of moderation, not out of laziness but as the best policy. But behind his benign air of fatherly concern lay an 'olympian egoism'. He knew that he would have to make concessions, and had in fact no desire for authoritarian rule; but what he understood in his bones was the *ancien régime*.

Those closest to him shared his certitudes. There was his headstrong younger brother, the Comte d'Artois, now known as Monsieur, who made very little effort to conceal his distaste for anything that touched on revolution. There was d'Artois's elder son, the Duc d'Angoulême, who though honest and sensible was handicapped by feelings of inferiority caused by his small size, poor sight, nervous tics, stutter and reported impotence. The Duchess, only surviving child of a queen many now chose to recall with nostalgia, could have won considerable sympathy and understanding. But the pale, trembling orphan of the Temple had become stiff and overbearing, with a grating voice and hard features, having inherited her father's brusqueness but not his bonhomie, and Marie Antoinette's pride but not her gracefulness. To charm the French, that left only the Duc de Berri, d'Artois's second son, another red-faced small man with short legs and no neck, but witty, generous and physically brave, if irascible.

For the moment, the imperfections of the restored Bourbons did not matter. Having refused to accept a Constitution proposed by the French Senate, Louis set about drafting a Charter of his own, coming up with 74 articles which promised equality before the law, freedom from arbitrary arrest, and reasonable liberty for the press, while retaining for the Crown the power to declare war or make peace. The King would be able neither to suspend laws nor dispense with them. A Chamber of Peers was set up, the revolution and the English system suggesting the need for one. All this was welcome to Lucie and Frédéric and allayed some of their fears about a restored autocratic monarchy. To Lucie's added pleasure, Frédéric, far from being punished for his years as Prefect, was rewarded with a peerage, for the wholehearted welcome he had extended to the new King. He now became the Marquis de la Tour du Pin Gouvernet. The second Chamber, that of the Deputies, was severely limited from the beginning by making only the 1 per cent of Frenchmen who paid over 300 francs a year in taxes eligible to vote.

The moderate tone of the Restoration was apparent in Louis XVIII's choice of ministers: Frédéric's friend, the elderly Malouet at the Ministry of the Marine, the Abbé de Montesquiou at the Ministry of the Interior, and Talleyrand back in Foreign Affairs, a post he had held, at various times, under the Directoire, the Consulat and the Empire. The Abbé de Montesquiou had resumed his ecclesiastical title, dress and manner, which Lucie, who could remember him in his defrocked days in a rose-coloured waistcoat laughing heartily at the theatre, found absurd. 'We must thank Providence,' remarked the new Director General of the Police, 'that we have a King made of dough of the finest constitutional flour', which sounded flattering until he added that it was all rather like a 'trompe l'oeil of divine right masquerading as a constitutional monarchy'. Frédéric and Lucie were not alone in feeling wary about the shape of things to come. With the end of ten years of Empire, and the very real changes brought about by the revolution, there was a profound question about how exactly the new France would be governed.

Fearing that any post in the new administration would be inferior to his two prefectships, Frédéric decided to return to diplomacy. He went to see Talleyrand, who rather to his disappointment offered him the embassy at The Hague. But he took comfort from the fact that as he left, Talleyrand added, 'Take that post for the time being.' When he talked it over with Lucie, they agreed that this undoubtedly meant that something better would be forthcoming.

The Allies had been most circumspect with the conquered French, their officers enforcing on their soldiers courtesy and respect for property. Though the 5,000 or so people whom Napoleon had rewarded with lands or revenues on foreign soil saw them confiscated, France was allowed to revert to the borders of 1792. Nor did it have to pay indemnities. It was also allowed to hold on to the stolen art in the Louvre.

Lucie arrived back in Paris, to an apartment in the Princesse d'Hénin's house in the rue de Varennes, to find Wellington preparing to take the job of British Ambassador. Claire, as wife to a senior courtier, was happily ensconced in the Tuileries, where the Duchesse d'Angoulême, now known as Madame, cast a chill over court life. The Duchess had barely been stopped from insisting that women return to the hoops of the 1780s. Just the same, only the chosen few were allowed to enter the throne room, the rest simply filing past the door, and Amédée de Duras, 'more duke than the late M. de Saint-Simon . . . graceless and scarcely polite', was busy concocting rules for precedence and hierarchy. For his part, the King appeared wrapped up in his own thoughts, eating gargantuan meals – he thought nothing of putting away a plate of cutlets as an hors d'oeuvre – before being wheeled back to his rooms to entertain his favourites with his erudite and elegant Latin quotations, aphorisms and metaphors. No restored monarch, it would be said, had ever treated those who restored him with such disdain: even the Tsar, arguably the most powerful ruler in Europe, had to present himself twice before he was received.

One evening Lucie was invited to a ball given by Prince Schwartzenberg. It was, she wrote later, not only 'the oddest spectacle to anyone given to reflection' to find herself surrounded by all the people, furniture and food that had so recently been those of Napoleon's court; it was also sad. Looking around her, listening to Claire talk about her good fortune, she reflected that 'not one of all the people there' was worthy to be Napoleon's conqueror, and that she was probably the only guest to feel shame at the speed with which Parisian society had gone over to the victors. Lucie's perception of events and people was unusually candid; never swayed by fashion or intrigue, it was as if she brought to her surroundings a curiously pure eye.

Even Josephine, despite her years as Empress, lost none of her friends and admirers with Napoleon's abdication. Malmaison continued to attract foreign visitors, come to look at the pictures and sculptures that made it more a museum than a house, and to stroll in the now famous hothouses where 184 new species had flowered for the first time in France. But late in May Josephine went riding with the Tsar and caught a cold, which turned into pneumonia. She died on the 29th, at the age of 50, just as the Allies were preparing to pull their soldiers out of Paris. She left no will, and her remarkable collections, divided by her children Hortense and Eugène, soon disappeared in sales and restitutions around the world.

* * *

Europe had been at war for more than 20 years. In those two decades, territories had changed hands, frontiers had vanished, dynasties had been overthrown and new monarchs brought to the throne. In September 1814, preparing for the Congress in Vienna which was to settle the affairs of Europe, Talleyrand wrote an analysis of what he hoped to achieve for France. Frédéric had been correct in suspecting that Talleyrand had plans for him: Lucie, calling on him one day, was informed that her husband was to prepare himself to leave immediately for Vienna as one of the plenipotentiaries at the talks. Talleyrand's own entourage included, not his wife, judged too blowsy and undistinguished, but his 21-year-old niece by marriage, the beautiful and accomplished Comtesse Dorothée de Périgord, who was to act as his hostess. Lucie longed to accompany them, but Humbert had joined the King's military Household, becoming a lieutenant in the Black Musketeers – so called after their black horses – and was looking for a wife. It was decided that she would remain in Paris with him, and that Auguste de Liederkerke would go with Frédéric to Vienna as his private secretary. Charlotte had given birth to a baby girl, Marie, and Lucie did not wish to be too far from them.

There had been considerable jealousy at the news of Frédéric's appointment, particularly when it became known that he would also retain his job as ambassador to Holland; Claire, in particular, was furious, having gone to great lengths to secure for Chateaubriand one of the postings to Vienna, and failed to do so. 'Can one love without suffering?' she wrote mournfully to a friend, fretting about Chateaubriand's future. 'To live is always to suffer.' Yet again, as in Brussels, there was something in Frédéric's uncompromising honesty that seemed to invite attack. At every turn Frédéric, like Lucie, was emerging as a figure strangely out of tune with the evasions and scheming of his age.

Vienna, in September 1814, was full not only of statesmen and their entourages, but of reigning royal families and courtiers from all over Europe. The German principalities, Italian states, Swiss cantons and the Catholic Church were all represented. At one moment, two emperors and two empresses, four kings, one queen, two heirs to the throne and three princes were all staying in the Royal Palace at the same time. Of the 100,000 foreigners gathered in Vienna for the Congress, it was said that 95,000 had come not to work but to be reminded of the grandeur and pleasures of 18th-century society. There were balls, banquets, hunting parties and even a medieval tournament held in the baroque hall of the

Imperial Riding School. The Congress, remarked the 80-year-old Prince de Ligne, who better than anyone could remember the splendour of Versailles, '*ne marche pas, mais il danse*'.*

Even so, the work done by Frédéric and his colleagues was both important and tricky, not least because of the real differences between the major powers, who in any case all wanted to exclude France from the preliminary talks. The Tsar wanted to emerge from the Congress with a kingdom taken from Poland; Prussia wished to swallow up Saxony; Metternich, the Austrian Minister for Foreign Affairs and the dominant statesman at the Congress, had no intention of allowing French soldiers to remain on Italian soil. Talleyrand, for his part, wanted to promote European equilibrium based not solely on military parity, but on principles of law and justice. As the soirées and receptions grew ever more fanciful and splendid, the relations between the powers grew steadily more tense. Out of the disagreements, Talleyrand soon drew a triumph for France, a secret treaty of mutual support with Austria and Britain, backed by some of the smaller states.

* * *

Lucie, having settled 8-year-old Aymar in Paris with a tutor in rue Notre-Dame-des-Champs, went to collect Charlotte and her baby daughter from Brussels, and installed them with her in the rue de Varennes. In the evenings they went out to call on the Princesse de Poix, who was once again drawing together the friends of her youth to talk about Voltaire and Montesquieu. They visited Claire in the Tuileries, and Lucie's stepmother Mme Dillon in the Faubourg Saint-Germain. In all these salons, Lucie was becoming uncomfortably conscious of undercurrents of intrigue running through Parisian social life. It reminded her of the uneasy months before Napoleon's coup of 18 brumaire.

Though Louis XVIII was personally popular – and showed Lucie particular marks of favour when she attended the court receptions – the rest of the royal family were already much disliked. The Duc de Berri's temper was offensive to many of his former supporters, while the Duchesse d'Angoulême's grim and despondent

---

* 'Doesn't work (literally: walk), but it dances'.

expression soured life at court. The excessive formality and etiquette, claimed the younger courtiers, were out of tune with the realities of the new France. It was no secret that discontent with the Bourbons was building up among Napoleon's former maréchals such as Ney, nor that his stepdaughter Queen Hortense's salon was often filled with soldiers openly regretting Napoleon's departure. What alarmed Lucie was the way that at court and in most royalist circles it had become fashionable to ridicule all talk of conspiracy. She was increasingly aware of whispering on all sides, odd glances, secret meetings.

Lucie might have worried more had she not been overwhelmed by more personal troubles. Her granddaughter Marie was nearly a year old and teething. One day she caught a fever: her temperature kept rising. There was nothing anyone could do. The baby died, while in Lucie's arms. 'I wept for her,' she wrote later, 'as if she had been my own child.' It was Humbert who attended to the funeral, while Lucie took the inconsolable Charlotte away to stay with the Princesse d'Hénin, then arranged for her to join Frédéric in Vienna. Lucie had now seen three infants – two of her own and a grandchild – die in the first few years of life.

Then came another crisis. Going to visit Aymar in his tutor's house, she discovered him in bed with a high temperature in the infirmary, a gloomy, north-facing room. Discovering that no doctor was due to call, she took her carriage in search of her own, a young man called Dr Auvily. They returned to find that Aymar was worse. Dr Auvily diagnosed pleurisy and told Lucie that if she wished to save his life, she should remove him instantly from the icy room. Aymar was wrapped in blankets and taken back to the rue de Varennes. On the sixth day, it was thought that he would not pull through and Humbert was told to prepare Lucie for his death.

But Dr Auvily refused to give up hope. Resorting to one of the more drastic remedies of early medicine, he had the little boy swaddled in a plaster jacket impregnated with cantharidin, used to raise blisters, leaving only his arms and feet bare, to which were applied mustard poultices. Every two minutes, Aymar was forced to swallow a teaspoon of warm liquid. Remarkably, he survived. Though the cantharidin had reduced his body to one large sore, his temperature fell. When he was able to leave the house again, Lucie asked Amédée to get her a special pass for the Louvre, and there, for the next six weeks, Aymar ran around in the warmth among the stolen Raphaels and Titians.

Even at the height of the Empire, Paris had not seen balls and receptions of such brilliance as those given in Paris during the winter of 1814. Lucie and Claire had decided to take Mme de Staël's 19-year-old daughter, Albertine, into society and to wean her off her mother's inelegant dresses and odd costumes in time for her marriage to the Duc de Broglie. Mme de Staël had been one of the first of Napoleon's exiled opponents to return to Paris, and was now making up for the wasted years of banishment with gatherings at which she drew together victors and vanquished, Bourbons and Bonapartistes. Somewhat stouter, but no less passionate, she continued to hold forth to admiring audiences on politics and literature. Her epitaph on Napoleon was unforgiving: a 'Condottiere without manners, without fatherland, without morality, an oriental despot, a new Attila, a warrior who knew only how to corrupt and annihilate'.

Lucie, who had continued to meet her from time to time during her ten years of exile, was in her drawing room in Paris one day when Wellington came to call. He had recently bought Pauline Borghese's house in the Faubourg Saint-Honoré for 820,000 francs. Wellington was not an altogether popular choice as British Ambassador, some of the French feeling that it had been tactless to appoint a man who had spent the previous six years engaged in defeating the French armies in Spain. Others complained that his dress was too casual and his open affair with a former mistress of Napoleon's too scandalous. But Lucie, who had known him as Arthur Wellesley since she was a child, met him again with pleasure.

With the Restoration, the English had once again poured across the Channel, delighted after a ten-year break to resume their grand tours of Europe, and to start them, as they had in the past, in Paris. Some 23,000 people crossed the Channel in 1815, to be charmed by the many new restaurants and cafés – there were said to be over 3,000 of each – by the new paintings by David's pupil Gros, by the ageing Talma, still playing his heroic roles, and by the wide tree-lined boulevards down which they wandered on fine afternoons. There was a new rhinoceros in the Jardin des Plantes, and at 188 rue Saint-Honoré could be seen the 'Venus Hottentote', a Botswana bushwoman with vast buttocks. The visitors gambled and danced – the cotillon, the polka, the waltz – at afternoon *thé dansants*.

The reaction of French society to their English visitors was somewhat disdainful. Deprived of French elegance for a decade, the English were said to have 'acquired the easy-going manners and customs of the tropics'. Cartoons in France portrayed the taller, plumper English girls as insipid, gawky and stiff. For their part, Englishwomen criticised their hosts for being 'pedantic and frivolous', overly conscious of social rank, and no better than 'amiable but thoughtless children' over matters of money. Lady Granville, soon to become British Ambassadress to Paris, admitted to feeling grudging admiration for the 'aplomb' of many Frenchwomen, but seriously doubted whether they were capable of the deep thoughts and feelings of Englishwomen.

Very quickly, salon life resumed; Parisian society, as Benjamin Constant observed, had no trouble jumping from 'one branch to another'. In the Tuileries, Claire de Duras vacillated between inviting ardent monarchists or liberal intellectuals, and allowed her salon to be dominated by the mournful Chateaubriand, who insisted on imposing his 'irritating, bitter and morose vanity' on all around. Mme de Récamier was also back in Paris, and it was in her salon that a new fashion was being tried out, that of putting out four or five little circles of chairs for female guests, between which were left corridors for the men – and herself – to circulate. To Claire's despair, Chateaubriand was beginning to show a marked interest in Mme Récamier.

Mme de Genlis had also weathered the jump from revolution through Directoire, Consulat and Empire to Restoration with exceptional ease. She had retired to the Convent of the Carmelites, 'disenchanted with the vanities of the world and the chimeras of celebrity'. Thin and pale but having lost none of her verve, she spent her days playing the harp and painting pictures of flowers. She was at work on a *Dictionnaire Critique et Raisonné* of etiquette and manners, in which she was trying to explain, to a generation who had never known it, the pleasures of *délicatesse, bon ton, politesse* and *douceur*. Before the revolution, she wrote, young women had been gentle and reserved, which was what they should be. They had since become bold and assertive, which made them seem prematurely aged. 'Gentleness' and 'submission', as prescribed by the evangelists, was what women should aim for.

Not everyone agreed: in the *Journal des Dames et des Modes*, women, portrayed sometimes as weak, sometimes as strong, but always as charming, were urged to heed the Comte de Saint-Simon's words: 'Rise up, Monsieur le Comte, you have great things to accomplish.' New manuals were appearing, on how to live, behave, dress and run a house, stressing thrift and orderliness, so that men could return in the evenings from public life into havens of domestic happiness.

Lucie was not alone in fearing the cross-currents of intrigue and discontent. Writing to Castlereagh in October, Wellington remarked that though Paris seemed tranquil, 'there exists a good deal of uncertainty and uneasiness in the mind of almost every individual that is in it'. The Comte d'Artois, surrounded by a group of ultra-royalists – the 'ultras' – made little secret of his desire to abrogate the Charter and restore royal absolutism. Much of the army, on the other hand, put on half-pay in an effort to reduce France's vast debts, mourned Napoleon's departure. Fouché, who had abandoned Napoleon, nonetheless spoke yearningly of a regency under Marie-Louise, or even of having the Duc d'Orléans on the throne instead of Louis XVIII. There were complaints on all sides of Louis's deafness to the mood of the country. In Queen Hortense's drawing room, which had become a meeting place for disaffected soldiers and courtiers, young men wore her favourite flower, a violet, in their buttonholes, as a sign of their attachment to the exiled Emperor. The corridors of the Tuileries were full of rumours about Napoleon.

Lucie had been right to feel apprehensive. One morning, couriers, having ridden day and night from the south, brought the news that Napoleon had landed with 900 men in Golfe Juan. He had been on Elba for just ten months. It was the first of his Hundred Days. 'Soldiers,' declared the deposed Emperor as he travelled north, unopposed and gathering entire garrisons along the way, 'in my exile I heard your voice.' Maréchal Ney, who had fought heroically during the Napoleonic wars, then been among the first generals to rally to Louis XVIII, was hastily despatched to check Napoleon's advance. Swearing before he left to return with the former emperor in an iron cage, he had a change of heart and joined Napoleon instead.

When word of Napoleon's return reached Vienna, where the Congress was still in session, the four Allied powers declared war, not on France, but on Napoleon personally. Bonaparte, said Talleyrand, was a 'threat to world peace'. Frédéric, declaring that his job as plenipotentiary 'now seems to me insignificant', asked Talleyrand whether he might return to France. Unlike Ney, Frédéric, having given his allegiance once again to the Bourbons, was not about to go back on his word. Talleyrand hesitated, then agreed to let Frédéric go south in order to raise support for the King among soldiers garrisoned in Toulon, and to convey to the Duc d'Angoulême, stationed in the Rhone valley, assurances of Allied commitment.

Travelling via Genoa and Nice, Frédéric reached Toulon, talked for four hours to the garrison, where he found the men extremely reluctant to support the Bourbons and eager to see Napoleon in power again, and went on to another garrison in Marseilles. He arrived to learn that the Duc d'Angoulême had already capitulated to Napoleon. Standing in the main square, Frédéric addressed the military and civilian authorities of Marseilles, as well as the *Gardes Nationales* and the people of the city, explaining the position of the Allies. It was his hope, he declared, that Louis XVIII would not lose his throne. The assembled crowds made it clear that it was not a hope they shared. After this, deciding that there was nothing more that he could do except 'give way before the storm', he took a boat to Genoa. Thrown off course by heavy winds, he reached Barcelona instead and from there made his way to Madrid and then Lisbon, where he found a ship bound for England. He stayed 24 hours in London, and dined with the Duchesse d'Angoulême, who had managed to escape and cross the Channel. He was certain, he assured her, that the Allies would do all they could to return Louis XVIII to the throne. With 'tears in her eyes', Frédéric reported later, the Duchess replied: 'I believe you. The King will return to France. But, oh unhappy France.' (The Duchess, remarked Napoleon later, was the 'only man of her family'.)

In Paris, news of Napoleon's approach had been greeted with extreme alarm. Princes, courtiers, ministers and their entourages piled their valuables into carriages and scattered north, west and east. For many of them it was their third or even fourth departure into exile. At midnight on 19 March, in pouring rain, Louis XVIII left for Lille, from where, after much uncertainty, he made his way to Ghent. Lucie, having decided to go to Brussels, went to the Ministry of the Interior to collect money owed to Frédéric. She was told to come back later for the money.

In the rue de Varennes, the Princesse d'Hénin and the now extremely stout Lally-Tollendal were frantically packing; they, too, had decided to make their way to Brussels. It was some hours before sufficiently strong horses could be found to drag their vast carriage, and it was not until late that the party finally set out. Lucie was delayed further by her banker, and by the time she had collected 12,000 francs in *napoléons*, there were very few available horses left in Paris. She spent the night at her window, anxiously watching soldiers file past, bunches of violets attached to their uniforms, sign of their allegiance to Napoleon, very visible by the light of the street lamps. At six next morning, two puny horses and a small barouche were produced and Lucie, Cécile, Aymar and a young Belgian maid set out for Brussels. Humbert had vanished somewhere with the Black Musketeers. The Faubourg Saint-Germain was deserted. It was Lucie's fourth flight into exile.

On 20 March, Napoleon arrived back in the Tuileries. The fleurs de lys were removed, the bees painted back on. Just a week before, thousands of people had lined the quays of the Seine to watch the first steam boat come up the river.

* * *

Lucie reached Brussels without trouble, and was soon installed in a rented apartment in the old city, where she met up with Princesse d'Hénin, Claire Duras and her mother and daughter, all once again refugees. It was many worrying days before she had news of Humbert or Frédéric. Then Charlotte arrived from Vienna, to describe how, with the news of Napoleon's advance, the Congress had dissolved and kings, statesmen and courtiers, instantly forgetting their differences, had scurried for home. Lucie was not greatly comforted to learn that Frédéric had insisted on going south in search of the Duc d'Angoulême. To Mme de Staël, who was back at Coppet and to whom she had become much closer in recent months, she wrote asking whether she had heard from him. 'If you only knew, my dearest, how acutely anxious people who feel deeply and are unable to take life lightly suffer at times like this, you would guess how I feel . . . This is the sort of situation that wears my mind down like a nail file . . .'

Brussels, she wrote, had become a military camp, full of cannons, drums and trumpets, with everyone fearful of a sudden invasion by Napoleon's troops. 'Goodbye my dear, write to me, love me as you do those for whom you care most deeply.' Then she added: 'If, at this moment of trial, God were to send skirts to all those in Brussels who are not real men, we would find ourselves in the midst of an enormous convent.'

One evening, when Lucie and Claire were together, a servant appeared to say that there was a 'gentleman' of their acquaintance who wished to speak to them but did not dare enter as he was not correctly dressed. Even in the middle of war, the manners of the court had to be observed. It was the Duc de Berri, who told them that a band of brigands had attacked his carriage and made off with everything he possessed. Lucie, who still had many friends in Brussels, arranged for a new wardrobe to be assembled. She also called on the Prince of Orange, who, after the defeat of Napoleon at Leipzig, had been made Prince Sovereign of Belgium and the United Provinces. The Prince received her warmly in the very rooms that had been hers when Frédéric was Prefect to the Dyle. 'In this salon,' he told her, 'I try to discover ways to be as well loved as your husband was.' ('Alas,' wrote Lucie later, 'the poor prince never succeeded.') Lucie mentioned her son-in-law Auguste to the Prince, who agreed to take the young man on to his staff. When it came to obtaining favours for those she loved, Lucie was ruthless.

In Ghent, a prosperous cotton- and paper-manufacturing town, Louis XVIII was lent a fine 18th-century mansion with a large hall, where he and those of his ministers and courtiers who had followed him met every day to discuss the day's news. The court in exile, presided over by the ever present de Blacas, remained a centre of intrigue, ultras and royalists, officers and diplomats, speculators, visitors and spies all spreading rumours and gossip. Chateaubriand, who, to his great satisfaction, had been appointed interim Minister of the Interior, had started a daily paper, *Le Journal Universel*, to which Lally-Tollendal contributed articles. Though there was a cautious feeling of optimism that Napoleon would soon be defeated, the foreign powers appearing to be unanimous in their intention to crush him, the émigrés lived frugally, remembering the money squandered at Koblenz.

The weather was fine and warm and there were occasional

outings to the surrounding countryside, to pause at inns to eat fish from the rivers, washed down with Louvain beer. On Sundays, the entire French community dressed up to accompany the King, in his uniform of pale blue silk embroidered in silver, to Mass in the cathedral. The loss of his kingdom had not diminished Louis's appetite. On the night of his arrival in Ghent, he polished off 100 oysters. Unlike Napoleon, who wished to get through his meals in under ten minutes, the King spent hours at table, savouring the sauces, trying out new dishes, tasting the Lafittes and the Tokais he particularly enjoyed, and the truffles in champagne, to which he was very partial. In Ghent, Louis's more moderate views were being constantly challenged by d'Artois and the ultras gathered round him, who insisted that Napoleon had been able to return precisely because the Charter had been so liberal, and talked of the punishments they would mete out, once the Bourbons were back in power, to those who had rallied to the former Emperor. Talleyrand, playing a waiting game, did not hurry to Ghent.

In Paris Napoleon's return had not proved as triumphal as he had hoped. Though large sections of the army had indeed gone over to him, drawn by his audacity, or disenchanted by the restored Bourbons, there were few cheers on the Champs de Mars when he proclaimed a new liberal republic. Parts of France had remained royalist, and he was having trouble raising men for the coming battles against the Allied powers. James Gallatin, a young American who had accompanied his father to Paris, noted that at the Opéra one night Napoleon looked 'fat, very dull, tired and bored'. Fanny, to her delight, was back in Paris, having returned with Madame Mère on a 74-gun ship of war sent by Murat to collect them from Elba.

Frédéric was eventually able to make his way to Brussels. For a while, the family was reunited, except for Humbert, who was with the Musketeers in Ghent. Humbert, Lucie wrote to her cousin Charlotte Beddingfield, 'at this moment of crisis is showing himself to be as noble, strong and manly as any loving mother could wish for'. Like Vienna at the height of the Congress, Brussels was immensely social. 'This is without exception,' wrote one young society lady to her mother in England, 'the most Gossiping Place I ever heard of.' Lucie complained wryly to Mme de Staël that she sometimes wished that God, who had given her the power of thought but not that of speech, had withheld speech as well, and made her deaf and dumb into the bargain, for she loathed 'chatter'.

Wellington had reached Brussels early in April; on 3 May he was appointed commander-in-chief of the Dutch and Belgian forces. Drawing up his men for the definitive battle against Napoleon, he had decided to intermingle veteran soldiers with fresh recruits, regular soldiers with militia; he complained that his army of 92,000 men was 'weak and ill-equipped' and his staff very inexperienced. To keep up morale, Wellington insisted on attending and giving parties, and he went to watch the English play cricket at Enghien; but on 8 June he warned the Duchess of Richmond not to organise a picnic too close to the border with France. In *L'Oracle*, Brussels's daily paper, Wellington was referred to as the 'hero of our age'. For her part, Lucie herself doubted that Napoleon would ever invade Belgium. The Emperor, she told Mme de Staël, was by all reports a 'changed man'. From all sides she had heard that his claws had been drawn and that he was now 'all moderation, sweetness and liberality'.

* * *

In 1814, waiting to be sent as Governor General to Canada, Charles, 4th Duke of Richmond and former Lord Lieutenant of Ireland, had moved to Brussels, taking over a house that had belonged to a carriage-maker in the rue de la Blanchisserie. Its former ballroom, used by the carriage-maker as a factory, had been papered in a pattern of roses and trellises, and it was here that on the night of 15 June the Duke and Duchess gave a ball. Most of the 222 guests were English, the men nearly all senior officers in Wellington's army, but Lucie and Frédéric were invited, together with Charlotte and Auguste. Wellington had received word that Napoleon had crossed the border, driven back a Prussian corps and occupied Charleroi, but he decided to let the ball go ahead so as not to spread alarm. As the evening wore on, the officers slipped away. Some had time to reach their barracks and change out of evening dress; others had arranged for servants to stand by with uniforms and horses. Among those who hastened away was Auguste, who went to join the Prince of Orange. But the dancing went on, and after supper was served at midnight, the bagpipes were played.

The Battle of Waterloo was fought between Napoleon and the Allied forces of Britain, Germany, Belgium, Holland and Prussia, with slightly more men and guns on the French side. At dawn, the outcome of the battle still uncertain, the inhabitants of Brussels, fearing the possible arrival of the French, began packing. From the city's ramparts, Lucie watched cart-loads of wounded men arriving from the battlefield. There was a constant booming from the cannons and heavy rain had turned the surrounding countryside into mud. Suddenly a troop of cavalry appeared out of the rain, and galloped through the streets, scattering the carts of wounded men and overturning carriages piled high with baggage, leaving a trail of cases and clothes. Rumours spread that the French were arriving. 'It was the most terrible sight I have ever seen,' Lucie later wrote to her English cousin Charlotte. 'Nothing can quite convey the idea of a town of 70,000 people seized by panic, all trying to flee at once. It seemed as if the end of the world had come.' It was some time before reassuring messages arrived, with the news that victory had gone not to Napoleon, but to the Allies.

As it gradually became clear, over the next few days, that Napoleon was retreating towards Paris, his army having virtually ceased to exist, people set out for the battlefields to see what they could do for the wounded. Brussels soon became a vast army hospital, 20,000 wounded men taken into churches, covered markets and private houses. Among the dead and dying were officers from the Duchess of Richmond's ball, some still in silk stockings, breeches and buckles, having not had time to change into their uniforms. On 22 July, Lucie wrote to Charlotte that she had come across her young cousin Browne, the son of Lord Kenmare, lying wounded in a shed by the river with a smashed thigh and broken leg; she had taken him home with her to care for.

Over the coming weeks, visitors to Brussels would get their coachmen to take them out to Quatre-Bras and Mont-Saint-Jean, to look at the skeletons of the fallen horses and collect musket balls as mementoes. 'Alas,' wrote Lucie to her cousin Charlotte, reviewing the events of the past few days with her clear and sceptical eye, 'I think that poor France is forever lost. I cannot believe that what has just taken place was the way to save her.'

CHAPTER FIFTEEN

Embarking on a Career of Grief

This time, the Allies were not quite as forgiving. During Napoleon's Hundred Days, much had been said and written about Louis XVIII's failure to govern, and the need to find other solutions for France, possibly in the shape of his cousin Louis-Philippe, son of the guillotined Duc d'Orléans. But the Battle of Waterloo, and Wellington's continued preference for the Bourbons, won the day. Louis XVIII was to be given back his crown. An armistice was signed on 3 July; the Allies entered Paris on 7 August.

A young officer called Alexandre Mercier, advancing towards the capital with the Allied troops, later described scenes of looting, destruction and drunkenness as châteaux were broken into, portraits sliced out of their frames with knives, furniture smashed, wallpapers stripped from walls, marble fireplaces shattered into splinters. In one château, Mercier came across a room 'which seemed to have been chosen as a place of execution for porcelain . . . The most ravishing Sèvres and Dresden vases, tea sets, all piled up in fragments on the floor.' In another, soldiers had dragged their bayonets across a fresco of a forest with nymphs playing in a pond. General Blücher had announced his intention of blowing up the Pont d'Iéna, named after a Prussian defeat in 1806, but the bridge was saved when Wellington posted a British sentry to stand guard.

In late June, Louis XVIII, travelling behind the Allies, met Talleyrand in Mons. Napoleon had abdicated once again and Fouché had been acting as provisional governor. Chateaubriand, who was at Saint-Denis when Fouché and Talleyrand came to greet the King, left a memorable description of the moment. He saw, he later wrote, 'passing in the shadows, Vice leaning on the arm of Crime', Fouché and Talleyrand, as 'the loyal regicide, kneeling, laid the hands which caused the head of Louis XVI to fall between the hands of the brother of the royal martyr; the renegade bishop stood surety for the oath'. Neither would receive the silver medal of loyalty, with its portrait of the King on one side and the word '*fidélité*' on a crown of laurel and oak on the other, given to those who had remained faithful to Louis and gone with him to Ghent.

Chateaubriand himself, hoping to be confirmed as Minister of the Interior, was blocked not only by Talleyrand, who mocked the idea of a poet becoming a politician, but by Louis himself, who considered Chateaubriand to be a difficult, egotistical and brooding presence. The 'large and tender' Lally-Tollendal, on the other hand, was rewarded with a peerage. And, once the Académie Française had been purged of Bonapartists and revolutionaries, he was given a seat where, freed from the Princesse d'Hénin's ruthless pruning of his clichés and platitudes, he became known for his interminable monologues about the past. Both he and the Princesse were in their late 60s, and in their rented house in Autcuil, which had a magnificent garden of cedars, they once again lit the flame of 18th-century conversation.

Lucie had not been looking forward to the return to Paris, fearing the ugly repercussions of Napoleon's second defeat. Frédéric was still ambassador to Holland and she was hoping that he might be posted to London instead. What she dreaded most was the idea that he could be recalled to court, 'the thought of which,' she told Mme de Staël, 'makes me feverish'. All that reconciled her to the idea of the second Restoration was the thought of seeing 'our dear and admirable Duke of Wellington again'. Soon after the Battle of Waterloo, Wellington had told Frédéric that he felt that he had 'bought the glory of saving France at not too high a price, but that it had personally cost him the lives of men he loved, who knew the secrets of his soul, and that he never wished for glory of this kind again'. To Charlotte, Lucie wrote: 'How one admires such goodness in so great a man.'

That autumn, while over a million Prussians, Austrians, Dutch, Russians, Bavarians, Germans, Scandinavians and even Swiss occupied large swathes of France, there were savage reprisals against those who had risen in support of Napoleon in the south. 'We have conquered France,' George Canning, who as British Prime Minister had helped orchestrate the war against Napoleon, said to Mme de Staël, 'and we want to crush her so profoundly that she will not stir again for ten years.' If the British troops were more or less kept in order by Wellington's insistence on discipline, the Austrians, and above all the Prussians, drank, looted and raped unchecked. In the Bois de Boulogne, the Allied cavalry was said not to have 'left wood enough to make a toothpick of'. 'Mercy,' observed Mme de Rémusat, 'is not fashionable this season.' Lady Jerningham, crossing the Channel once again to visit Lucie and the French Dillons, remarked that the Faubourg Saint-Germain had become 'very dismal' now that it was overrun with foreign soldiers.

One of Frédéric's first duties in the new Chamber of Peers was to try Maréchal Ney, the soldier who had fought so bravely in Russia for the Emperor, then switched allegiance to Louis XVIII, only to join forces with Napoleon once again. Ney could very easily have escaped, but he allowed himself to be arrested, asking only that he be tried, as was his right, by the peers. All but one of the 162 present that day voted him guilty of high treason. Frédéric was one of the 139 who also voted for his death, but who at the same time appealed to the King to show mercy. Louis chose not to. Ney's wife, Aglae, frantically trying to save her husband's life, begged Wellington to intervene. He refused, saying that he was merely a servant of the Allies. Maréchal Ney was a much-loved and very brave man; indeed he was known as 'the bravest of the brave'. He was shot in front of a wall in the Luxembourg Gardens, having himself given the order to fire.

All over France, investigations were being launched into how individual people had behaved during the Hundred Days; almost a quarter of all civil servants would lose their jobs in the '*épurations*', the purges, that followed. Frédéric was nearly one of them. Metternich, who had long disliked and mistrusted him, suggested that someone with a less equivocal past should be posted as ambassador to Holland; but this Louis refused to do. It was a measure of Frédéric's perceived honesty, his transparent desire to do the right thing, and to admit freely to mistakes and lapses of judgement, that his years of service to Napoleon were not held against him. He did not apologise for them, but he made it plain that he now intended to give his full support to the restored Bourbons, and that his belief in monarchy was absolute. It was the lack of fuss and speed with which both he and Lucie had rallied to Louis that sometimes made them appear unprincipled.

It was Britain which now emerged from the Napoleonic Wars as master of the world's largest empire. The final act of the Congress, signed on 9 June 1815, would keep the peace in Europe for almost another 30 years. Sovereignty was seen to lie no longer in the legitimacy of their dynastic ruling houses but in the legally defined states themselves.

In Paris, a new elective Chamber of Deputies – men who were 'pure but moderate' – with the numbers increased from 262 to 402 and the age of voters reduced from 30 to 21, brought to power an unexpected majority of royalists. The casualties of Napoleon's Hundred Days included the two most powerful intriguers of their day, Talleyrand and Fouché, the mood of France having swung against apostate bishops and regicides. Talleyrand's fall, however, was softened when he was offered the position of Grand Chamberlain, which brought with it 100,000 francs a year and inclusion at every important occasion. As Lucie had long remarked, Talleyrand was a 'wily old fox'. The Duc de Richelieu, grandson of the famous Maréchal de Richelieu, who had fled marriage with Lucie's friend, the hunchback Rosalie-Sabine de Rochechouart, and served for many years in Russia, was invited to form a new government. Richelieu was seen as uncontaminated by recent events. Sensitive, modest and loyal, he soon won respect from the Allies, but his inexperience led to a feeble team of ministers. His first task, to restore stability to France, was made all the more difficult by the terms of the peace deal: northern and eastern France were to be occupied by 150,000 Allied soldiers for at least three years, their costs borne by France, and 700 million francs were to be paid in war indemnities. There was further humiliation when the four great powers formed a pact, the Holy Alliance, effectively placing France under supervision.

After the first Restoration, the Allies had somewhat surprisingly left their looted art in Paris. The great gallery of the Louvre, nearly a quarter of a mile in length, where Aymar had played during his convalescence, remained full of stolen masterpieces. In 1815 the Allies were not disposed to be so generous. Parisians, looking out of their windows or sullenly lining the streets, watched as foreign soldiers, helped by porters with barrows, ladders and ropes, began to remove, pack up and carry away paintings and sculptures, leaving the walls of the Louvre every day a little more denuded. The worst moment came when a troop of Austrian cavalry, keeping at bay an angry crowd, climbed to the top of the Arc de Triomphe and started to lower the famous horses, looted by Napoleon from St Mark's in Venice. Louis XVIII chose to leave the Tuileries for the day rather than witness their departure. That evening four carts, each bearing a horse lying on its side, set out for Italy accompanied by cavalry and infantry, to the sound of drums. The sculptor Canova, sent from Rome by Pius VII to oversee the return of the looted Italian art, was ostracised by the Parisian art world; Antoine-Jean Gros, painter of heroic Napoleonic scenes, cut him dead.

* * *

Frédéric and Lucie now prepared to divide their time between The Hague, where King William had his court, Brussels, which remained administratively important, and Paris. Humbert, still looking for a wife, had been appointed aide-de-camp to Maréchal Victor.

In spite of frantic protests, tears, and and even threats that she would throw herself overboard, Lucie's sister Fanny had once again been forced to follow Napoleon into exile. This time it was to a British island lying 4,000 miles from France and 1,200 miles from the coast of Africa. The journey by sea took over two months. The first glimpse of St Helena, wrote Fanny's husband Bertrand later, was of a rock, more 'like a huge dark-coloured ark lying at anchor . . . than a land intended for the habitation and support of human beings'. Writing to her daughter Charlotte, Lady Jerningham compared Lucie and Fanny to figures on a weather vane: as the fortunes of one rose, those of the other sank. 'Mme de la T du P is again got to "fine". The other poor Thing leaves for St Helena. I fear for her state of agony.' Fanny wrote a forlorn letter to Lord Dillon, asking whether he might be able to send her a piano and some French and Italian songs for a mezzo-soprano: 'I am so unhappy! I am the unhappiest of Women!'

St Helena, with some 5,000 inhabitants, was extremely hot in summer; bare rocks rose above green valleys full of guava, mango and prickly pears. It rained incessantly and there was constant fog and mist. The island was infested by enormous flies and a great number of rats. The Bertrands had their own house, not far from Napoleon's sombre and gloomy Longwood. Napoleon had brought with him two horses, and a barouche was ordered from the Cape for Fanny's use. At Longwood, where water dripped down the walls, and rats scampered over the parquet floors, etiquette was observed as inflexibly as in the Tuileries. Six sailors from the *Northumberland*, the British ship that had conveyed them to the island, were put into imperial livery and turned into footmen. Courtiers, in full uniform, stood when in Napoleon's presence. A *pâtissier*, M. Pierron, brought from Paris, spun amber palaces in sugar. But Sir Hudson Lowe, the narrow-minded, jealous major-general appointed to watch over Napoleon, insisted on treating his captive as a Corsican adventurer, refused to address him as anything but 'General', and made his life miserable with petty restrictions. News, visitors, post and contacts were all rationed.

Lowe, who held the devoted Bertrand responsible for some of Napoleon's demands, longed to get rid of him and used minor acts of tyranny against Fanny, hoping to reduce her to such despair that she would force her husband to return to France. Fanny had her three children, Napoleon, Hortense and Henri with her, and a fourth was on the way. She had no patience as a teacher, and as restrictions against her grew harsher, and she could no longer leave the grounds without being stopped by sentries, so she took to spending all her time indoors, quarrelling with Bertrand. Napoleon became very fond of Fanny's children.

* * *

There were times when a streak of remorseless misfortune seemed to run through Lucie's life, when even her almost perfect courage and determination failed her. The next blow to strike the family concerned Humbert. He was now 26, a good-natured young man even if somewhat spoiled by his parents. Like Frédéric, Humbert was truthful and honourable. The day he took up his appointment on Maréchal Victor's staff, 27 January 1816, he went to the aides-de-camp' mess to meet the other new officers. One of these was a Major Malandin, a rough, brave but prickly man who had risen through the ranks under Napoleon. Seeing Humbert and perceiving his aristocratic manners, the Major made a number of vulgar and offensive remarks. Prevented from replying by the sudden arrival of the Maréchal, Humbert went home and asked Frédéric what a young officer should do in these circumstances. He was, he told his father, referring not to himself but to a good friend. Frédéric believed him. 'Challenge the aggressor,' was his advice. And what, continued Humbert, should his friend do if an apology were offered? 'Refuse it,' replied Frédéric. 'Your companion should be all the more zealous in defence of his good name in that he, unlike the man who insulted him, has not had to pay in blood for the insignia of his rank.'

Despite attempts to prevent it, duelling remained a popular way for military officers – and indeed civilians – to settle their quarrels. Humbert returned to barracks and demanded that the Major offer him satisfaction with weapons. The Major was a renowned shot, but he was also a decent man. He offered to apologise. Humbert refused to accept his apology. Seconds were found who, knowing of the Major's skills, suggested sabres or swords. Humbert insisted on pistols. Maréchal Victor, appalled as were all the officers, felt he could not intervene. Humbert spent the evening at home, telling his parents that he was going out riding early next day. Lucie's cousin Mme de Boigne, not always a totally reliable witness, wrote later that when Humbert set out next morning he appeared troubled, pausing to kiss Lucie fondly and to leave a lock of hair that Cécile had asked for in her sewing basket.

At the arranged meeting place in the Bois de Boulogne, Major Malandin made one more attempt to apologise. Humbert was seen to cross the clearing; the assembled officers sighed with relief, thinking that he had decided to accept the apology. Instead, Humbert struck the Major across the forehead with his pistol. 'Monsieur,' he said, 'I think that now you will not refuse to fight.' The Major was a quick-tempered man. With what Aymar, describing the scene as it had been told to him many years later, called 'the startled bewilderment of a hen suddenly spat upon by a gazelle', he drew himself up and announced: 'He is a dead man.'

The rules of the duel were that the two men were to stand 25 paces apart and that they would continue to exchange shots until one of them was too badly wounded to continue. The signal to fire was given. Humbert fired first. The Major raised his pistol and as he fired was heard to say, 'Poor child – and his poor mother.' Humbert, shot in the heart, spun round and fell face downwards on the ground.

Frédéric was walking in the Champs-Elysées when a friend told him of Humbert's death. It was only now that Frédéric realised the part that he had unwittingly played in his own son's killing. For many weeks, the salons that surrounded Lucie and her family – those of Claire de Duras, the Princesse d'Hénin, the Princesse de Poix, Mme de Boigne – went into mourning. Frédéric, remarked his friends, appeared profoundly altered. 'What shall I tell you about myself?' Lucie wrote to her goddaughter Félicie from The Hague in May. 'I am just the same, I will never be consoled, I will never forget, my heart is still at precisely the same spot, like a clock which has stopped.' Relations with Claire were evidently bad again, as were those between Claire and Félicie. Lucie continued in the same bleak tone: 'It is perfectly simple. I don't hold it against your mother, I wasn't made to be her friend, it was an honour to have had her friendship for a while, it vanished. At my age, if one still has a heart that is young, one no longer has time for pretence, and one is no longer fooled by appearances.'

* * *

The Hague, where Lucie and Frédéric moved soon after Humbert's death, was known to its inhabitants as a 'village' but it was a village of 35,000 people, magnificent palaces and fine streets, connected to Delft and Leyden by a regular boat service. Its only drawback, noted a visitor, were its 'green and stagnant canals, which too often emit an almost pestilential stench'. After the Battle of Waterloo, the 'intimate and complete' union of Belgium and Holland had been proclaimed. Remarking that the new single state had not been greeted with much enthusiasm and that 'it is doubtful that either side will be happy', Frédéric was struggling to keep track of former revolutionaries and Bonapartists banished from France after the Second Restoration. In his reports, he discussed his constant worries about 'spiteful troublemakers' and 'incendiary writers'. 'All the bandits of Europe,' he warned, 'are finding homes here.' One of these was the painter David, whose studio became a meeting place for former 'regicides'. Frédéric's official letters were full of digressions on human nature. He was thinking of trying to convince the Dutch King that it would be far better to expel the troublesome émigrés, rather than to impose censorship, along the lines of the English Aliens Bill.

In September, Cécile, who was not yet 17, became engaged to a soldier called Charles de Mercy-Argenteau. He was ten years older, but devoted and rich. Lucie praised the young man warmly, and commented sharply on criticisms reported to Lady Jerningham by 'malicious groups' in Brussels. 'I despise malice,' she wrote, 'and I do not fear it.' The date for the wedding had been set for 15 December, and it was to take place in the chapel owned by their friend the Duc d'Ursel in Brussels, who had been the host at Charlotte's wedding four years earlier. Lucie, gathering together her much-loved Cécile's trousseau, remarked that it needed to be very plain, because the Dutch and Belgian people among whom Cécile would find herself were 'simple, without pretensions, generally agreeable and well brought up'.

She had very little time to enjoy her daughter's happiness. By December, Cécile was ill and the marriage was postponed. Lucie and Frédéric took her to Nice, hoping that the warmer climate would cure her. They watched, waited, kept hoping. On 20 March 1817, Cécile died.

'You must not leave this world which so admires you,' Lucie wrote to Mme de Staël, herself not well:

> It is to me, my dear, that death should come. What is there left for me? I had put all my pride, my joy, my tenderness, my hopes in those two children . . . After losses like these it is not possible ever to rise again, everything is over for me now, the world and its distractions disgust me; I have embarked on a career of grief and I shall never leave it again.

She had seen two of her grown children die in a little over a year; only the married Charlotte and Aymar, aged 10, were left. As with all her other losses, Lucie now fell silent. She wrote little and waited, as she had so often before, for time to make it bearable.

Frédéric was in such despair, his mood almost wild, that there were fears he might not rally. In The Hague, where he had left the embassy in the care of his chargé d'affaires, there were rumours that he had been dismissed. The mood in Brussels was uneasy, with fears of food shortages. But he did finally return wearily to his battle against the 'revolutionary germs' menacing the new country. The Hague, he said, had become a rallying point for the 'demagogues and rabble rousers', and the town was a nest of intrigue feeding on rumours and scandals, dimly perceived 'through the veils of politeness and politics'. Both the Dutch King and his son, who disagreed over everything else, were united in wishing France ill.

Lucie, planning to spend the summer with Charlotte in a rented house between Berne and Thun, was hoping that she might find peace from her constant grieving in the countryside. To Mme de Staël, she wrote miserably: 'My life is without purpose. The gap left by my adorable child can never be filled.'

Mme de Staël, too, was dying. Earlier in the year, leaving a party, she had fainted. She was now back in Paris, clinging to the gathering of friends she so loved and needed, arriving at receptions exhausted but then, buoyed up by the conversation and the company, talking as brilliantly as she had ever talked. Louis XVIII had finally repaid the 1.9 million francs owed to her father. But as the days passed, so she found it harder to leave her sofa. When friends came to see her they found her skin blotchy, but they remarked on the way that, without rising, she could still hold the entire room spellbound. To Chateaubriand, who often visited her, she said: 'I have always been the same, lively and sad. I have loved God, my father and liberty.' She died on 14 July 1817, the anniversary of the fall of the Bastille.

* * *

The terms negotiated by Richelieu with the Allies in the Second Treaty of Paris had again been surprisingly mild. In Paris, with the foreign soldiers gone, three distinct political movements, grouped around different men, ideas and newspapers, were taking shape. There were the 'ultras', who continued to wish to reverse everything the Enlightenment and the revolution had stood for and who spoke of the Charter as a 'work of folly and shadows': these were the 'green cabinet', green being the colour adopted by the Comte d'Artois backed by some of the restored senior clergy, by the secret monarchist and Catholic Chevaliers de la Foi, by men like Chateaubriand, Mathieu de Montmorency and the Duc de Fitz-James. Then there was the 'constitutional' party, the group of intellectuals who opposed the ultras and gathered around Mme de Staël's son-in-law, the Duc de Broglie and the historian Guizot; and a third party of independents, lying somewhere in the middle, a loose gathering of republicans, Bonapartists and Orléanists, many of them Freemasons.

Of all Louis's ministers, Richelieu, the dry, seemingly cold Président du Conseil, with his ironic glance and his air of gloom, was the most sensible and moderate; but the King did not much care for him. On the other hand, Louis XVIII did like Décazes, his Minister for Police, a charming, intelligent and good-looking royalist, whom Talleyrand said reminded him of a wig-maker. Décazes was appalled by the violence and extremism of the ultras, and worried about the very real threat that they posed to the stability of France. On 5 September 1816, the Chamber was dissolved; in new elections the number of ultras in power shrank to 92. Louis had always said that what he desired was '*repos*', a restful and peaceful France, and that disobedient and reactionary subjects were fatal to the health of the country. And, for a while, '*repos*' did indeed come to France. Though Richelieu fell, and the new ministry, dominated by Décazes, carried France further towards the left, the country became both peaceful and prosperous enough for the war indemnities to be paid off quicker than expected. That autumn, the grape harvest all over France was exceptionally good: wine-makers referred to it as the 'wine of departure', naming it after the foreigners' withdrawal from French soil.

In September 1818, soon after Charlotte had given birth to a strong and healthy daughter – she was so large that for a while Lucie thought it must be twins – Frédéric was sent to negotiate the terms of the early withdrawal of the Allied forces at the Congress of Aix-la-Chapelle. Charlotte called her new baby Cécile, but since her sister's death was still so raw in their minds, she was at first known as Séraphine; though Séraphine too was the name of another dead child, Lucie's first daughter.

Louis XVIII desired not only *repos* but luxury. While the last of Napoleon's bees were turned into fleurs de lys, he set about enjoying his great court, his gilded royal palaces, and his delicious food: pheasants and partridges from the royal parks, pears from the royal gardens, truffles from Piedmont, oysters from the Channel. Wanting to be seen by his people, Louis drove out every afternoon, surrounded by his *Gardes de Corps* in their tall crested helmets. In the Tuileries, where chairs had been strengthened to bear the weight of his great bulk, he was surrounded by innumerable footmen, in his red, blue and silver livery, and by the friends who had remained consistently loyal to the Bourbons. The atmosphere of deferential silence in which he liked to live – though in private his anecdotes were said to be so lewd that one diarist, the Comte de Saint-Chamans, refused to record them – was made easier by the whole web of customs and etiquette, the pomp and majesty of traditional Bourbon monarchy. In the King's presence, not even members of his family addressed each other as '*tu*'.

For the first time in over a century, Paris became a truly royal city. Louis opened his official bedroom and state apartments to the public, appointed Cherubini to compose rousing choral music, and instructed his court to turn out in their most magnificent costumes for Mass in the chapel of the Tuileries on Sundays. He ordered carpets woven with garlands, and appointed the neo-classical painter Gérard *Premier Peintre du Roi*. Enormous, red faced, smiling, with his three chins and his 'penetrating, lynx-like, look', Louis also found time to enjoy a new favourite, the 35-year-old Zoe Talon, Comtesse de Cayla, a former pupil of Mme Campan's academy. Mme de Cayla was an agreeable, plump, blonde, seductive woman, with pock-marked skin and bad teeth, for whom the King would build a pavilion of marble and acacia, with an orangerie and stables, later home to the sheep with long silky coats to which she would give her name. Mme de Cayla's husband was said to have gone mad as a result of her infidelities. Growing older, Louis spoke of the 'exhausting glories' of royal life.

Though Louis did not himself care for salons, Parisian society had never lost its taste for the intimate, subtle groupings of friends, come together to talk of politics, or literature, or about each other. When Mme de Staël died, Claire de Duras, deciding that her friend's mantle had fallen on her, invited to her new salon in the rue de Varennes foreigners, artists, ultras and politicians. Though fashionable, Claire's salon was regarded as not very relaxed, due to her tendency to lecture people rather than talk to them. It lacked, observed the guests, that guiding spirit of the 18th-century salons, where strong women like Madame du Deffand kept the tone high and the talk witty.

Claire herself had suffered a blow when her daughter Félicie, who had been left a widow at 17, had first chosen to remain with her mother-in-law, the Princesse de Talmont, and then married again, against her wishes. Auguste de la Rochejaquelin was another hero of the Vendée, whose brother Louis had died fighting the revolutionary army. Bitter about the rejection, feeling betrayed by the daughter she preferred, Claire refused to attend the wedding, though she had been present when, two weeks earlier, Clara had made what she considered a good match, to Henri de Chastellux, the King permitting him to take the old Duras title of Duc de Rauzun, there being no male Duras descendants. Claire continued to scheme on behalf of the constantly dissatisfied Chateaubriand, but the *chère soeur* was having to fight terrible jealousy towards Mme Récamier, who had become decidedly close to him. An earlier love of Chateaubriand's, Claire's cousin Nathalie, had recently lost her wits; as had his sister not long before. 'Ah my God! Poor Nathalie!' Chateaubriand wrote to Claire, with the self-regard that Lucie so deplored in him. 'How the fates pursue me! Didn't I tell you that everyone I have ever loved, known, spent time with, had gone mad?' Lucie's distaste for Chateaubriand had not been lost on the poet, who carefully erased her presence from his published memoirs.

* * *

The King's *repos*, and that of France itself, was now shattered by a former royal saddler called Louis-Pierre Louvel. In May 1816, the small, fair-haired, pigeon-toed but extremely attractive granddaughter of Ferdinand I of the Two Sicilies, the Bourbon Marie-Caroline of Naples, had arrived in Marseilles in a golden barge, with 24 oarsmen in white satin and blue-and-gold sashes, to marry the wild and quick-tempered Duc de Berri. As with Marie Antoinette and Louis XVI, her first meeting with d'Artois's younger son took place in a clearing in the Bois de Fontainebleau. The marriage, in Notre Dame, was exceptionally magnificent and the fountains of Paris flowed all day with wine. Though the new Duchess was barely literate, and showed no interest in the tutors provided for her, she was charming and high-spirited and soon lightened the leaden atmosphere of the stuffy court. She was also the Bourbons' one hope for an heir.

On 13 February, soon after Frédéric's return from Aix-en-Chapelle, a masked ball was held in Paris. It was midnight when word reached the revellers that the Duc de Berri had been stabbed by Louvel, a solitary nationalist fanatic intent on avenging Waterloo and exterminating the Bourbons. Without changing, harlequins, shepherdesses and paladins hastened to the Tuileries for news. The Duc de Berri died six hours later, but not before he had told the King that his wife was pregnant. They already had one daughter, but had lost a son and another daughter. When, not long afterwards a boy was born – the Duchess refusing to allow the umbilical cord to be cut before the witnesses appointed by the King had seen the baby attached to her – there was widespread rejoicing that a Bourbon heir had finally appeared. The King, come to see the baby, wet his lips with wine from Turançon, as was the custom.

But even while rejoicing went on over the new Duc de Bordeaux, as he had been named, a storm of anti-liberal feeling had been released by the murder of his father. While Louvel went to the guillotine, Décazes, whose moderate leanings and conciliatory attitude towards former revolutionaries were blamed for de Berri's death, was forced to resign. And though Richelieu was brought back as Président du Conseil, only to be forced out again, the ultras were gaining in power, censorship of the press was tightened up, and emergency laws passed allowing for detention, without trial, for those suspected of plotting against the state. A dry, charmless, provincial noble from Toulouse called the Comte de Villèle, who in 1814 had advocated returning to the *ancien régime* and was widely seen as one of the leaders of the ultras, was rising to power. De Berri's violent death had the effect of dissolving the centrists, and leaving just two parties, ultras and liberals, who viewed each other as bitter enemies.

Frédéric, still Ambassador to Holland and sending back regular reports to Paris on the state of disaffection of the Dutch and the Belgians, was named by Louis Ambassador Extraordinary to Spain, in order to press the Spanish King to adopt a Charter along the lines of the French one. But the British were unwilling to see further French influence in the Spanish peninsula and managed to intervene; and even before he got there, Frédéric was recalled. Italy, a mosaic of small states, in some of which revolutionary ideas appeared to be on the verge of triumphing over conservative rule, was now regarded as one of the most volatile countries of Europe. Frédéric, whose performance in Holland had pleased the King, was asked whether he would like to go as the new Ambassador to Turin.

Lucie, having once again fought back unbearable grief and despair, had resolved to take up a new occupation. She would write her memoirs. They would be not a book, for that might suggest publication in her lifetime, nor confessions, nor an essay on her opinions, nor 'the journal of my heart', but merely 'a few facts from a troubled and restless life'. Confessing that her thoughts rambled, that her memory was poor and that she had a tendency to be led astray by her imagination, she set to work on 1 January 1820. 'Let us make the most,' she wrote, 'of the warmth that is left to me, and which may at any moment be chilled by the infirmities of age.' Lucie was 50. Over the next 30 years she would write one of the finest memoirs of the age.

CHAPTER SIXTEEN

A Pocket Tyranny

By 1820, with Europe at peace after many years of revolutionary upheavals, travellers were again making their way south across Europe and into Italy. They came to see for themselves what had happened to the places they knew from the Grand Tours of the 18th century, to spend the winter months in hotels in Florence and Lucca, to explore the excavations at Pompeii, and to follow Childe Harold through the ruins of Rome. Napoleon had opened the Corniche along the French Riviera, but for Frédéric, who travelled on ahead of the others, as for most of those coming from the north, the route to Italy remained over the Mont Cenis pass to Susa and then crossing the Duchy of Savoy. Officially informed of his new posting in April, it was August when he set out. Frédéric was evidently much in favour with Louis XVIII, who wrote to his dear 'brother and cousin' Victor Emmanuel I, that he was sending him a man of 'zeal, prudence and great loyalty'.

Lucie and Aymar, pausing on their way to see Charlotte in Berne, where Auguste had been made Minister for Holland, joined him in the early autumn. Lucie was extremely, and anxiously, attached to her last surviving daughter. In her cool, clear-eyed way, she described her as not exactly witty or loving or even of irreproachable behaviour, but '*divertissante*', fun to be with, which was immensely important to her since she found so many people tedious. Their journey over the Alps took three days. Travellers compared the French stagecoaches to slave ships or the Black Hole of Calcutta for heat, stuffiness and bad smells, six people crammed inside an airless box, with a wicker basket like a hen coop above their heads, stuffed with coats and bags.

The diligence left Lyons at seven in the evening. Until midnight it climbed slowly. Lucie and Aymar woke to farms, thatched cottages and the sloping valleys of Savoy, and stopped for breakfast and a thorough customs search at the border town of Pont de Beauvoisin, where travellers often forfeited books considered inflammatory by the police. At Les Echelles, they exchanged their slender French horses for six large, strong-boned animals with high haunches. And, as night fell, they entered a tunnel at Saint-Christophe-la-Grotte, emerging into 'an entire kingdom of mountains', with snowy precipices and sunless valleys. From the top of the pass, looking down, could be seen waterfalls and, far below, horses and carts labouring up the track towards them, 'like crows or flies'. When the diligence finally reached the plains of Lombardy, Lucie found them to be covered in mulberry trees, Indian corn and vines of Muscat grapes, strung in festoons between the trees. The long, straight road leading to Turin, where they encountered minuscule carts, drawn by large dogs, was shaded by elm trees, and in the distance, rising above the plain like a pointed cone tipped with snow, could be seen Monte Viso, whose thick forests sheltered wild boar, renowned for their size and fierceness.

Both Frédéric and Lucie were delighted by Turin. They took a house on the edge of the city, on a hillside overlooking the Po, with views across the plains to the Alps where snow lay on the peaks all year round. Their 'panorama', Lucie told Félicie, was 'the finest that I have ever seen'. A Roman colony under Augustus, for centuries destroyed and rebuilt, deserted and repeopled, Turin, the smallest royal capital of Europe, was quite unlike other Italian cities. Built on a grid, its streets wide and absolutely straight, intersecting each other at right angles and running in direct lines from gate to gate through large and imposing squares with arcades, it had none of the charm and lightness of the Tuscan towns. Its predominant colour was neither yellow nor ochre but grey; its feel was northern, even somewhat French. But Lucie liked its handsome houses, with their white or striped awnings over the windows to keep out the summer heat, and the way that the well-paved streets were kept clean by sluices of crystal-clear water. She felt relieved by its sense of sobriety and orderliness.

On the ground floor of their rented house were the hall, kitchen, offices and dining room, where Frédéric hung the portrait of Louis XVIII by Gérard, a personal present from the King; above, there was a drawing room and bedrooms for themselves, the children and guests. There was also a small chapel attached to the house.

Lucie soon settled into a routine. She rose at 7.30 and went to Mass; then came a lengthy, slow toilette, before breakfast at 10, a substantial meal of kidneys and eggs, taken on her white-and-blue English plates. While Frédéric went into Turin to pay visits – having been instructed by the French Foreign Ministry to present France in a grand and powerful light – Lucie worked with Aymar on his lessons, sewed, painted and wrote letters. Her plan, she explained, was to write at least three times a week to Charlotte. From time to time, she worked on her memoirs.

At two o'clock, one of the footmen came to offer lunch; she had soup or some fruit and sometimes a small glass of vermouth, which she considered rather bitter. In the evening, when Turin came alive, there were expeditions into the city, to receptions or to the opera, where the *castrato* Giovanni Battista Velutti and Domenico Donzetti, a '*tenore robusto*', 'sing like angels, but to whom I alone listen', as she wrote to Félicie. The opera, she explained, was the centre of social life, but everyone chattered and paid visits to each other's boxes, pausing only to listen to their favourite arias. The stage of the Teatro Regio was enormous, large enough to accommodate, as happened with the more spectacular performances, an entire troop of cavalry. Turin in winter, Lucie soon discovered, was very cold, and often foggy, thick mists rising from the river, but she felt well. The air was clean, and in spite of eating red partridges – a delicacy she was dubious about – she suffered from no stomach upsets. 'We are very happy here,' she noted. 'All I ask is that we be allowed to remain.'

For a French ambassador in 1820, Turin was an important posting. The patchwork of states that was Italy was simmering with cross-currents of revolt. In 1720, the then Duke of Savoy in Piedmont had acquired half-wild Sardinia, and because Sardinia was a kingdom, the dukes were allowed to incorporate a royal title. In 1796, Piedmont had been one of the first places to fall to Napoleon, who, along with knocking down Turin's ramparts, had eventually installed his sister Pauline and her husband Prince Borghese as governors, ordering them to replicate the manners of the French court and enforce the Code Napoléon. When, in 1815, France was defeated, the Congress of Vienna, in which Frédéric had played a part, reneging on an earlier promise of independence made to Genoa, ceded not only Genoa, but also a strip of the Riviera, to the Kingdom of Sardinia.

King Victor Emmanuel I, returning from his long exile, immediately set aside the Napoleonic Code, reintroduced pre-revolutionary laws, re-established feudalism, sacked Napoleon's teachers and put education back in the hands of the Jesuits. He returned Jews to their ghettos. Long after much of Europe had abandoned them, Victor Emmanuel wore wigs. On learning that the King of Bavaria had become a liberal, and that the King of Prussia had promised his people a Constitution, he declared: '*Io solo sono veramente re*' – 'I alone am truly King'. Piedmont was again the least enlightened part of Italy. 'Of all the little despotisms of Italy,' remarked Lady Morgan, continuing on her travels around Europe, 'Piedmont seems the most complete, perfect and compact; in a word, a "*despotisme de poche*",' a pocket tyranny.

Frédéric, quickly detecting an undercurrent of rebellion, forwarded to Paris a recently published pamphlet in which the people of Piedmont 'supplicate their King not to . . . treat them as the people least worthy of liberty of all the nations of Europe'. Frédéric wrote often and at great length, letters that were frequently full of passion and seldom very diplomatic. France, he insisted, needed to take a leading role in Italian affairs, and not stand by while others did so.

The idea of a united Italy, free of foreign interference, had been around for a long time; both Dante and Machiavelli had visions of what a liberated Italy might look like. Yet foreign rule had not always been unpopular, not least because it checked the absolute power of local tyrants. There were many Italians who had welcomed Napoleon's invasions, seeing in him a force against the Church and feudalism, and the harbinger of a new age of equality. The Congress of Vienna, which left Austria powerfully entrenched all over Italy, had by contrast pleased some of these absolutist rulers, whose ancient privileges the Austrians upheld. But as the tensions triggered by the suppression of Napoleon's more liberal policies began to spread among the existing secret revolutionary societies, the Carbonari, so the Austrians became more repressive in those parts of the country where they held sway: Tuscany, Parma, Modena and the Italian kingdom, which ran from Venice to Ancona.

Early in 1820, some six months before Frédéric's arrival in Italy, a revolution had broken out in Spain and a liberal Constitution had been successfully imposed on the King, the Bourbon

Ferdinand VII. Inspired by this, the Neapolitan Carbonari unleashed a similar insurrection against the Bourbon Kingdom of the Two Sicilies; the people of Naples rose up and joined them. 'Will they defend themselves or not?' wrote Lucie to her goddaughter Félicie who, since Cécile's death and Charlotte's marriage, had become like a daughter to her. 'For my part I believe that they will be subjugated but they will not be submissive.' A Constitution, on the Spanish model, was proposed. But Ferdinand, while publicly accepting the Constitution, privately called on Metternich and the Austrians to come to his assistance. With the help of Austrian troops, the rebellion was quickly crushed, the Constitution was abandoned and absolutist rule restored.

Frédéric, keeping a close eye on these events from Turin, continued to insist, in sometimes intemperate language, that the French should take a stronger position. To the fever of liberalism abroad in Europe, the Russians and the Austrians proposed military crackdowns. The French, pushed this way and that by the ultras and the liberals, vacillated. France, observed the Tsar, now inspired 'neither fear in her enemies nor confidence in her friends'. After Naples, Frédéric warned his superiors in Paris, there was a very real danger that revolt would spread to other parts of Italy, and particularly to Piedmont where, he said, the Austrians were much hated. 'A volcano has been lit,' Lucie remarked, and it would not be easy to put out. In his reports, Frédéric continued to complain bitterly that he was receiving no orders from Paris.

In Piedmont, as Frédéric was discovering, secret societies had become fashionable, not only among students, but among some members of the Savoy aristocracy and the army. On 10 March 1821, students in Alessandria, some 40 miles east of Turin, joined forces with young soldiers and rose up against the King. Victor Emmanuel put up little resistance. Very late on the night of 13 March, Frédéric was summoned, together with the entire diplomatic corps, to the palace, to be told that the King had decided to abdicate. By five next morning Victor Emmanuel was already on his way to Nice, having handed the crown to his brother, Charles-Felix, Duke of Genoa, and the temporary regency to his 22-year-old nephew, Prince Charles-Albert, Duc de Carrignan. Frédéric liked the young prince and knew that his sympathies lay partly with the Carbonari. He requested, and was granted, an audience with Carrignan; the young man assured him that he would do nothing to disgrace his ancestors, and that he was ready, if need be, to die. 'It is not a question of dying,' Frédéric told him, 'but of living and ruling.' In his letters to Paris, Frédéric boldly recommended taking the initiative in promoting a united Italy. If France played its cards right, he added, it might even gain Savoy and Nice for itself. Was it conceivable, he asked, that France was really prepared to sit by and see Austria reign all the way from Vesuvius to Mont Cenis?

It seemed that France was. Frédéric continued to receive no orders. Carrignan, leant on heavily by the Russian minister, resigned the regency and left Turin. Charles-Felix followed the example set by the Bourbon King of Naples and called on the Austrians to suppress the revolt. A skirmish at Novaro gave an easy victory to royalist and Austrian troops. Frédéric, saying that he was witness to much bad faith and many intrigues, wrote that he feared that the Austrians and Russians between them would 'devour this admirable and unhappy country'. Three hundred students continued to hold out in a citadel, and a regiment which had gone over to them still held 37 pieces of artillery. It would indeed all be laughable, he wrote, were it not for the fact that it was bound to end in bloodshed. As, indeed, it did: Austrian and Italian troops loyal to Charles-Felix attacked, the students were routed and a number of ringleaders were executed. A company of Grenadiers, noted Frédéric sourly, had marched 'against a fistful of children'.

At some risk to himself, and disregarding the orders of the Foreign Ministry in Paris, Frédéric provided a number of the rebels with French passports, enabling them to escape. Charles-Felix arrived in Turin to assume his throne, having become more than ever a devoted ally of Austria, which announced that it would leave a garrison of 10,000 men in Piedmont. When the French Foreign Ministry finally pronounced on the uprising, it was to uphold Austria. Frédéric's role, he was instructed by Paris, was simply to convey to Charles-Felix that Louis XVIII would support no one 'who defies the authority of the legitimate sovereign'. All that was left to Frédéric was to snipe against the Austrian Ambassador, Baron Franz von Binder, referring to him as a 'real madman'; for his part, von Binder wrote to his own superiors that Frédéric was a man of 'well below average abilities'. Walpole's friend Miss Berry, passing through Turin on her travels, went to lunch with

Lucie and Frédéric, and heard accounts of the revolt and the return to feudalism. 'To hear many persons talk here,' she observed, 'one would suppose oneself in the 13th century.' For her part, Lucie wrote sadly to Félicie that she was exhausted by all the 'acts of cruelty and vengeance' that she had been forced to witness. 'Most men are ugly creatures. I much prefer outright villains to sly traitors: the former horrify me but the latter disgust me.'

Frédéric's original appointment to Turin had come as something of a surprise to those who knew how much he was mistrusted by Metternich – now Chancellor of Austria – and the French ultras. After his outspokenness over the insurrection, it was widely thought that he would be recalled to Paris. Metternich lost no chance to run him down. Reports sent back by the Austrian officials in Piedmont described Frédéric as 'lacking in character' and much influenced by his domineering wife. But Frédéric held on; when it was suggested to him that he would do well to request a transfer, he did nothing.

Paris now appeared to lose all interest in Piedmont, leaving Frédéric for weeks on end without replies to his letters. His position was made all the weaker by the fact that both the court in Piedmont and the ministers had turned against him. Even so, he continued to speak out, to complain, to criticise and to send back reports full of disquisitions on the nature of diplomacy and the wickedness of the Austrians. There was an admirable stubbornness in him, a refusal to be silenced. 'The state needs the truth,' he declared, 'it is for this that we have been appointed to our posts.' But it did not make either his or Lucie's life easy. And it was perfectly true that Lucie seldom held back from voicing her own opinions, strongly held and forcefully expressed. Neither one of them was diplomatic by nature; whether in Brussels, Amiens or Turin, they stood apart for the reckless ease with which they challenged political decisions they considered to be lacking in morality or common sense. And neither minded greatly what anyone said about them.

* * *

Once the novelty of Piedmont wore off, Lucie found Turin extremely dull. 'There is very little wit and very few ideas in this country, as far as I can see,' she wrote to Félicie, telling her that all the Savoyards did was to moan about their health and warn each other to avoid sun and wind and to keep the windows shut. It was, she said, the 'nec plus ultra of monotony and boredom'. Though French remained the one common language of Piedmont, and all visitors were struck by the French tone of the court, where women dressed in French fashions, danced French quadrilles, ate French food and had salons where they discussed poetry, the city soon seemed to her stultifying.

After the carnival, when boxes at the Teatro Regio were furiously fought over, the city slumbered under a stifling routine of strolls or rides up and down the Via del Po, the ambassadors and courtiers bowing to each other from their carriages. There were starchy visits and games of whist. Lady Morgan noted that while the ladies of Turin engaged in the 'pleasant persiflage' of the French, the older women had a disagreeable tendency to show off too much of their 'tanned and prematurely withered necks'. Houses, magnificent on the outside, were chilly and sombre within; floors were made of stone or marble and there was seldom any heating. Lucie begged Félicie to send her large, cheerful lamps, and wall brackets for candles, in order to bring some proper light into the gloomy rooms of her house, made darker by their heavy damask curtains and mahogany furniture.

In June 1821, came the news that Napoleon had died. The Emperor had lived for five and a half years on St Helena, with the same etiquette and formality as on Elba, telling his secretary Emmanuel Las Cases, to whom he was dictating his memoirs, 'I closed the gaping abyss of anarchy, and I unscrambled chaos.' Las Cases, forced to leave St Helena and return to France, had been spreading word of Napoleon's martyrdom around a France already somewhat nostalgic for their heroic Emperor. On St Helena, Napoleon had become increasingly unwell, apparently suffering from stomach ulcers and a liver complaint. He had seen very little of Lucie's half-sister in recent months, but Fanny was with him, with her children and all his courtiers, when he died on the afternoon of 5 May, and she kissed his hand. There was some talk of poison, but a malignant tumour was more widely thought to have been the cause of death. His heart was removed and placed in a silver box; Napoleon had requested that it be sent to his wife, Marie-Louise, but Sir Hudson Lowe, dogmatic to the last, refused. Fanny's 13-year-old son Napoleon walked in the funeral cortège with his father Bertrand, holding one corner of the cloth draped over the coffin.

Bertrand, constantly nagged by Fanny to leave St Helena, where, she said, the children were growing up wild and uneducated among the soldiers, had managed to stay there until the end. With Napoleon dead, the family hastened back to Europe, arriving in Portsmouth to be welcomed by Lady Jerningham, who found her niece very thin in her dark mourning clothes, and somewhat stooped. But Fanny, as she told her daughter Charlotte, had an '*air distingué* and is very agreeable, both in French and English'. The devoted Bertrand was clearly much upset by Napoleon's death. The Bertrands rented a house in the Edgware Road, while overtures were made to the Duc de Fitz-James, Fanny's brother-in-law and a man with some influence at the French court, to get the sentence of death imposed on him after Waterloo lifted, so that they could return to France. Lucie's sister brought back with them from St Helena an exquisite bird cage, built for Napoleon by Chinese artisans, without nails, each wooden piece fitting perfectly into the next, and constructed with three separate floors, one for each of three kinds of bird. The cage had originally been delivered with an eagle perched on the top, but Napoleon had asked for this to be removed, saying that it looked absurd.

* * *

The fourth anniversary of Cécile's death fell on 20 March 1821. Charles de Mercy-Argenteau, the young man she was to have married, came to Turin to stay with Frédéric and Lucie, who were both devoted to him. Like Lucie, Charles loved music, and together they went to hear Velutti sing. In July Charlotte arrived from Switzerland to join them, bringing with her Hadelin, now 5, and 3-year-old Cécile. For a while, Lucie again felt herself to be surrounded by all the pleasures of family life she so enjoyed. They rode out in the cool of the evenings, with Charles and Charlotte on horseback, and Lucie and the children in her barouche. At night, the young people played whist with Frédéric while Lucie did her petit-point. 'I revel in being loved by you, by my excellent husband, by my daughter and by my dear Charles,' she wrote to Félicie, adding that Charles was like a 'tender and devoted son' to her.

But soon, Lucie detected a change in her only remaining

daughter. The once lively and irreverent Charlotte was becoming thin and listless. She and the children stayed on in Turin all through the winter of 1821. She was often feverish and shivery, and to Lucie she seemed to be growing weaker. 'I find her terrifyingly pale and extremely frail,' she told Félicie, saying that she was now being 'devoured by a mortal despair', and that no two days passed without 'worry and torment'. She had once again forgotten what it was like to feel safe. Fretting constantly about the cold damp air, Lucie longed for the spring to come, so that she could take Charlotte up into the mountains; when the weather became warmer, she was sure, Charlotte would get better.

However, Charlotte, like Cécile before her, grew weaker. In the summer of 1822, she returned to Switzerland, accompanied by Lucie and the children. It was there, in the Château de Farblanc near Evian, that she died, on 1 September. She was 26. Not long before, Charlotte had written in her diary:

> I have often felt the desire to write, as if I were being pushed by some unknown force. I have started twenty novels in my head, but I hate novels, apart from a few exceptions, so how could I possibly write books of the kind that I usually find so reprehensible and miserable? At other times, I think I would have liked to discuss morality, but what a ridiculous idea at my age! I don't believe that one should tackle such a subject until the age of passions has passed and one can be the proof of what one wants to say.

She did not have the chance to find out.

With Charlotte's death, Lucie and Frédéric had lost five of their six surviving children; only Aymar was left. Even for the age, it was an exceptionally cruel number of losses. For the rest of that year, Lucie was extremely low. She seemed to shrink into herself, silent and enduring. Her letters to Félicie ceased. When she did at last write it was to say that she now clung to her goddaughter as her 'dearest confidante' on whom she would depend 'all the days left to me in this sad life'. She did her tapestry and struggled on, so depressed that she felt unable to help anyone around her, even those she loved.

But then, early in 1823, Charlotte's husband Auguste came to stay, bringing with him Hadelin and little Cécile, and Lucie, who in the past had often found her son-in-law remote and difficult, drew closer to him. When he set off back for Switzerland, he took Hadelin with him, but asked if he might leave Cécile for Lucie to bring up. Very slowly, warmed by the presence of the little girl, she began to revive, telling Félicie that Cécile was almost too intelligent but that she could be sulky when crossed. Even with the child by her, Lucie still never left the house without seeing the world through Charlotte's eyes. 'There is a deep layer of grief in my heart,' she wrote, 'which colours everything.'

When, not very long afterwards, the Princesse d'Hénin died, she left Le Bouilh – which she had earlier bought in order to help Frédéric and Lucie with some of their financial difficulties – and all her money to Aymar. Even if the older woman had never liked her, Lucie felt that she would now miss her. For Frédéric, his aunt's death had severed the last surviving link with his childhood.

For a while, the family's finances seemed to improve, and both she and Frédéric felt intense relief that at least Aymar's future was assured. But the Princesse's lawyer fled to England, taking with him all Mme d'Hénin's fortune, and leaving debts and legacies that could be paid only by selling Le Bouilh. His elderly cousin, Mme de Maurville, to whom Mme d'Hénin had left some money, was again penniless, and Cécile's legacy of 50,000 francs had disappeared. Lucie and Frédéric would now face ever-increasing financial anxieties, made worse by Frédéric's extreme impracticality when it came to money. Nor were their worries lessened by the fact that constant attempts were being made to remove him from his post, most of them not by the Ministry for Foreign Affairs, but by envious and scheming people in Paris, anxious to secure the Turin embassy for relatives or friends.

In May 1824, Lucie and Frédéric celebrated their 37th wedding anniversary. 'I never stop to wonder whether my husband is more or less intelligent, knowledgeable or capable,' Lucie wrote to Félicie. 'But not a single day goes by when I do not detect, by some word or glance, how much goodness, greatness of character and nobility there is in him; and I hope that he can say the same for me.' It had been, and remained, an exceptionally happy marriage.

* * *

By early 1824, it was clear that Louis XVIII was dying. He had gangrene in his right foot and up his spine and he fell asleep in public, his head lolling, his clothes hanging loose over his shrinking bulk. 'The King,' wrote Heine later, 'rotted on his throne.' Soon he no longer left the Tuileries. 'A King is permitted to die,' he told his courtiers, 'but he is never permitted to be ill.' During the summer he struggled on, receiving his ministers with his head propped up on a cushion on his desk. On 12 September, orders were given to close the stock exchange and the theatres. It was his companion, Mme de Cayla, who managed to persuade the dying King to take the last sacraments. A large, silent crowd collected outside the Tuileries. On the afternoon of 16 September, in the presence of ministers, ambassadors, courtiers and deputies, Louis XVIII died, his body decomposed and smelling strongly. Dressed in a coat of mail, with gauntlets and spurs, his embalmed corpse was displayed for a month before being lowered into its tomb at Saint-Denis.

The reign of Louis XVIII, which had lasted, apart from the Hundred Days of Napoleon's return, for just over 10 years, had been surprisingly prosperous. For all the bitter rivalries between ultras and liberals, the shrewd and subtle Villèle, with his long nose, pock-marked skin and nasal accent, had steered a deft course between them with caution and subterfuge. In 1822, Villèle had been ennobled and made President of the Council. He was neither eloquent, nor imaginative, nor brave, and he knew nothing at all about the arts or the sciences, but he was tenacious, instinctively attuned to politics, and he was not easily rattled. Though Pozzo di Borgo, the Tsar's powerful Ambassador to France, insisted on talking about the 'decrepitude of an ageing monarchy', France had in fact grown steadily richer and more stable. After Britain, it was now the most active economy in the world.

All this was about to change, and the changes would ultimately bring devastating consequences to Frédéric and Lucie; though not quite yet. A new political world was emerging, heavily influenced by the right-wing ultras, and it was, once again, one that neither of them found sympathetic. The ultras came in many shapes – moderate, passionate, clerical, mystical, intransigent – but what they had in common was that all longed for the *ancien régime*. As the deputy Louis de Bonald put it, the *ancien régime* had been the most moral, spiritual and perfect age, and all the years since then had been mired in the 'spirits of the shadows'. The revolution had been a time of madness, a monster 'feeding on cadavers'; the Charter had been a foreign perversion and had to be repealed. What was needed was careful censorship (to stop peasants acquiring inflammatory ideas), judicious education (by priests), a return to a powerful landed aristocracy and above all the crushing of the new meddling, ambitious middle class. These views were anathema to Frédéric and Lucie.

As brother to the King, the Comte d'Artois had been content to bide his time, gathering around him ultras. Crowned Charles X in Rheims, in full panoply of regal splendour, he intended to return France to a better age of chivalry, and to rule by divine right, according to religion and its truths. Lazy, ignorant, stubborn, but neither cruel nor unjust, Charles X wished to govern as a powerful monarch, through the intermediary of his loyal servants, men who, like him, regarded kingship and Catholicism as inseparable, and the emerging middle class as nothing but a bothersome Third Estate. For himself, he preferred to go hunting. Though 67 years old, with a long face, short white whiskers and a slightly bovine expression, he was both fit and sprightly.

Because he liked to please, Charles did not go in for confrontation; because he preferred ease and comfort to formality, he surrounded himself with charming disorder, and made himself available to everyone, even regicides, defrocked bishops and Napoleon's former maréchals. At the coronation, Talleyrand was again much in evidence, in a new carriage with his quarterings of three crowned lions, on hand as Grand Chamberlain to slip the King's feet into their purple velvet boots. But behind the new King's ease lay a clear purpose: it was to eradicate the last vestiges of revolutionary spirit, to cast Voltaire, Rousseau and the Enlightenment *philosophes* into the outer darkness, to clean up the godless universities, and to wage war on all the lawyers, teachers, entrepreneurs and industrialists who had risen up and made their fortunes. Natural rights, warned the Comte de Frémilly, who had watched the sans-culottes parade the heads of his friends on pikes around the streets of Paris in 1789, were a chimera. The new Directeur des Beaux Arts ordered that the naked statues in the Tuileries be dressed, and told actresses and dancers at the Opéra that if they wished to please him, they needed 'loose trousers, and some morals'.

From his embassy in Turin, Frédéric had seen French foreign ministers come and go; none had shown much interest in the affairs of Piedmont. After Claire de Duras had worked to secure the Foreign Ministry for Chateaubriand, he had passed through Turin to see Frédéric and Lucie. She wrote to Félicie to say that he had spent many hours talking and that she felt exhausted at the thought of the turmoil into which Claire would be plunged once Chateaubriand was actually in power. 'I have always been horrified by the spectacle of women meddling in the affairs of state,' she wrote, in a letter that said more about herself than about Claire, 'because it is my belief that they spoil everything, inflame everyone, and contribute only small insights and small passions.' Her words were disingenuous: though in public Lucie was scrupulous in her deference to Frédéric, behind the scenes she remained forceful and not above a little meddling herself.

With Chateaubriand at the Foreign Ministry, Frédéric's position was for a while relatively safe, but when, in August 1824, the arch conservative Duc de Damas was brought in, there were fresh calls for his dismissal, the ultras in Paris pressing for their relations to be appointed in his place. Though, wrote Lucie, 'I cannot think why as [Turin] could please only people who are as sad, solitary and unhappy as we are'. Frédéric wanted to stay on in Turin until the beginning of 1829, when he would be 70, but he remained unpopular, not least because he still never missed a chance to complain about the Austrians. When he asked de Damas whether he might appoint Aymar, who had just turned 18 and was, according to Lucie, 'spirited, sensible, tactful and possessed of beautiful and noble manners', as second secretary in the embassy, he was informed curtly that his son would have to join the queue at the bottom.

With Cécile, Lucie was rediscovering the pleasures of teaching; she found the little girl sweet-natured and gentle and took care to organise her life very carefully, saying that she believed that 'method and precision are both useful and necessary to women'. These were virtues she had held to all her life. Lucie had developed rheumatism in her knee but refused to go to one of the many spas that had become fashionable, insisting that she loathed such places. Bored by the company of the two young men attached as secretaries to the embassy, she preferred to spend her time with the Mother Superior of Les Dames du Sacré-Coeur, a woman who reminded her of 'my poor beloved Charlotte', or with the clever, witty, Comtesse Valpergue, who had a château in the Val d'Aosta. From time to time, Frédéric was obliged to accompany Charles-Felix and the court to Genoa. Left with one of the attachés, M. de Marcieu, a talkative and amiable young man, but very pedantic, Lucie wrote to Félicie that he reminded her of someone who had a great many clothes in his suitcase but who chose to go about naked. 'Perhaps I am a bit like that myself? I am full of ideas, but I do not dress myself up to show them to all the world.'

Describing herself at around this time, writing with candour and humour, Lucie noted that her hips were rather wide – 'of huge dimensions' – that her waist was disproportionally narrow, and that, because her lower back was weak, she was obliged to wear a corset. Since she felt uncomfortable having anything tight around her neck or shoulders, what would suit her perfectly, she concluded, was to go around with an armoured girdle below her waist, and to remain naked above. She chose, now, to dress only in black, without ornament of any kind, having decided that grey tended sooner or later to go to lilac and look shabby. All vanity, however, had not deserted her: to Félicie she wrote somewhat smugly saying that her figure was still that of a woman 20 years younger, but that she preferred to keep it hidden. When Adrien de Laval, who had been suggested as a possible husband for her in 1787, came to Turin and they met for the first time in many years, she wrote that 'we each found the other to be old: for my part, consenting to be so, while he does all he can to keep age at bay'. She was 55. A portrait showed her smiling slightly, with a rueful, quizzical expression, in the all-enveloping black dress she believed appropriate for her advanced age.

By the summer of 1825, Frédéric had been ambassador in Turin for five years. Though the city was not totally without its pleasures, particularly at carnival time, when Lucie had one of the much coveted boxes at the Teatro Regio, both of them longed for a break in their routine. Charles de Mercy-Argenteau, to whom they had remained extremely attached, had taken holy orders and they very much wanted to attend his first Mass in Rome. They also wished to spend a winter away from the snow and the icy winds of Turin, which, Lucie complained, made the city seem more like Russia than Europe. They had been saddened by the recent death of one of the young attachés in the embassy, whom they had nursed through typhus, but had not managed to save. The Duc de Damas, whom Lucie regularly referred to as a 'monster' or an '*idéophobe*', someone allergic to ideas, when asked for his permission, said that they could take a long holiday only on half-pay, something their constant financial worries made impossible.

Aymar went off to Rome on his own, and Lucie spent five weeks on the lake at Evian with Auguste and the children, where, she wrote sadly, everything reminded her of Charlotte. But in Rome Aymar was introduced to the Princesse d'Esterhazy, who was close to the French court, and when the Princess wrote to Paris describing Lucie's desire to visit Rome, de Damas felt obliged to offer Frédéric leave, on full pay. Early in October, they set out for the south.

* * *

For many travellers in the 1820s, reared on the classics and Goethe's *Italian Journey*, Rome remained the symbol of the Grand Tour and the city they most wished to explore. Since the journey from Paris took 28 days by public diligence, visitors tended to stay for several months, crossing the Alps before the snows came and spending Christmas and Easter in Rome. The Holy City was said to be 'pestilent with English . . . wishing to be at once cheap and magnificent'. The Piazza di Spagna was widely known as the Ghetto degli Inglesi, and the word *inglese*, 'English', had become synonymous with 'foreigner'. They came, the English, French, Germans and Russians, to sketch, to look at churches and art galleries, and above all to wander in the ruins, absorbing what the Marquis de Custine called '*le parfum de classicisme*'. The Romantics too were discovering Rome, and visitors had ceased to regard the Romans solely as custodians of past splendours but rather saw them as a people struggling for unification.

Though Napoleon had outraged the Romans by his removal of the Pope and annexation of their city, and appalled them with his overhaul of laws, prisons and the postal service, the men he had sent as governors had proved conciliatory and astute. During the years of French domination more had been done to embellish the city than had been done by other governors in a century. Avenues of trees had been planted, monuments cleaned and restored, and the Pincio gardens laid out.

Lucie was charmed by Rome. After the severe right angles of Turin, she loved the 'complete disorder of the buildings', the way the narrow streets wound around the houses, the fact that no two were alike, and the soft and changing colours of the city in the southern light. The contrast was so striking, she wrote, that everywhere she walked, she felt surprise and pleasure. Adrien de Laval had been appointed Ambassador to Rome and it was with him that they met Mme d'Esterhazy and her six children, and Lucie was soon dreaming wistfully that one might make a good wife for Aymar, though the ages did not quite fit and there was no fortune on either side.

After Turin, Rome was extremely lively. In 1815, Pope Pius VII had offered a refuge to Mme Mère and the Imperial family, and Napoleon's sister Pauline, Princess Borghese, had made Rome her home, holding court in the Villa Paolina by Porta Pia until her money, health and looks ran out and she retired to the better air of Florence to die. Mme Mère, until her own death in 1836, lived in splendour in Piazza Venezia, where she was said to sit at her first-floor window to watch the crowds strolling below. Shortly before Lucie's arrival, the ex-queen Hortense, in whose house Lucie had arranged Fanny's wedding, had arrived and opened a salon to which every new visitor sought an entrée. And in October 1823, made miserable by Chateaubriand's dalliances with two new favourites, Mme Récamier had come to settle at 65 Via Babuino, in a house opposite the Greek church. She was still very beautiful, in her white dresses and blue shawls, still graceful as she undulated her hips in the shuttered light of her salon; in comparison, said David's pupil, Etienne Delécluze, Italian women looked like 'little savages'. After the starchy receptions in the dark palazzos of Turin, the light and frivolity of Rome were intoxicating to Lucie. There were also expeditions to Frascati, Albano and Tivoli, though visitors were warned against the *mal'aria* of the Campagna. At a ball given in the Palazzo Torlonia, 16 English aristocrats came in matching outfits and plumes and danced a *contredanse*. Writing to Félicie, Lucie reported that the gossip of the city was that by the early 1820s both Pauline and Hortense had decided that they wished after all to live with their husbands, neither of whom could abide them.

Charles de Mercy-Argenteau had been made a chamberlain at the Vatican by Leo XII, elected Pope in 1823. When Lucie and

Frédéric were forced, reluctantly, to return to Turin in the middle of February 1826, they left Aymar with him. Lucie continued to hope for a diplomatic posting for her last remaining child, having decided that his particular mixture of curiosity and reserve, and his good knowledge of foreign languages, made him the perfect candidate. There had been a brief hope that Frédéric might exchange the embassy in Turin for that of Rome – much promoted by Lucie – but it came to nothing. When, not long afterwards, Chateaubriand was appointed ambassador instead, there were complaints that the once bustling French Embassy became sombre and silent. Mme Chateaubriand was said to be retiring and pious, while the poet was an uneasy host and more interested in himself than in his guests. Before Lucie left Rome, she sent Félicie a piece of stone as a paperweight, saying that she had picked it up on the path that Cato had walked when visiting Cicero. She looked on her and loved her as her own daughter, she said, then added: 'But I hardly dare say this to you, because coming from me the name of daughter has a ring of death.'

* * *

One, perhaps inevitable, result of Lucie's long and close marriage to Frédéric, of her absorption in her children and the tenderness she felt for Félicie, was that she had never devoted much time to exploring close friendships. Born early enough to have observed the dying days of the great 18th-century salons, to have heard for herself conversation elevated to an art form and used as an expression of affection and intimacy between people who wanted to please and amuse their friends and who delighted in one another's company, she considered most of the people she had met since pale reflections of that subtle and lost world.

Circumscribed by formality and manners, seldom settled for long in a place she regarded as home, repeatedly battered by the loss of the children she loved, she had formed few deep and lasting attachments with either men or women with whom she might have developed trust and understanding. Her insights into character, her obvious enjoyment of wit and intelligence, her humour and occasional sharp tongue, and her very real generosity of spirit would have made her a most rewarding friend; but true friends had not come her way. The one exception, someone who had exasperated and tormented her for years, whom she criticised constantly yet could not quite bear to let go of, was Claire de Duras.

For Claire, the Restoration had brought worldly success, money and, through her husband's position at court, power. But it had not brought happiness. Chateaubriand, who was probably incapable of real love, had eluded her; Félicie, the daughter for whom she felt most attachment, preferred her mother-in-law, the Princesse de Talmont. And her ceaseless intriguing on behalf of friends and relations had ended in ridicule. Having nagged Chateaubriand incessantly to promote Clara's husband to the post of head of a political department, for which he had neither the skills nor the experience, she was obliged to see him ignominiously dismissed.

As early as 1813, when she declined to attend Félicie's second wedding on the grounds of ill-health, Claire had taken refuge in being unwell. Whether her physical ailments were real, or the product of her frantic and tormented mind, it is impossible to know. But they caused her painful self-doubt, wounded pride and the sort of tricky changes of mood that she visited on Lucie, along with unbearable, all-devouring envy. The fact that Félicie and Lucie had become so close, driven into each other's arms by Claire's intriguing, must have been extremely hard to bear.

By the middle of the 1820s, Claire and Lucie had not seen each other for more than six years. As Lucie noted sadly, 'You cannot command friendship.' Lucie's sympathy for her old friend remained both sincere and shrewd. Hearing how acutely jealous Claire had become of her daughter's mother-in-law, Mme de Talmont, she wrote to Félicie: 'Your mother is much to be pitied; this illness of the heart and the mind is even harder for her than it is for other people, for they can escape or avoid it, while she, poor woman, cannot escape herself.'

But, sometime in 1818 or 1819, Claire had sought refuge in fiction. Her first novel, *Ourika*, about a black girl dying of unrequited love in an aristocratic French family, was enormously successful, in part because it touched on racial questions that were being talked about. Paris had lost none of its pleasure in turning events into fashions: there was an Ourika shawl, an Ourika colour and Gérard painted a celebrated picture of a dreamy black girl. Embroidering a rug for Félicie when she heard of *Ourika*'s success, Lucie observed wryly: 'How pathetic it is that all I can give birth to is a carpet.' Claire had sent a copy of the book, not to her, but to Frédéric. When Lucie read it, she remarked that she particularly disliked its heedless mixture of truth and invention, 'one spoiling the other'. She had little patience with lack of clarity.

However, as Lucie quickly pointed out, Claire was not really writing fiction at all. Ourika was the 'little captive' brought home from Senegal before the revolution by the Chevalier de Boufflers for his aunt, Mme de Beauvau. And in Ourika's mournful lament – 'I no longer felt pity for anyone but myself' – Claire was really writing about herself and Chateaubriand.

Claire's next novel, *Édouard*, another tale of love and unhappiness, was a similar success. This time, Lucie herself appeared as a character, in an unexpectedly sympathetic light, so thinly disguised that she might as well have been named. In the novel, a group of people are sitting together one evening talking. A Mme de Nevers is asked why she has no intimate friend. I had a friend, she replies,

> one who was very dear to me . . . We had been friends since childhood, but I fear that we have now been estranged for many years . . . The Marquis of C., her husband, is Minister in Holland. I feel her loss keenly. No one else has ever been as necessary to me . . . She is my conscience, and I have never sought to find another to replace her; now that I am alone, I can never make up my mind to anything.

Mme de C., Mme de Nevers goes on, had once tried very hard to persuade her to struggle against certain feelings that she believed to be wrong and misplaced; she had indeed tried to flee from them, but had failed, and Mme de C. had travelled all the way from Holland 'to pull me back from the abyss into which I was about to fall'. The words were the very ones written by Lucie to Claire in 1812 about Chateaubriand. But what was striking was Claire's evident sadness about their lost friendship. This barely disguised account of their estrangement touched Lucie.

All through the spring of 1826, Claire was ill. She had rheumatism in her neck, was short of breath and slept badly. There were days when she could neither read nor write. 'Death terrifies me,' she wrote to a friend, 'but I must accustom myself to the idea of it, because I feel it to be approaching with great strides.' She became depressed and dreaded seeing her friends; like a child, she wept. She begged for news of Chateaubriand.

Lucie had gone to Le Bouilh – still looking for a buyer – taking with her Aymar. Though angry that the tenants who had been living in the château had left it in a terrible mess, she was pleased to find herself back in France and among people she was fond of. The house reminded her poignantly of happier days when Claire 'used to be so fond of me'. Félicie was to have visited her, but instead went to Paris to her mother, whose health seemed to be growing worse. Lucie, telling Félicie that she should not be too upset if her mother rebuffed her, wrote: 'You must fill your heart with charitable thoughts. Just because she was not a good mother, that is no reason for you not to be a good daughter.' Then she added, again revealing more about herself than about Claire: 'How she must feel the emptiness of all that worldly success . . . all that noise that prevented her from hearing the true voice of her own heart or enjoying those natural and gentle feelings which should be enough in the life of any woman.'

In August, Claire lost the sight in one eye; half her face became paralysed. Moods of silence and self-obsession alternated with periods of tenderness and generosity. She told a friend that she felt herself to be somewhere between life and death and that she was determined to face the end with courage. 'In vain I seek to be happy: but I can no longer do it; I have suffered too much.' Now that she was clearly dying, Chateaubriand came often to see her (but was unable to refrain from writing to a friend: 'I am menaced by great unhappiness. Mme de Duras is dying.'). From Turin, Lucie offered to send packets of grissini, particularly good, she said, for invalids, in with the weekly consignment of truffles that travelled regularly from Piedmont to the King's table in Paris.

Still Claire hung on. There was a rapprochement with Félicie, which pleased Lucie though she felt sad on her own behalf. 'If I were with her,' she wrote, 'I would look after her night and day . . . But friendship is something else. Friendship, once lost, cannot be recovered.' But then the day came when Claire seemed well enough to travel south. Lucie decided to make one last effort. She went to Switzerland and, on Lake Maggiore, the two old friends met. But the encounter did not go well. Claire was silent, cold and tearful and Lucie returned early to Turin. 'I was appalled, crushed,' she wrote to Félicie.

At the end of September 1827, Claire went to stay in Nice,

taking Félicie and Clara with her. The weather was wet and very windy. Frédéric and Lucie crossed the Col de Tende to see her and this time, the meeting went very well. Claire seemed extremely touched that they had made the journey, and said to Lucie: 'My dear friend, it is as it once was between us.' It was life imitating art: in *Édouard*, Claire had written that when Mme de Nevers was ill, her friend had arrived and held her in her arms. Lucie returned exuberant to Turin. She had, she said, 'refound' her friend at last. On the evening of 16 January 1828, Claire told her two daughters that she felt grateful to God for allowing her to die slowly, for she had always feared sudden death. Barely able to speak for ulcers on her tongue, she asked for the last rites; and then she died, smiling and holding their hands.

With Claire's death disappeared one of Lucie's last links with the past, and what had been, for all its flaws and estrangements, a true friendship. Increasingly now she would devote herself to Félicie, telling her again and again, in letter after letter, that she felt for her all the love she had felt for her own daughters. 'Your friendship,' she wrote, 'is necessary for my heart.' In return, Félicie, whose nature was elusive and adventurous, was affectionate; but she kept her distance, letting months pass without replying to Lucie's many letters and very seldom making plans to visit her.

* * *

By the summer of 1827, Villèle's government in Paris was in trouble. Under siege from both the right and the left, Charles X decided to create 76 new peers and to dissolve the Chamber of Deputies. Frédéric, as outspoken as ever, wrote furiously to the ministry to say that he personally felt that nothing since the revolution had done more harm to the French aristocracy, to France itself or to the monarchy: 'You have gone against all that is natural, shaken everything up, made compromises.' Villèle's government fell. A new government was sworn in, and an attractive and likeable man, M. de Martignac, an ultra, but not a fanatic, was invited to become Minister of the Interior. To Frédéric's relief, the Ministry of Foreign Affairs went to an old friend and colleague, the Comte de la Ferronays, former Ambassador to Denmark and Russia.

The new government had no real leader, though Martignac was regarded as more or less in charge. Charles X and his close friends, it was clear, were really looking to the day when the governance of France might be returned into the hands of men of the far right, '*ultras exaltés*', who believed that France's future lay in a strong aristocracy and a powerful Church. Even Lucie, in her own way, was dismissive of the middle class that had risen to positions of power since the Restoration, saying that she would like to see noble families restored to their ancestral seats, and that it was a shame that only men she called 'industrialists' could afford them now. 'It will take more than steam engines,' she wrote, 'to create a genealogy.' When it came to leadership, all Lucie's years in America, her very real commitment to political reform, had not shaken her fundamental belief in the superiority of the aristocracy.

In 1829, Frédéric turned 70. He was now the doyen of Turin's diplomatic corps. With de la Ferronays in the Foreign Ministry, he was no longer talking of retiring, but planned to ask for six months' leave to accompany Aymar to Paris, where he hoped to find him a wife. Lucie had not been well and had been bled copiously, then had leeches applied to her head which made her look, she said, like a 'gorgon'. That summer, they went to Rivalta, not far from Turin, to a château with a moat and battlements covered in ivy; there was a river nearby, its banks lined with willow trees, and views of the Alps. Lucie had thought of letting Frédéric and Aymar go alone to Paris, while she and Cécile stayed at Le Bouilh, and she wrote to Auguste's mother, the Comtesse de Liederkerke Beaufort, urging her to visit her granddaughter 'whose upbringing is the whole purpose and consolation of our lives'.

In the event, reflecting on Frédéric's age, she consented to accompany them to Paris, stipulating only that they should take rooms, not in Saint-Germain, which Cécile might find deserted and sad, but somewhere more cheerful near the Tuileries or by the Seine. 'I want to be able to see something from my windows,' she wrote, 'I like the quays, the squares, somewhere I can walk.' Félicie was asked to find them something suitable, with rooms for the family, a maid and a valet, with an open fire over which Lucie could cook, and a good supply of wood nearby. It should also have *lieux à l'anglaise*, an English bathroom. But, she announced, she had no intention whatsoever of going out into society, to hear people exclaim, 'Ah! how she has aged!'

CHAPTER SEVENTEEN

A Warm Heart

Frédéric, Lucie, Aymar and Cécile reached Paris soon after Lucie's 60th birthday. It was their first visit for almost ten years. The winter was exceptionally cold; the Seine had frozen over at the end of November, and the thaw did not start until late in February. Sledges sped along the streets, with women dressed in thick furs. At balls and receptions, bare shoulders and arms were enveloped in shawls. There was an acute shortage of bread and at the Bal des Indigents, referred to by some as 'the ball of the rich for the poor', held at the Opéra in aid of those starving on the streets, ticket holders remarked on the brilliance of the light cast by the new gas lamps. '*La vie élégante*', with its new aristocracy of money, journalism and politics, had taken over, said visitors to France, from the old '*vie aristocratique*' of the *ancien régime*.

Spring was starting as they arrived. On the first fine days, Paris was as crowded, noisy, busy and foul-smelling from the 'malodorous muck' of the streets as it had been during Lucie's childhood. The Seine had almost disappeared beneath the number of boats transporting food, wine and charcoal to the city; along the quays were anchored bathing establishments, offering every variety of bath, from Russian to Turkish or Chinese, hot or cold, scented, with or without a massage or attendants. In the Jardin des Plantes, a giraffe, sent as a gift to the King by the Pacha of Egypt, was taken from its enclosure by its African keeper for a walk around the gardens, dressed in a wool coat. It was said to be very partial to rose petals. In Le Rocher de Cancale, Paris's 'supreme temple of gastronomy' on the corner of the rue Mandar and the rue Montorgueil, there were 112 fish dishes on the menu. Brillat-Savarin had inadvertently invented steaming, when the enormous turbot he was cooking proved too big for his largest pot and he borrowed a cauldron from a laundress, placing his fish surrounded by shallots and herbs in a wicker hamper above the boiling water.

Talma, the great heroic actor, was dead, but Mlle Mars, 'young under the Directoire, beautiful under the Empire, glorious under the Restoration', continued to draw crowds to the Comédie-Française, even if a marked taste for the ghoulish drew crowds to watch the guillotine at work, beheading dogs, at 20 francs a ticket. A 'Combustible Spaniard' was entertaining Parisians by sitting for 14 minutes in a hot oven, alongside a chicken, which cooked, after which he ate it. 'Here pleasure is a luxury, a delight,' noted one traveller, 'and not a laborious affair, as in England . . . not a slow moving canal, but a wild and bubbling brook in which the French, like corks, dance for joy.'

At court, Charles X continued to pass his days hunting, travelling between Saint-Cloud and Paris in a cloud of soldiers, bodyguards and liveried attendants, while his courtiers went about their duties, impeccably uniformed. Their wives spent a great deal of time in church. 'The good Lord,' remarked a visitor, 'is much in fashion these days.' The Duchesse de Maille, who refused to attend court on the grounds that it was too dull, maintained that to please the King one had to be 'ultra and stupid', and to please the Duchesse d'Angoulême, 'ultra and devout'. It was only the young Duchesse de Berri who brought life to the Tuileries, with her costume balls on Turkish, Persian and Scottish themes, and her new passion for swimming in the sea at Dieppe, something that few people had thought to do before. In mid-July, the Duchess, swathed in a long dark wool shift, wool trousers and a waxed taffeta cap, with boots to keep the crabs at bay, was escorted to the water's edge by a doctor. She was then led into the sea by a '*guide baigneur*' in a special uniform, watched by spectators with opera glasses from the sea front. A cannon was fired to herald the first swim of the year. In this outfit, observed one of her ladies-in-waiting sourly, even the most beautiful woman in the world looked like a 'monstrosity' as she emerged from the water enveloped in clinging, sagging wool.

There was, however, a rival court in the Palais-Royal, home once again to the Duc d'Orléans after his return from exile in 1817. Louis-Philippe was now 56, a portly, courteous figure, married to Marie-Amélie, one of Ferdinand IV's 18 children, a tall, blue-eyed woman with a long face and a long neck, who described herself as having an 'air of modest but imposing nobility'. Together, they had cleared the Palais-Royal of the rubbish that had accumulated during the Duke's exile, and turned it into a salon for writers and intellectuals. 'The Duc d'Orléans,' his cousin Louis XVIII shrewdly once remarked, 'remains absolutely still, but nevertheless I notice that he is moving forwards. What does one do to stop a man who does not move?' To the Parisians, Louis-Philippe seemed not just appealing, with his open, easy-going manner, but approachable, and he was said to play with his children even when people came to dinner.

Lucie and Frédéric reached Paris too late for Mme de Genlis's 85th birthday, a musical reception held by her niece Pulchérie de Valance, but, given Lucie's feelings about fashionable society, this was not a hardship. As she said, 'I love "home"' – using the word in English – 'wherever I find it . . . and I have no need of strangers in my life.' During the ten years of Frédéric's absence in Turin, the foreign embassies had turned into showcases for the countries they represented, the Austrians vying with the British for grandeur. 'What,' asked Lady Granville, wife of the British Ambassador, 'would the *parlez-vous* do without us?' Lady Granville, whose sharp tongue enlivened many gatherings, said that the American Ambassadress, Mrs Brown, when asked how she was, replied that she was in 'foine spirits and very hoppy'.

It was soon after their return that Victor Hugo's *Ernani*, which had been rehearsing in the arctic cold of the winter, opened in Paris. Its first night had been sold out for many months. *Ernani*, a tale of youthful love, fidelity and self-sacrifice, was not a particularly fine play but it hit a chord with the largely young audience who had queued for tickets. The Romantic school seemed to come of age; the day of the perfect meter and classical austerity was over. Voltaire and the subtle ironies of the 18th century were 'so much debris and ancient ruins'. 'Writers,' announced Hugo, 'have the right to take risks, to become daring, to create, to invent their style and not cling too hard to grammar.' *Ernani* ran, to enormous excitement and much popular debate, for 45 performances, during which the Classicists and the Romantics in the audience came to blows. And it was not only in literature that Romanticism flourished. The '*beau idéal*' that had inspired David and his disciples, with their heroes and their scenes from classical mythology, had been replaced by Delacroix's taste for colour and the exotic, and soon by Corot and the new landscape painters. In music, Berlioz's *Symphonie fantastique* was welcomed with delight. The composers who became popular after 1830 – Chopin, Bellini, Meyerbeer – would have practically nothing in common with pre-revolutionary France. After 15 largely stagnant years, Paris was once again becoming the cultural heart of Europe.

Uneasy in Paris, profoundly disliking what they saw of the ultra government, Frédéric and Lucie had moved to Versailles. Frédéric decided that the moment had come for him to retire from the diplomatic service and settle at Le Bouilh – still unsold and heavily in debt. He was 71, and he was tired of intrigues. Versailles, they discovered, had become a melancholy and deserted town, the great palace given over to workmen preparing to turn its immense rooms into a museum. Lucie attended Mass in the Royal Chapel, remembering the last time she had been there, 42 years earlier. 'Then,' she wrote, 'people found me beautiful and very fair-skinned, in my pink dress, wearing diamonds worth six million francs.' They were invited to a reception at which Charles X – whom Lucie had known since he was a child – was present, but she insisted on remaining hidden in the background, saying that she preferred now to leave the bustle of society to others. They spent their days showing Cécile the sights and visiting a cousin of Lucie's who had settled in France; she had nine children with whom Cécile played. In the evenings, in their lodgings, they read aloud from the diaries of Saint-Simon: they had reached the 12th volume. Lucie was content, saying that she felt fortunate in that she did not fret for the things she no longer had. 'I am happy at heart,' she noted, 'and my peace of mind allows me to enjoy what I have left.' It was an aspect of her nature that had served her well all her life.

* * *

Villèle's cautious, repressive leadership had, by 1830, alienated even the right-wing Catholics. At court, Charles X, surrounded by complacent ultras, continuing to believe that the future for France lay in autocratic religious government, had become ever more remote, not only from the audiences who cheered Victor Hugo, but from the Chamber of Deputies. On 16 May 1830, fearing the winds of insurrection, the King abruptly dissolved Parliament; the Chambers were prorogued. The new electoral laws, it was announced, would effectively limit franchise still further. In the Palais-Royal, Louis-Philippe gave a magnificent ball for his brother-in-law, the King of Naples. 'A Neapolitan night,' remarked a deputy, listening to the shouts of '*à bas les aristocrats*' coming from the street, 'and as at Naples, we are dancing on a volcano.' The words were almost exactly those used by Gouverneur Morris in 1792. On 26 July, Paris woke to learn that freedom of the press had been suppressed.

In Paris, journalists wrote angry pieces about the censorship of the press and editors, in defiance of the ban, printed them. The weather turned very hot. Printers and journalists gathered in the streets. Trees were chopped down and turned into barricades; pavements were ripped up. At Saint-Cloud, the King went out hunting and the courtiers played whist. On 29 July came an attack on the Louvre: some of the troops defected to the insurgents, the elderly Lafayette again at the centre of the rebellion. In the rue Florentin, Talleyrand, ever in tune with the currents of disaffection, paused to dictate a sentence to his secretary: 'At five minutes past twelve,' he recorded, 'the elder branch of the Bourbons has ceased to reign.'

Next day, the weather still very hot, Paris was under siege: the banks, the stock exchange and the theatres stayed shut. Bands of young men gathered in the Champs-Elysées; young women attached tricolour cockades to their hats and belts. The name of Louis-Philippe was being openly discussed as a possible saviour of monarchy, a man who could both guarantee the rights of property and stability for the bourgeoisie and promise reform for the liberals. In the Tuileries, ministers found themselves isolated and powerless. From the windows of their lodgings in Versailles, Lucie and Frédéric listened for the noise of cannon or guns which might herald some kind of coup, much as they had once listened from Chantilly for the sounds that might have announced that Louis XVI had been rescued. It was the third time, Lucie reflected, that they had seen turmoil surround a king of France: Louis XVI, Louis XVIII and now Charles X. It was as if her entire adult life had been spent watching the rise and eclipse of kings.

Though it was now clear that the reign of Charles X was over, there were still hopes among the ultras that the monarchy might be kept in Bourbon hands. But the Dauphin, the small, awkward, stammering Duc d'Angoulême – of whom a courtier had written 'he is not a Prince, and even less so a man: he's a nothing, a human envelope, that's all' – had neither the presence nor the courage for the role.

On the night of 30 July, the third day of what quickly became known as '*Les Trois Glorieuses*', the three glorious days, the royal family left Saint-Cloud in a cortège of carriages on which the fleur de lys had been painted over, surrounded by bodyguards, cadets and horsemen. The Duchesse de Berri travelled on horseback, in men's clothes, with two pistols at her waist.

They had reached Rambouillet, where they planned to sit out the troubles, when they learnt that Louis-Philippe had agreed to become Lieutenant General of the kingdom. On 2 August, Charles X formally abdicated, stating that he wished that the 9-year-old Duc de Bordeaux, the Duchesse de Berri's son, be crowned in his place, as Henri V. But it was all too late: the insurrection was gathering pace. The royal party took the road for Cherbourg, escorted by 800 men of the household. At one stop, there being no square table of the kind that etiquette demanded the King sit at, the sides of a fine round acacia one were chopped off to make it square. The heat was appalling. The cortège rumbled slowly on towards the coast in clouds of dust. On 16 August, with crowds lining the docks in total silence, the former Charles X and his family boarded two British boats for Cowes and another exile.

When, in Versailles, Frédéric had learnt of the abdication he had not hesitated. Leaving Lucie and Cécile – Aymar was away hunting in the mountains – to make their way to Le Bouilh, he had set off immediately for Orléans, thinking that he might find Charles X there, preparing for a possible counter-attack. As with Louis XVI, 41 years before, he was ready, even at the age of 71, to die defending his king. Hearing that Charles was in fact on his way back to exile in England, Frédéric made his own way to Bordeaux.

In Paris, there had been violent confrontations between loyal troops and insurrectionaries, leaving 150 soldiers and 600 civilians dead. On 9 August, Louis-Philippe accepted the crown. And after many discussions and debates about press freedom and individual liberties, about education and religion, about the electoral laws and franchise, a new France was hammered out, one that would combine the cross-currents of Romanticism with those of industrial expansion and capitalist enterprise. The July revolution brought to power an intellectual and liberal elite, whose model was England; the Duc de Broglie, Mme de Staël's son-in-law, was appointed to the Ministry of Education, François Guizot, translator of Gibbon and Shakespeare, to the Ministry of the Interior. The country they confronted had an empty treasury, a disordered capital, a disaffected civil service and divided and suspicious European allies. But Louis-Philippe, anxious to please and to be liked, was determined to keep France at peace, and though there would now be eight months of social disruption, savage anti-clericalism, bankruptcies and unemployment, and fears that the bellicosity of the emotional left might lead to war with the rest of Europe, bit by bit order was restored.

Talleyrand, the supreme conciliator, taking with him his niece Dorothée, set out for London, from which he had been expelled 36 years before, and where he would prove to be an exceptional ambassador. Though many of the old nobility regarded Louis-Philippe as a usurper, Mme de Boigne, Lucie's cousin, considered him a giant among pygmies. And if writers like Hugo continued to think nostalgically of the lost *douceur de vivre*, the subtleties and nuances of another age, when sociability and worldliness were perceived as fundamentally important in themselves, they had also come to see them as insubstantial, even a little irritating.

Frédéric, as ever, had acted impetuously and with honour, but with little regard for the consequences. From Le Bouilh, he had written to the Chamber of Peers a letter later reprinted in *Le Moniteur*, stating that having sworn an oath of loyalty to Charles X, he could not now in all conscience swear another to the King who had taken his place. Like Lucie – only more adamantly – he believed that the Bourbons alone were the legitimate kings of France, and that the Orléans branch could never be anything but pretenders. The consequences were not long in coming. Banished from the Chamber of Peers, he forfeited his 12,000 francs annual salary. Having resigned from the embassy of Turin, he lost his diplomatic wages. All that was left was a modest pension and the very small revenues from Le Bouilh and Tesson. Debts of 300,000 francs were still outstanding, and all hopes of holding on to either of the two properties were abandoned. Nor had there been any compensation for the lands and wealth lost during the revolutionary years.

It was not in Lucie's nature either to be reproachful or to complain. With winter coming, there were no immediate buyers for Le Bouilh, which, before a bridge was built over the Dordogne at Bordeaux, lay inconveniently far from the city. They had found the château very shabby, Lucie writing to Félicie: 'Give me warning when you plan to come to stay, so that the only pair of sheets that is not covered in fleas can be washed.' But were it not for their poverty, Lucie added, she would be perfectly content. She rose at six, or earlier, did lessons in history, geography and mathematics with Cécile, read, played music and discussed politics with Frédéric, and felt busy and useful.

She had bought *Le Cuisinier Royal* – it was characteristic that it did not occur to her to lower her standards – and with it was teaching both herself and their single elderly servant to cook, dish by dish. In order to save money they ate only what Le Bouilh and its farm produced. Dinner was at six, after which the family gathered around the lamp to sew and play piquet, while Frédéric read aloud, as he had always done. Since the Princesse d'Hénin's legacy had been stolen, Frédéric's cousin, Mme de Maurville, was again living with them, but she had lost her memory and was often bad-tempered. Cécile, now 11, sewed beautifully. In October, in perfect soft autumn weather, peasants from all around Saint-André-en-Cubzac came to help pick the grapes, gathering twice a day in the courtyard of Le Bouilh to eat soup, meat and bread. 'We are as peaceful as doves,' wrote Lucie.

Cholera, brought by the Russian army to Poland, had reached France. Its first cases in Paris had come soon after Lent, when some of the revellers, in their fancy dress, had fallen dying in the streets. There had been talk of witchcraft and poison, and the cholera, for which there was no known cure, was moving to other parts of the country. 'My one wish,' Lucie wrote to her grandson Hadelin, 'is to die the last in my family, so that I may nurse and take care of those I love, before I join them in a better world.' She meant precisely what she said: her love for her family had always been and remained the strongest and most important thing in her life.

* * *

It was not cholera, but misfortune of a different kind, one which they had not yet experienced, that hit them now.

The revolution of 1830, which had brought the liberals to power, had been disastrous for those referred to as the Bourbon royalists, men who, like Frédéric, could not tolerate the idea of the Orléans branch on the throne. Of the 75 prominent generals under Charles X, 65 lost their jobs. The royal bodyguards were dismissed, along with many civil servants, administrators and deputies.

No sooner had Charles X reached England than he made contact with those who remained loyal to him in France, and who were angry and apprehensive about what the July revolution might bring. After 40 years of swings of fortune, the royalists were masters of subversion and clandestinity. The west and south-west and the Vendée remained areas of known Bourbon support, with networks of secret societies and agents.

Among those who had accompanied the deposed King to England was Auguste de la Rochejacquelein, Félicie's husband, whose two brothers had perished in earlier Vendéen insurrections and who was himself a soldier. By birth and by inclination, Auguste was an '*ultra exalté*'. Félicie, for her part, still childless at 31, had grown up wild and boyish, her imagination fed on tales of Vendéen heroism, and, in spite of all Claire's efforts to turn her into an elegant courtier, had preferred to learn to shoot, ride bareback and break in horses. Drawn to the Duchesse de Berri, whose lady-in-waiting she became, she had come to believe that the future for France lay in the sole remaining Bourbon heir, the young Duc de Bordeaux. Together, to the disapproval of the court, the two women often dressed up in men's clothes and went out hunting; but though they frowned on such antics, the courtiers were also charmed by the Duchesse's silky fair hair, and the energy that she put into creating pleasure for herself and those around her.

After the Duchesse's departure for London with Auguste and the court, Félicie retired to live in the Château de Laudebaudière in the Vendée, surrounding herself with a coterie of ardent royalists. Many of the young women, noted the local Prefect, M. de Sainte-Hermine, who had been ordered to keep an eye on them, were to be seen in men's clothes. 'An impenetrable air of secrecy reigns over the Château,' he reported, 'and only the initiates are party to its mysteries.' He was right to be alarmed: Félicie and her friends were plotting.

The Duchesse de Berri, in one of her male disguises.

In England, Auguste and the Duchesse de Berri were busy making plans for an insurrection, convinced that if the legitimate monarchy, in the form of the 9-year-old Duc de Bordeaux, could be restored to France, then it would follow that legitimacy would triumph, not only in France, but throughout Europe. Neither Charles X, nor de Blacas, who had accompanied him to exile in England as he had once accompanied his elder brother, were quite as keen on the idea of a military uprising. But Auguste and the Duchess continued to make their plans, drawing comfort from the uneasy state of France in the first months of Louis-Philippe's reign. In Holland, Mme de Cayla, Louis XVIII's favourite, and M. Ouvrard, the rich financier with whom Thérésia Tallien had once lived, were both prepared to raise money for the cause. Military leaders were appointed from among the secret ultras to co-ordinate the insurrection, which was to be triggered by the Duchesse's return to France when, so they believed, disaffected people all over France would rise and sweep the young Duc de Bordeaux to the throne. Since Auguste remained abroad, it was

Félicie who took his place as commander of one section of the Vendée. She was a natural soldier, if somewhat severe and brusque, devoted to her followers, competent, obedient to superior orders and apparently fearless. From England, the Duchesse de Berri sent her a lock of her hair.

It was at this point that Aymar entered the story. In 1830, Aymar was 24, an excitable, affectionate young man, whose passion for hunting had filled him with nostalgia for the romance and heroism of the Vendée. Lucie, for whom her sole remaining child possessed 'a soul as pure as the purest mountain crystal', admitted that even so he could be both unsophisticated and rough. Not long before, Aymar had met Félicie for the first time; he had been captivated by her high spirits and daredevil ways. Towards the end of 1830, with very little to keep him occupied at Le Bouilh, and hearing constant rumours about this possible uprising, he set out for Laudebaudière. There he found not only Félicie, but her friend, the sculptress Félicie de Fauveau, a manly-looking young woman whose uncompromising monarchist views and taste for the medieval and chivalric had won her fame in Paris. The château was full of young Bourbon royalists looking for adventure. Aymar joined the cause.

In January, as he and one of the other young men were making their way to Bordeaux, they were stopped by police and questioned. Aymar, as his mother acknowledged, was capable of arrogance. Accused of criticising the new government and praising the Vendéen rebels, he was sent for trial to the assizes in Niort in May. A letter to Frédéric had been found on him in which he had written: 'There will be war, I think; I don't quite dare say, I hope; but that would be the truth, because it would be very hard to imagine the people any more unhappy than they are . . .' At Niort, he was sentenced to three months in prison; though the sentence was quashed, 2,000 francs of the fast-dwindling de la Tour du Pins' fortunes had gone in lawyer's fees. At this stage, however, Lucie was still speaking of 'this stupid little affair of my son's'.

By now, the Vendée had been split into separate sections to prepare for the uprising. But in September, when the different leaders – of whom Félicie was one – met to discuss tactics, it was agreed that it was too soon to act. To Auguste, Félicie wrote long letters calling for more guns, and more money, having by now gone through most of her own. Aymar had returned to the château of Laudebaudière and the little band of royalists passed their days training, recruiting men from among the local peasants and farmers and assembling weapons.

Early in August 1831, Aymar was sent by Félicie to carry a message hidden in the leaves of an album to Auguste in Spain, where he was trying to raise men and money for the cause. Pausing in Bordeaux to get a passport, Aymar was spotted by police and arrested. Frédéric, who saw him taken away, hurried back to the inn where Aymar had been staying and took away the album and several kilograms of ammunition. Aymar himself, left alone in a room in which a fire was burning, seized the chance to burn the bundle of papers he was carrying. Just the same, he was conducted, in handcuffs, back to the Vendée, where police questioned him about his friends in the château of Laudebaudière. What he did not know was that the château had been searched, that crates of guns, sabres, pistols and ammunition had been discovered, and that most of the others were already in detention. Félicie, dressed as a servant, had escaped.

Aymar was moved to the prison of La Roche-sur-Yon to await trial. Frédéric went with him, and took a room in the town, so that he could spend two hours every day with his son; Lucie stayed at Le Bouilh with Cécile. Three months later, Aymar was released for lack of evidence. To Félicie, Lucie wrote that their money was running very low, and that no buyer could be found for Le Bouilh. Everything that could be sold had already gone. 'There are green leaves just appearing on the trees, but I, my dear child, am not growing green again, I am becoming old and obsolete and the vicissitudes of life no longer amuse me as they once did.' Only her heart, she said, still felt warm, which was as well, for there was nothing in the world she so disliked as a cold heart.

The Vendée, to the royalist leaders, at last felt ready for insurrection, the people stirred out of the lethargy into which the July revolution had plunged them by the brutality of the local gendarmerie. The Vendée was good guerrilla country, with thick forests of oak and chestnut, fast-running rivers and narrow roads winding between deep ditches which offered perfect cover. Scores of men, trained and armed, were believed to be waiting for the Duchesse's signal. In April 1832, having wasted a certain amount of time wandering around Europe, the Duchesse de Berri landed in Marseilles. She was surprised to find very little support or excitement, and pushed on to Toulouse and then to Bordeaux, where she spent two nights at Le Bouilh with Lucie and Frédéric. 'Henri V,' she told her followers, referring to her son, 'summons you to arms. His mother, regent of France, is committed to your well-being. Long live the King, long live Henri V.'

The Duchess travelled on to the Vendée. But still the uprising did not come. Supporters shrank away, back to their farms and villages, saying that their weapons, left so long hidden in damp places, had rusted. In Paris, the staunch Bourbon royalists, Chateaubriand and Fitz-James, said that they could do nothing as they were being watched. Support that might have come from the Tsar of Russia or Metternich did not materialise.

Still the Duchess pressed on, using the code name of Petit Pierre, losing men in skirmishes with police and soldiers, making forced marches by night, sometimes on foot, sometimes on horseback, struggling on through driving rain over rocky ground. She was dressed as a peasant, her face smeared with mud. Aymar, terrified lest he miss out on the fighting, set out to join her. Hidden in Nantes by an elderly royalist, he met up with other supporters and they walked all through vineyards and along river banks, guided by loyalists.

But the moment had been lost; all element of surprise had vanished. When small bands of men did rise up and attack police stations and barracks they were easily rebuffed. The weather had turned cold and very wet: the rain poured down, soaking the rebels and their ammunition. The government soldiers, meanwhile, well fed, well equipped, and in far greater numbers, had no difficulty defeating them. When Aymar and his men attacked a village, burnt the tricolour flag and rang the church bells, it was all too late. The insurrection had collapsed.

The Duchesse de Berri, Félicie, Mlle de Fauveau and Aymar were among those who managed to escape capture. They made their way into Nantes, where some 400 of them were hidden in the cellars and attics of loyalists, while the soldiers searched the town and its surroundings. For almost four months, Aymar lived in hiding with two elderly sisters; under Mlle de Fauveau's tuition, he began to paint a Gothic missal.

On the night of 6 November, the Duchesse de Berri was betrayed by one of her followers. The house in which she was hiding was surrounded by soldiers; she had just enough time to slip into a secret chamber behind the chimney. All that night, the men ransacked the house, finding nothing. Next morning, two young soldiers, left on guard and feeling cold, decided to light a fire in the chimney. The Duchess held out as long as she could, then burst out, her dress smouldering, into the room. She was taken off to prison at Blaye.

Her followers, meanwhile, were rounded up. Young men who had never set eyes on the Duc de Bordeaux now went to the scaffold in his name. In some parts of France the feelings against the rebels, inflamed by the liberal press, were so violent that the carts bearing them to trial were stoned. Aymar was lucky. On 26 November, disguised as an artisan and travelling on false papers, he managed to board a boat at Nantes bound for Jersey. Félicie, too, was lucky. She came out of hiding, caught a diligence for Paris, lay low for several months, then got out to Switzerland.

But for Lucie and Frédéric, the story of the Duchesse de Berri's uprising was only just beginning. When an article had appeared in the government paper, *L'Indicateur*, about a 'Carlist brigand, Aymar La Tour du Pin' and the armed thefts he was carrying out in the countryside, Frédéric, with his customary impetuousness, had written a sharp letter of protest. Whatever else, he wrote, posterity would regard his son and the other young rebels as victims of their great 'loyalty and devotion'. *L'Indicateur* refused to carry the letter, but the legitimist *Journal de Guyenne* agreed to run it. Frédéric was arrested, charged with incitation to civil war and sent for trial before the assizes of the Gironde, together with the editor of the paper.

In court, on 15 December 1832, Frédéric insisted on reading out a statement. His one regret, he said, was that, on account of his age, which was 73, he had been unable to stand at his son's side among the insurgents. He felt nothing but shame, he declared, that he was not one of the 'sainted victims'. Though Maître Saint-Marc, his highly regarded lawyer, put up an eloquent defence, pointing out that four members of Frédéric's family had gone to the guillotine for the French monarchy, Frédéric was given three months in jail and a fine of 1,000 francs. It was considered very lenient, out of respect for Frédéric's evident anguish and his long years of service to his country.

The Fort du Hâ where Frédéric was a prisoner and where Lucie chose to join him.

On 19 December, 1832, Frédéric entered Bordeaux's notorious prison, the Fort du Hâ, where their friend M. de Chambeau had spent several months during the Terror. Lucie insisted on going with him.

* * *

The Fort du Hâ, built in 1456 as the residence of the King's representative in the south, was a forbidding rectangular fortress, with towers, a moat and a dungeon. Used during the revolution of 1789 to house suspects, it was damp and semi-derelict. It had underground cells 6 inches deep in water, and child prisoners were said to be enclosed in its dark, airless dormitories. When, during the Directoire, Thérésia Tallien was asked why she wore rings on her toes, she replied that they helped conceal the scars of the rat bites she had received as a prisoner there.

Lucie and Frédéric, however, were not unhappy. Frédéric was always cheerful when he believed that what he was doing was right, and it was in Lucie's nature to make the best of everything. Soon after their arrival, Frédéric wrote to Aymar that they had been given an airy room, overlooking the courtyard where the 'criminals' took their exercise. Though it saddened him to see them, he said that he greatly preferred the spectacle of men who had committed crimes through necessity, rather than that of many men in society, whose corruption and cowardice revolted him. Frédéric, who referred to Lucie as 'your incomparable mother', was already making plans for his release, when they would spend a few days putting order into their affairs, before meeting Aymar somewhere in Switzerland.

At the bottom of the letter, Lucie added a few words. 'I feel myself to be in a palace,' she wrote, 'and the thought that I can be really helpful and agreeable to your father is unquestionably the nicest thing that has ever filled my heart.' Since the weather outside was so wretched, she said that she did not mind not going out. It was carrying selflessness to an extreme degree. Neither she nor Frédéric uttered a single note of reproach. On the contrary, Frédéric insisted that Aymar had brought nothing but 'honour to my white hair'. More surprising, perhaps, Aymar seemed to feel no remorse for where his escapade had led them; later, he would write that the insurrection and his part in it had transformed his life, which might otherwise have been spent in idleness and indecision.

Lucie, who was not a prisoner, was allowed to come and go. To Hadelin, Cécile's brother, living in Rome with Auguste, who had been appointed Minister to the Vatican, she described their cell. It was sunny, she wrote, with two clean beds, several tables, a dresser with plates, a small cupboard for Frédéric and another for herself. In one corner, which they called the kitchen, there was a basket for wood, two pitchers of water, and several brooms 'because I like it all to be clean'. In the 'sitting room' there was a comfortable chair for Frédéric and a wicker one for herself by the window, where she could see to sew. On the floor, there was a fox-fur carpet given to her by Frédéric. Lucie rose at 6.30, lit the fire, got dressed, helped Frédéric to get up, then made him a cup of hot chocolate that she had left melting by the fire. At nine o'clock arrived a maid, with the *Gazette*, who spent two hours cleaning and making lunch. At 5, Lucie lit the lamp and prepared supper: either two meat dishes, or one of meat and one of fish, stewed fruit, and a small glass of Médoc. A second daily paper was delivered between 7 and 8; after reading it and discussing the news, Frédéric went to bed. Lucie herself stayed up later. As prison, it was not harsh.

Five afternoons a week, between 1 and 2.30, Lucie went to

visit her granddaughter Cécile, whom they had placed in a convent, walking the half-hour to the Sacré-Cœur and back for exercise, except if it was raining, when she took a carriage. On Wednesdays and Saturdays, Cécile came to spend the day in their cell. The one drawback that Lucie would admit to was the noise, both from the 'common criminals', some of them small children, sleeping on straw in rooms off the courtyard, and from what she referred to as '*les prisonniers un peu messieurs*', the better class of prisoner, who argued loudly over their meals at a communal table. In her letters, she sought constantly for humorous details.

Under the terms of his sentence, Frédéric was permitted six regular visitors, who did not need passes, and occasional additional ones, who were vetted by the obliging prison governor. Two old friends, both of them lawyers and peers who, like Frédéric had resigned after the July revolution, came every day. When, towards the end of February, word came that the Duchesse de Berri was pregnant, as the result, she claimed, of a secret marriage that had taken place while in Italy on her way back to France, Lucie was outraged. She wrote to Aymar that there had never been a 'more dishonourable act in the history of the world. It's appalling! Miserable woman! How she deceived everyone!' And when, shortly afterwards, word came the Duchess had announced that she had given up all pretensions to the throne for her son and that she was retiring to Sicily, the little group sitting around the fire in the cell during visiting hours in the Fort du Hâ agreed that she was 'the most despicable of people'. What puzzled them was the identity of the child's father, since the Duchess had been in the Vendée, and not in Italy, at the time of the supposed conception. For a while, they talked of little else.

Aymar was still in Jersey. At the beginning of March came the news that he had been condemned to death *in absentia*. Félicie was reported to be in Portugal, where she and Auguste were embroiled in trying to restore another dethroned legitimist ruler. What preoccupied Frédéric and Lucie most was not only where they would live, given that Aymar could not return to France, but what money they would have to live on. 'We must resign ourselves, devote ourselves, to uncertainty,' Frédéric told Aymar. They talked of Savoy, but agreed that it would have to be in a village or small town, since cities and spas were beyond their means. A cousin, Louis de la Tour du Pin, offered financial help, but they both found it very hard to think of accepting charity. Advertisements for the sale of Le Bouilh and Tesson had gone into the Paris papers. They were haunted by the thought that they might die leaving Aymar destitute. Frédéric came down with a chest infection and worried Lucie intensely by spitting blood. Leeches were applied; slowly, he recovered.

On 20 March 1833, Frédéric was released from the Fort du Hâ. By June they were in Nice – still part of Italy – reunited with Aymar; Auguste brought Hadelin to stay with them. Lucie noted that Auguste had lost his hair and that his sight was poor and she was irritated by the way he took so little notice of his daughter Cécile, whom he had not seen for four years. At 18, Hadelin, she wrote to Félicie, was tall, with a weak chest, and a rather unfeeling manner, probably 'because he had been brought up by a father who hated intimacy'. She thought constantly and with longing of her tender, loving Charlotte. Life in Nice was cheap, but they soon realised that they would have to move to a smaller apartment and get rid of their cook. She saw her life, Lucie told Félicie, as a series of drawers, in which she placed what talents she possessed. 'When those of a lady and an ambassadress were called for, I closed that of the housewife; now I know exactly where to look for what I shall need in my new situation, and I have completely forgotten all the other drawers, without experiencing the least vestiges of regret or complaint.' She felt, she said, not just resigned but cheerful.

For a while, Lucie, Frédéric, Aymar and Cécile wandered, spending a few months here and there, usually in hotels, settling for a few months in Pinerolo, not too far from Turin, where Frédéric was obliged to go every three months to renew the certificate for his small pension. Wherever they were, they saw few people, spending their days reading, drawing, playing music; Cécile did her lessons, Lucie her tapestry. She had taken to writing stern letters to Hadelin, who was to study law in Paris. 'I consider it important,' she instructed him, 'for you to move in a "high circle"', adding, interestingly, 'because I am not a liberal', a reflection of her unchanging belief in the values of the aristocracy. It was also essential that he improve his spelling, grammar and handwriting. 'A sales assistant would be embarrassed to write so badly ... Beware of sentences that everyone uses without realising what they are saying.'

In the autumn of 1834, leaving Aymar in Italy, they returned to Le Bouilh, where the house remained unsold and their debts unpaid. While Cécile went on with her lessons and her drawing, Lucie embroidered a pair of slippers for Félicie. She missed her goddaughter, and she wrote to tell her so repeatedly, begging her to pay them a visit. From now until the end of her life, her longings for Félicie's company would be a refrain that would run through all her letters.

In May 1833, Auguste, like Aymar, had been condemned to death *in absentia* for his part in the insurrection, but he had insisted on returning to France for an appeal, arguing that he personally had not been in the Vendée at the time, and he was acquitted. Félicie, who had been banished, also decided to appeal and returned for her hearing. Receiving no news, Lucie became frantic with worry, until she read in the *Gazette* that Félicie had conducted herself magnificently and quoted Joan of Arc. She, too, was acquitted. By contrast with her clever and imaginative goddaughter, Lucie herself, she wrote, using the words in English, was a 'matter-of-fact person', and incapable of being other than she was. And, she added ruefully, she suspected that she was losing, day by day, what little intelligence she had ever possessed. 'Poverty shrinks the mind, I know this only too well.' They were planning to spend a few days in Paris, but Lucie was adamant that this time she would see no one. 'I know what it is to be old and poor,' she wrote, 'and to hear people whisper: "Who is that old woman over there?"' But with Félicie, she added wistfully, she would not mind being poor. Félicie wrote, from time to time; but she did not come.

* * *

Somewhat to her surprise, Lucie enjoyed Paris. They took rooms in a hotel in the rue de Tournon and were immediately called on by old friends, even if Lucie remarked that they had only come to inspect what prison had done to them. She herself refused to pay visits, announcing that she did not intend to 'show my old nose in society'. Amédée de Duras, who had since remarried, told her that he was astonished to see that she still had her teeth and that she had not become 'decrepit'. She was touched by his words, but also a little impatient. 'I suffer only in my heart, from not being able to see those whom I love,' she wrote, 'and not on account of worldly losses.'

The walls of Paris were covered in cartoons and caricatures of Louis-Philippe, whom Lucie continued to refer to as an 'animal'. Thérésia Tallien had just died, at the age of 62, after a long and happy marriage to the Prince de Chimay, having had 11 children by four different men. Pulchérie de Valance brought them the gossip of Paris, Lucie observing that she would have preferred to talk politics, even if hers were of the wrong kind, and that she was no longer used to such egotistical, hypocritical chatter. Just the same, she was flattered to find herself '*en vogue*' among fashionable Parisians. After a brief pause to digest the political tone of the new monarchy, some of the salons had opened again. Her cousin, Mme de Boigne, whom Lucie had never cared for, received politicians; Clara, Félicie's sister, whom she did not care for much either, writers. On all sides, Lucie heard complaints that the court of Louis-Philippe was full of greedy shopkeepers. Mme Récamier was still receiving, and still wearing white, entertaining her guests in her rooms at L'Abbaye-aux-Blois, where she gave readings from Chateaubriand's unpublished memoirs. In some ways, Paris had changed astonishingly little; as Talleyrand had once said, talking about the émigrés: 'They have learnt nothing and forgotten nothing.'

Frédéric and Lucie returned to Le Bouilh in time to complete the sale on the property to a merchant from L'Isle de France for 160,000 francs, all of which would have to go towards paying off their debts. Lucie hated having to show the new owner around the château, discussing what they would leave and what they would take. 'It's rather like a foretaste of death,' she wrote, 'to see your ancestral roof taken from you.' She had been happy at Le Bouilh. Many of her best years, when Humbert, Charlotte and Cécile were all alive, had been spent in those cavernous rooms. Its loss struck her like a blow to the heart with a double-edged sword, from whose wounds 'one recovers only through perfect resignation to the will of God'. She had struggled for perfect resignation all her life.

There was one more visit to Paris, late that summer, a far easier journey than in her childhood: 72 hours, in a large diligence with separate compartments, drawn by eight horses, with stops at post-houses along the way. Even Bordeaux now had its first horse-drawn omnibus. Louis-Philippe had weathered five years of submerged plotting, revolutionary unrest and several attempts on his life, aided by France's growing prosperity and his own tolerant and domestic nature. The King, said Victor Hugo, combined something of Charlemagne and something of a country solicitor. Mme d'Agoult, a leading society hostess and historian, complained that 'Anglo-American habits . . . le club, le sport, le cigare' had dealt the old salons a death blow and that 'that innate gift which for two centuries made the Frenchwoman queen of everything most elegant in Europe' had finally disappeared. The women who had taken her place, she said, were coarse, shrill and over-familiar, and knew nothing of the 'discreet intimacies and delicate gallantry' of the past.

Lucie was still in Paris when, on 28 July 1835, a '*machine infernale*' exploded as the King rode out to review his troops on the Boulevard du Temple. He was unhurt, but 41 people were killed. From her hotel window, Lucie watched the funerals, a vast, silent crowd filing slowly past. The cartoons and caricatures disappeared from the streets. There were soldiers everywhere and much talk of press censorship. 'Laws,' she wrote, 'will now slip through like honey.' Before leaving Paris, she dined one last time with Amédée in Versailles. 'We philosophised,' she wrote, 'on human affairs in this town of so many misfortunes.'

That summer they received word that Aymar was to be banned not only from France but from Piedmont. Lucie was worried chiefly about Cécile, whom she had come to love as her own daughter, and who was growing up to be charming, affectionate and strong-minded, and appeared determined to accept no husband other than one she chose herself. 'I am a completely hopeless grandmother,' she wrote sadly, 'good for nothing at all, without money, position or contacts. I have nothing left to offer in this business of life.' It was not, she added, that she felt cowardly or despairing; simply that 'I no longer wish to swim against the current, because the world does not seem to me to be worth it'. She dreaded that a day might come when Auguste took her granddaughter to Brussels, in order to find her a husband.

So tender with those she loved, Lucie had lost none of her sharpness towards those she did not. She had met Fanny again briefly in 1827, but recorded nothing then about her feelings for the half-sister she had not seen for 16 years. In March 1836, she heard that Fanny had died of cancer, at the age of 51, and had been buried alongside her mother in the cemetery of Père Lachaise in Paris, leaving four sons and a daughter. Lucie's first reaction was to worry for the children, for, she said, Bertrand was 'three quarters mad and a fool'. But then she felt guilty that she seemed to feel so little affection for them, because it was against nature not to love the grandchildren of a father she herself had loved so much.

She and Frédéric were, once again, searching for a home. The epidemic of cholera that had raged across France – killing over 19,000 people in Paris alone, at times so virulent that the city seemed deserted except for those carrying litters of the dead – had spread to Germany and Italy. Italians were fleeing north over the borders into Switzerland and France. Félicie, who had bought a house in Lausanne, offered it to them when she was at last able to return to Ussé with Auguste. Lucie would have much preferred Italy, despite the cholera, being, as Frédéric said, 'of a totally fatalistic nature', but they were frightened for Aymar and Cécile. Lucie also worried that if she left Lausanne, she might never see Félicie again, for her goddaughter had never liked Italy. 'I have only one desire,' she wrote, 'and that is to see you.' What she really hoped was that she might spend the last years of her life living with Félicie; but she was coming to understand that this would never happen. Félicie was too restless, too volatile. 'There are things in life that one should neither analyse nor go on and on about,' she wrote sadly. 'They are as they are. One must bear them. Absence is one of these.'

In the autumn of 1835 they moved into Félicie's house, the Villa Sainte-Lucie; it was warm and pleasant and it had a large terrace with fruit trees. Lucie, however, was bored. 'This has never happened to me before,' she wrote. 'I look at the lake and see it as a mirage in the desert.' They had an elderly French visitor who could not bear to see the family so apparently cheerful with so little and whose sense of discontent was such that it was like 'Vesuvius erupting'.

Early in December, Frédéric had severe pains in his stomach; a little later he was struck down with 'gout of the head', for which he was given morphine. He now spent much of his time in his room, reading and writing to Hadelin, long letters mulling over his own life and urging the young man to study, to think on serious matters, to develop a taste for reflection. He should turn, he wrote, towards 'the vast questions of humanity: there you will find true riches'. More than anything Frédéric ever wrote, these letters to his grandson revealed a thoughtful and liberal man, intelligent, full of fears and doubts about the future, and intensely clear about the nature of responsibility. It was in history, he told Hadelin, that he should seek to find ways of understanding the world, and to learn how to make his mark on it; for it was to history that 'one must look to discover motives and judgements, the source of ideas, the proof of theories too often imaginary and vague'. Reflection, he added, was 'the intellectual crutch on which the traveller must lean on his road to knowledge'.

Some time towards the end of 1836, Félicie at last came to visit them. She was still there when, on 26 February 1837, Frédéric died. He was 78. As a diplomat, he had been proud of the grandeur of France; as a prefect he had been a liberal and a reformer, never shying away from speaking out, and never afraid to admit to making mistakes; and he had spent his entire life striving for morality. If he had sacrificed both himself and those around him to his ideals, he had also loved and looked after them. It had been a singularly happy marriage. He and Lucie had been together almost exactly 50 years.

* * *

Lucie now needed all the courage and the resignation she had ever possessed. She reacted to Frédéric's death as she had always reacted to loss: she retreated into herself, wrote little, endured, let the time pass. This was the first blow she had had to survive alone.

When Félicie departed again for Ussé, she suggested that they continue to live in the little house in Lausanne. Lucie always found parting from Félicie painful; it was even worse now. 'The moment you left,' she wrote to her, 'I sank into a sea of sadness and felt that I was drowning . . . I pray to God to give me strength. I am fighting as hard as I can against despair and hopelessness.'

Very slowly, the days and then the weeks passed; Lucie recovered some of her confidence. She talked of completing her memoirs, started almost 20 years before and continued in a desultory way since then. She travelled a little with Cécile, around Switzerland, to spas and mountain villages, and loved its sense of peace and equality, in comparison with which France seemed to her filled with 'vainglorious castles and sad hovels', and 'wicked and despicable' men. When a priest in Lausanne proposed saying a Mass for Frédéric, she shrank back, saying that she could not bear to have people looking at her or intruding on the intimacy of her grief. She was not lonely, sitting in the little garden, under the lilac and the flowering chestnut tree; Félicie had left her a cat. Very occasionally, she and Cécile went out rowing on the lake, stopping to eat in an inn on the waterside, after which she would paint or draw. 'I feel myself to be an old tree,' she wrote, 'from which every day someone cuts off a branch; the trunk, which once sustained it, no longer exists; all that is left are a few faded leaves. Ah! how terrible it is to become old.' Sometimes she spoke of wishing that she had died with Frédéric. She could not get used to the fact that there was no longer the person about whom she could say: 'This is another me.' In April 1838 she had a fall, which left her bruised all over and with a black eye. The doctor put her on a strict diet which, she said, with a touch of her old humour and self-mockery, made her wrinkles look more pronounced and her nose bigger. She was now always enveloped in black, but her expression remained wry and quizzical.

Lucie worried constantly about Cécile, whose health was often poor, with headaches and sudden attacks of anxiety, and about Aymar, whose life in Lausanne seemed to have been reduced to playing whist and falling asleep after dinner, 'squatting on his wicker chair like a chicken on a perch', but this did not bother her as she loved him 'as much asleep as awake'. Hearing of the death of Talleyrand, she doubted that anyone would mourn him; she reflected that he was after all not much older than she was, but that to her he had always seemed to belong to another generation. As much as she was able, she forced herself to go out, pay visits, receive guests, in order that Cécile would meet people and have some kind of life. She avoided the Vendéen exiles who had settled along the shores of the lake and whose company she found profoundly boring. Their sighs, she said, 'would turn a windmill'. The small lives of a small society had always filled her with dread. When her granddaughter left, she said, she would become like a bear, and go nowhere.

In the spring of 1841, Cécile was 23, a thoughtful young woman with strong views of her own. The few young men who had asked for her hand had been politely but firmly turned down. Auguste was still in Rome, complaining of ill-health, and Lucie railed against his meanness towards his daughter. To make a little money, Cécile embroidered tobacco pouches and Aymar drew. But in May Auguste suddenly announced that he would be taking Cécile back to Belgium, to live in the Château de Vêves near Dinant with his father and sister. The days before her departure were agony for them both, Cécile silent and fighting back tears, Lucie so distraught that she longed for her to leave, so that the terrible moment of parting would be over. 'At my age,' she wrote, 'separations are serious things. This little one has become a friend, a companion, someone to whom I can talk of anything. I am like a cat: when they suffer, they go away alone to a corner of the attic.'

At first, Cécile's daily letters from Belgium were despairing. Her grandfather was 'at the last point of decrepitude', while her aunt would allow her neither to be sad nor to be ill. But bit by bit her spirits improved, and she became close to Hadelin. In October Lucie travelled to Brussels, where Cécile had just announced her engagement to a man she had chosen for herself, the younger son of the Baron de Beckman, who looked more Spanish than Belgian, with an 'aquiline nose, black hair and a small mouth, deep-set dark eyes and a lovely smile'. Ferdinand, noted Lucie, would bring her granddaughter not just money and security but 'love'. She was delighted to see Cécile so happy. She stayed for the wedding, though she was irritated by Auguste's arguments over the trousseau.

For a while, she agreed to settle at Noisy, thinking that she should cease to be a burden on Aymar. Hadelin had also married and was living not far away, and Lucie continued to worry that he was too like his father, cold and weak, and she regarded weakness as 'the greatest cause of danger and unhappiness'. But she hated the grey skies of the Ardennes, and the good circulating library in Brussels was not enough to keep loneliness at bay. She missed Aymar, as she always missed Félicie; alone, she spent the days remembering, mourning the past. When Auguste fussily told her that the arrangement was not to his liking, she was delighted to return to Lausanne.

That autumn, Cécile and Ferdinand spent several months with her in Switzerland. Cécile was four months pregnant and Lucie found her much softened. They read Shakespeare – Lucie did not care for the comedies – and Dickens. She was now hard at work on her memoirs, and had just reached the march on Versailles in October 1789, when she had stood at Mme d'Hénin's windows and watched the women of Paris advance under the pouring rain. It seemed to her an unimaginably long time ago. Though she longed for Félicie to visit her, she was not discontented, or would not allow herself to be. 'I feel all right here,' she wrote. 'I am very afraid of having to leave and die far from the person from whom I was never apart for so many long years.' She liked the hills and crooked streets of Lausanne, the sailing boats that looked like feluccas on the water, and her very occasional outings on the *William Tell*, the new steamboat that went up and down the lake, though she said that she now went out so little that a trip to Geneva was 'like going to China'. With each new spring, she was charmed again by the greenery and the buds. Not long before his death, Frédéric had written to Félicie about Lucie: 'My wife's bottomless reserves of courage will always serve you well. Ah! how admirable it is to be so completely buffeted by storms, yet to remain so fundamentally unbroken.'

CHAPTER EIGHTEEN

The Rhapsodies of Life

Lucie lived for another 11 years. At 72, she remained very upright and her hair was still fair and thick. Rheumatism in her hands and knees sometimes made writing and walking difficult. Her sight was not always good. She wore nothing but black.

Taking stock of her life, she wondered whether it might not be best for her to retire to a convent, in order to leave Aymar not just free but in possession of the whole of Frédéric's 3,000 francs pension, which was all they had between them. But she did not think she could quite bear it. 'The word sacrifice,' she wrote to Félicie, 'does not seem to be in my dictionary.' She was being modest rather than disingenuous. More than that, she could not contemplate parting from Aymar: the day that he wished to leave her would be that of her 'spiritual death' if not her physical one. Félicie wanted them to stay on in her house in Lausanne, but Lucie felt stubborn about their poverty. She and Aymar discussed settling somewhere in Italy, where everything would be cheaper, and where she was drawn by the thought of the yellow light and the warm skies. It had to be in a town: she needed to be able to watch people walking in the street.

In early November 1842 they hired a barouche and, taking a maid called Claudine with them, set out for the walled town of Lucca, where Napoleon's eldest sister Princess Elisa had ruled with ruthless efficiency, leaving a legacy of schools, libraries and artistic foundations. They took lodgings and walked around the ramparts; in the evenings Lucie sewed and read in their landlord's well-stocked library. But Lucca was full of foreigners, Swiss, English, Russians and Germans, come to spend the winter looking at Tuscany, and the only apartments still empty were expensive. At the end of the year they moved on to Pisa, having heard that it was less fashionable and that the cholera epidemic of the 1830s had left many houses empty. In 'The Tower of Famine' Shelley had written of a desolate place 'which was the cradle and is now the grave / Of an extinguished people', but even so he considered that no sight matched that of the sunset in Pisa, seen from along the sweep of the Arno. It was in Pisa that Byron, living in his menagerie of dogs, monkeys and exotic birds, had written part of *Don Juan*.

Their first glimpse of the town, approached across a gentle plain lying between the Apennines and the sea, was of the great dome of the cathedral, with its leaning tower nearby, then of houses, walls and an aqueduct. Lucie and Aymar found rooms at 717 Lung'Arno, in front of the charming black-and-white-striped, 13th-century, Gothic Church of Santa Maria della Spina, with its spires and statues and its dark mullioned windows. They had a large sitting room, with an open fire and an alcove with a writing table by the window, where Lucie sat writing her memoirs and her letters looking out over the river. There was a green carpet and a desk, on which she placed a picture of Félicie. She was pleased with Pisa and the grass that grew among the cobbles along the deserted streets, despite its air of 'melancholy stillness', observing that the climate of Italy was for her like putting oil in a rusty old lock. At eight every morning, she crossed the road for early Mass in Santa Maria della Spina, sometimes going on to walk in the Campo Santo, where, she said, she hoped she might one day be buried. In the evenings, now that Frédéric was dead, it was Lucie who read aloud, while Aymar carved and painted small items of furniture which he sold. When, late at night, he went to smoke a last cigar by the river, she sewed socks.

Though she found Pisa's circulating library to be poor, she liked the sense of equality in Italian churches, and she made friends with the local priest. She was scornful, however, about Pisan society, dismissing the evening *passeggiata* – stroll – along the banks of the Arno as being 'what passes here for people of distinction'; Lucie had not lost her taste for distinction. She was hard at work copying out her memoirs into red leather notebooks, carefully numbering the pages, and filling every inch of the paper in her small, neat, sloping hand, never altering or crossing anything out. They were not intended for publication, she said, but as a record, for Aymar. She was resolved neither to romanticise her own past nor that of France, but to look back, with a precise, cool eye, on the world that she had known; and there was something dutiful in her determination to set things down as she had seen them, without embellishment or apology. 'I am busy describing the rhapsodies of my life,' she wrote, with all her old self-mockery, to Félicie. 'Old age has so far got me only by the heels, which ache (gout I think, or rheumatism).' She felt, she added, happy and at peace. It was a refrain she had sung all her life, as if the words themselves kept sadness and disorder at bay.

They lived frugally. They walked everywhere, very occasionally hiring a carriage to take them to the pinewoods by the sea, where they saw deer and came back with bags full of pine cones for the fire, and sometimes with violets. Though a new train service ran between Pisa and Livorno, some 15 miles away, they never took it. Sometimes there was no money for tapestry wool and Lucie was forced to abandon a pleasure and a habit of a lifetime. At one point, overcoming her deep revulsion for begging, she wrote to Lord Clifford's heirs, asking whether she might have the £300 he had promised her in his will after she had had his son living with them in Turin for almost two years. But she received no reply. Lucie and Aymar moved into cheaper, smaller lodgings, away from the river.

But there were frequent letters and occasional visits from Cécile and Hadelin, both happy and with children of their own, and visits from other foreigners settled in Pisa and old friends from Lucie's distant past. Only once in all those years, in 1847, did Félicie come, still young-looking in her mid-forties, though her thick fair hair had gone completely white.

Like Félicie, Lucie retained a streak of restlessness, and her curiosity had never left her. If anyone proposed a trip to America, India, Tahiti or Sydney, she announced, she would accept it in a second. Meanwhile she was content to spend her life looking after Aymar, anticipating his every wish, cooking him the meals he liked, just as she had done for his father. 'I do not want to fail in this duty,' she wrote, 'which is also the happiness of my last years.' But she had lost neither her wit nor her sharpness, remarking when she read Chateaubriand's *La Vie de Rance*, that there was nothing more pitiful than the 'decrepitude of genius'. When asked by Hadelin to be godmother to his new baby, she wrote to say that he was to tell the godfather, a man in his 20s, that though Lucie was three times his age, 'sixty years ago I was one of the most fashionable *lionnes*' – lionesses – 'in the whole of Paris'.

The year 1848 was one of revolution in Europe. Hunger, unemployment and economic depression, spreading through Italy, Germany, France and Austria, fuelled liberal and nationalist movements and brought down governments. In Vienna, Metternich, after 27 years as Chancellor, fell. The new Pope, Pius IX, promised reforms; the rulers of Naples, Tuscany and even Piedmont, where Charles Albert had proved an even more repressive ruler than Charles Felix, were forced to grant constitutions. For a while, Louis-Philippe in France hung on to rigid conservatism. 'Our civilisation is very ill,' wrote Count Molé, a man who had served almost as many rulers as Talleyrand, 'and nothing would be less surprising than a good cataclysm which would put an end to it all.' He did not have long to wait. The opposition, gathering strength from the rising price of bread, panic among speculators, issues of electoral reform and clashes between the army and the National Guard, drove Louis-Philippe off the throne and to a second exile in England. The Bourbons and the Orléans had gone: there would never be another king of France. A mob entered and ransacked the Tuileries, as it had 56 years before, spilling so much wine that several revellers were rumoured to have drowned. The poet Lamartine, conservative turned republican, was named head of a provisional government, only to be ousted in the struggles that followed the setting up of the Second Republic.

Napoleon's only son had died in 1832 of tuberculosis at the age of 21. His nephew, Louis-Napoleon, who had been in prison for his part in two attempted coups, returned from exile to run against General Cavaignac, former Minister for War, winning a landslide as the first elected President of France. In 1852, after a *coup d'état*, he became Emperor Napoleon III; liberals were banished, sent to penal colonies or fled, Victor Hugo among them. Napoleon I had never quite carried out all his plans to make Paris the most magnificent city in Europe: his nephew and Baron Haussmann would now sweep away narrow, dark streets and bring lighting, pavements and space. The Champs-Elysées were still green fields; but they would not be so for long.

The France into which Lucie was born, in the spring of 1770, was no more. Versailles had become a museum. Steam, the telegraph, trains, gas lighting, the smokestacks of industry had between them transformed the landscape of her childhood into a world she would no longer recognise. Charles Darwin was soon to publish his work on the origin of the species. Queen Victoria had been on the throne for 16 years, and a million people had died in the Irish potato famine and a million more had emigrated, many of them to the United States. Slavery had finally been abolished, though in France it had taken a revolution, an empire, two restorations, a bourgeois monarchy and a republic to end it. The word 'tourist' had entered the vocabulary, and some said that it was invented by Stendhal. The people who had filled Lucie's world, the friends, the statesmen, the soldiers, the famous women who had opened their salons to all the most brilliant people of their age, had gone, and it would never again be possible to believe in the divine right of kings. Even Bertrand was dead, buried in the Invalides near the man he had served so faithfully, and Mme Récamier, who had in her lifetime been painted by David, Gérard and Isabey. The graves of Lucie's children lay scattered, Humbert in Paris, Charlotte – with Frédéric – in Lausanne, Séraphine in Albany, Edward in Richmond, Cécile in Nice.

Pisa remained untroubled, a quiet, peaceful backwater, the news of Europe's turbulence only distant sounds. On 2 April 1853, soon after her 83rd birthday, Lucie died. She was buried, not in the Campo Santo as she had hoped, but in a cemetery on the outskirts of Pisa, in the walls of a vaulted grey arcade, among the cypresses. In an age when rivers mattered, when life unfolded along their waterways and banks, she had lived on the Seine, the Hudson, the Thames and the Garonne, and she died by the Arno. 'The days pass by like instants,' she had written not long before. 'I miss nothing that vanity once might have caused me to regret. I no longer dream of all those footmen in livery, those horses, those carriages, that excellent cook . . . All that is now so far from me that it is as if I had never known it.'

Afterword

The year after Lucie's death, Aymar married and had a son; but he in turn had no male descendants. The Gouvernet de la Tour du Pin branch of the family died out. Félicie de la Rochejacquelein lived another 30 years. Cécile, Lucie's much-loved granddaughter, died in 1893, having had three sons; Hadelin, who predeceased her, had four.

Lucie's papers and her red leather notebooks made their way to the Château de Vêves in Belgium, home to the Liederkerke-Beaufort family, where she had spent several unhappy months after Cécile's marriage. They included the many volumes of memoirs, covering the years 1770 to 1814, but stopping with Louis XVIII's accession to the throne; though she never explained her decision to write nothing about the 40 years that followed. What made it possible to document those years were the seven boxes of letters between Lucie and Félicie, starting in 1821 when Lucie was 51 and her goddaughter 23, and continuing until not long before Lucie's death. Only a very few of these many hundreds of letters, as full and as detailed as her memoirs, have ever been published. There were also her letters to Hadelin, to his father Auguste, to Mme de Staël and to various friends, and Frédéric's own papers and letters.

Hanging on the walls of the château at Vêves are also portraits of Lucie's grandchildren and of the Princesse d'Hénin; and at Le Bouilh, still in the hands of the family who bought the château from Frédéric, are pictures of Lucie, Frédéric and their children.

In 1907, 54 years after Lucie's death, Hadelin's son Aymar-Marie-Ferdinand decided to edit his great-grandmother's memoirs. They were published under the title *Le Journal d'une femme de cinquante ans*. Quickly recognised as one of the most exceptional portraits of an exceptional age, it was soon translated into English and German. It has seldom been out of print since then, and it has provided scholars and readers with a rich fund of detail on France during a long and uniquely troubled period of its history.

Acknowledgements

My first thanks go to M. le Comte de Liederkerke-Beaufort for allowing me to read in the family archives in the Château de Vêves, near Dinant, in Belgium. While working there, I was greatly helped by Mme Rouard, and M. Frédéric Rouard. Beatrix de Blacas, descendant of Mme de Duras's daughter Clara, kindly arranged for me to visit the Château of Ussé in the Loire, and provided me with a family picture of Lucie de la Tour du Pin.

In Bordeaux, I was able to visit Le Bouilh with the kind assistance of Charles Pelletier-Doisy and Guy de Feuilhade. I thank them both. Julien and Michelle Sapori took time to show me around Hautefontaine and provided me with information about the early life of the house.

The descendants from Lucie de la Tour du Pin's English family, Isabel and Alec Cobbe, generously allowed me to see their family papers, as did Teresa Waugh, whose own work on her great-great-uncle, Archbishop Dillon, was invaluable. I thank them all very much.

For documents and papers, and in some cases permission to quote from them, I would like to thank the following institutions and their staffs: the Bibliothèque Nationale; the Service des Archives, Ministère des Affaires Étrangères; the Bibliothèque Historique de la Ville de Paris; the Archives Nationales de France; the Bibliothèque Municipale de Bordeaux; the Bibliothèque Municipale d'Amiens; the Musée Carnavalet; the National Archives of Brussels; the British Library; the London Library; the National Archives at Kew; the Library of Congress; the New York Public Library; the manuscript division of Albany State Library; the Massachusetts Historical Society.

I am greatly indebted to Colin Jones and Anne Chisholm for

reading the manuscript, to Philip Mansel for his kind help and advice, and to Elfrieda Pownall, with whom I talked over the idea for the book. The Wingate Foundation very generously gave me a scholarship, which allowed me to travel to the many places in which Lucie lived at some point in her long life. Without its support I should not have been able to explore them all.

And I should like to thank the companions of my many journeys: Christopher Balfour, David Bernstein, Anne Chisholm, Monnie Curzon, Virginia Duigan, the late Alfred Gellhorn, Miles Morland, and my son, Daniel Swift. The trips were made all the more pleasurable through their company.

Once again, I would like warmly to thank my editors, Jennifer Barth and Penelope Hoare, my agent, Clare Alexander, and Douglas Matthews, for his index.

C. M.
London
October 2008

Bibliography

Primary Sources

The most important sources for this book are to be found in the published and unpublished papers, letters and journals of Lucie Dillon, Marquise de la Tour du Pin, and of her husband, Frédéric de Gouvernet, Marquis de la Tour du Pin. The unpublished letters and manuscripts are in the Château de Vêves, near Dinant in Belgium, in the possession of M. le Comte de Liedekerke-Beaufort. These files include:

Dossier 316. Diplomatic dispatches from M. de la Tour du Pin from The Hague (1792 and 1818–19) and from Turin (1820–8).

Dossiers 324–39: Correspondence between Lucie Dillon and Félicie de la Rochejacquelein.

Aymar de la Tour du Pin: Unpublished memoir.

Various letters to and from Mme de Staël and Lady Bedingfield.

Manuscript Sources

Important manuscript sources are to be found in the Archives Nationales in Paris (Series F1, F2 and F3); in the Ministère des Affaires Etrangères in Paris (files on The Hague and Sardinia); in the Archives Municipales in Bordeaux (files on La Tour du Pin and Le Bouilh); in the National Archives in London (files on refugees from the French Revolution 1789–1800). Contemporary newspapers are to be found in the State Library of New York State in Albany; in the Library of Congress, Washington; in the Archives Municipales at Amiens; in the British Library's Newspaper Library, Colindale Avenue, London; and in the Bibliothèque Nationale in Paris.

I translated the quotations from the French and Italian myself.

Select Bibliography

The 18th century in France, the revolution, the Directoire, the Consulat, the Empire and the two Restorations are periods of history that have been extensively written about: in memoirs, novels, scholarly histories and academic journals. The following is a brief selection of some of those most frequently consulted for this book.

1) France in the 18th and 19th centuries

Alméras, Henri d', *La Vie Parisienne sous le Consulat et l'Empire*. Paris.
Ariès, Philippe and Duby, Georges (eds), *Histoire de la vie privée*, 4 vols. Paris 1986.
Baldensperger, Fernand, *Le Mouvement des idées dans l'émigration française: 1789–1815*. Paris 1924.
Bertaut, Jules, *Le Faubourg Saint-Germain sous la Restauration*. Paris 1935.
Bertier de Sauvigny, G. de, *La Restauration*. Paris 1955.
Bertier de Sauvigny, G. de, *Metternich et la France après le Congrès de Vienne*. Paris 1968.
Braudel, Ferdinand, *The Structures of Everyday Life: The Limits of the Possible*. London 1981.
Cabanis, José, *Charles X, roi ultra*. Paris 1972.
Carpenter, Kirsty and Mansel, Philip (eds), *The French Émigrés in Europe and the Struggle against Revolution*. London 1999.
Childs, Frances Sergeant, *Refugee Life in the United States 1790–1800*. Baltimore 1940.
Cobb, Richard, *Terreur et subsistances: 1793–1795*. Paris 1965.
Cobb, Richard, *Reaction to the French Revolution*. London 1972.
Cobb, Richard, *Paris and its Provinces: 1792–1802*. London 1975.
Cooper, Duff, *Talleyrand*. London 1932.
Craveri, Benedetta, *The Age of Conversation*. New York 2004.
Darnton, Robert, *The Great Cat Massacre and Other Episodes in French Cultural History*. New York 1984.
Darnton, Robert and Roche, Daniel (eds), *Revolution in Print: the Press in France: 1775–1800*. Berkeley 1989.
Englund, Steven, *Napoleon: A Political Life*. New York 2004.
Fairweather, Maria, *Madame de Staël*. London 2005.
Forneron, J. *Histoire Générale des émigrés pendant la révolution française*, 3 vols. Paris 1884.
Forster, Robert and Ranum, Orest, *Food and Drink in History: Selection from Annales, Vol. 5*. Baltimore 1979.
Forster, Robert and Ranum, Orest, *Medicine and Society in France: Selection from Annales, Vol. 6*. Baltimore 1980.
Fraser, Antonia, *Marie Antoinette*. London 2001.

Furet, François and Ozouf, Mona, *A Critical Dictionary of the French Revolution*. London 1989.

Godechot, Jacques, *La Vie quotidienne en France sous le Directoire*. Paris 1977.

Goncourt, Edmond de and Goncourt, Jules de, *Histoire de la Société Française pendant le Directoire*. Paris 1840.

Goncourt, Edmond de and Goncourt, Jules de, *Histoire de la Société Française pendant la révolution*. Paris 1889.

Herald, J. Christopher, *The Age of Napoleon*. London 1963.

Hesse, Carla, *The Other Enlightenment*. Princeton 2001.

Hibbert, Christopher, *The French Revolution*. London 1980.

Hibbert, Christopher, *The English: A Social History: 1066–1945*. London 1987.

Hobsbawm, Eric, *The Age of Revolution: Europe 1789–1848*. London 1962.

Hufton, Olwen H., *The Poor of Eighteenth-Century France*. Oxford 1974.

Hufton, Olwen H., *Women and the Limits of Citizenship in the French Revolution*. Toronto 1992.

Jones, Colin, *The Great Nation: France from Louis XV to Napoleon*. London 2002.

Ketcham Wheaton, Barbara, *Savouring the Past: The French Kitchen and Table from 1300 to 1789*. London 1983.

Lacroix, Paul, *Directoire, Consulat et Empire: mœurs et usages, lettres, sciences et arts en France 1795–1815*. Paris 1884.

Mansel, Philip, *Louis XVIII*. London 1981.

Mansel, Philip, *The Court of France: 1789–1830*. Cambridge 1988.

Mansel, Philip, *Paris between Empires: 1814–1832*. London 2001.

Mansel, Philip, *Dressed to Rule*. Yale 2005.

Mercier, Louis-Sebastien, *Tableau de Paris*, 12 vols. Geneva 1979.

Mercier, Louis-Sebastien, *Les Rues de Paris au XVIIIième siècle*. Paris 1999.

Poniatowski, Michel, *Talleyrand aux Etats Unis 1794–1796*. Paris 1967.

Poniatowski, Michel, *Talleyrand et l'ancienne France 1754–1789*. Paris 1988.

Poniatowski, Michel, *Talleyrand et les années occultées 1789–1792*. Paris 1995.

Ribiero, Aileen, *Fashion in the French Revolution*. London 1988.

Ribiero, Aileen, *Dress in 18th-Century Europe 1715–1789*. New Haven 2002.

Rudé, George, *Europe in the 18th Century: Aristocracy and the Bourgeois Challenge*. London 1972.

Schama, Simon, *Citizens*. London 1989.

Schieff, Stacy, *A Great Improvisation: Franklin, France and the Birth of America*. New York 2005.

Taine, Hippolyte, *Origines de la France contemporaine*, 5 vols. Paris 1882–8.

Tombs, Robert and Tombs, Isabelle, *That Sweet Enemy: The French and the British from the Sun King to the Present*. London 2006.

Verlet, Pierre, *French Furniture and Interior Decoration of the 18th Century*. Fribourg 1967.

Weber, William, 'Musical Tastes in 18th century France', *Past and Present*. November 1890.

2) Memoirs

Agoult, Comtesse de, *Mémoires, souvenirs et journaux*, 2 vols. Paris 1990.

Boigne, Adèle de, *Mémoires: Souvenirs d'une tante*, 4 vols. Paris 1908.

Chastenay, Mme de, *Mémoires: 1771–1815*. Paris 1987.

Coigny, Aimée de, *Mémoires*. Paris 1902.

Cradock, Mme, *La Vie Française à la veille de la Révolution: 1783–1786: Journal inédit*. Paris 1911.

Desbassayns, Henry-Paulin Panon, *Voyage à Paris pendant la révolution 1790–1792*. Paris 1985.

Ducrest, Georgette, *Mémoires sur l'Impératrice Josephine*. Paris 1828.

d'Espinchal, Comte, *Journal*. London 1912.

Gallatin, James, *Diary*. London 1914.

Genlis, Mme de, *Mémoires*. Paris 2004.

Hesdin, Raoul, *The Journal of a Spy in Paris during the Revolution*. London 1985.

Hézecques, Comte d', *Page à la cour de Louis XVI*. Paris 1987.

Kotzebue, August von, *Travels from Berlin through Switzerland to Paris in the year 1804*, 2 vols. Paris 1850.

Meister, Henri, *Souvenirs de mon dernier voyage à Paris: 1795*. Paris 1910.

Millingen, J. G., *Recollections of Republican France between 1790–1801*. London 1848.

Moody, C. L. (ed.), *By a Lady. A Sketch of Modern France in a Series of Letters to a Lady of Fashion, written in the years 1796 and 1797 during a Tour through France*. London 1798.

Moré, Comtesse de, *Mémoires 1758–1837*. Paris 1898.

Morris, Gouverneur, *A Diary of the French Revolution*. London 1939.

Reichardt, J. F., *Un Hiver à Paris sous le Consulat: 1802–1803*. Paris 1896.

Rémusat, Mme de, *Mémoires*, 3 vols. Paris 1880.

Rossi, Henri, *Mémoires aristocratiques feminins 1789–1848*. Paris 1998.

Staël, Mme de, *Dix années d'exil*. Paris 1887.

Vigée-Lebrun, Elisabeth, *Souvenirs*, 2 vols. Paris 1984.

3) The Dillon and de la Tour du Pin families

Aston, Nigel, *The End of an Elite: The French Bishops and the Coming of the Revolution*. Oxford 1992.

BIBLIOGRAPHY

Audibert, Louis, *Le dernier Président des Etats Généraux du Languedoc: Monseigneur Arthur Richard Dillon*. Bordeaux 1868.

Berthelot, Michel, *Bertrand, Grand Maréchal du Palais*. Châteauroux 1996.

Bertrand, General, *Lettres à Fanny*, ed. Suzanne de la Vaissière-Orfila. Paris 1978.

Clully, Lucien de, *La Tour du Pin*. Paris 1909.

Gaissart, B. de, *La Naissance, le mariage et la mort de Fanny Dillon, Comtesse Bertrand*. Paris 1967.

The Jerningham Letters, 1780–1843. ed. Egerton Castle. London 1896.

McManners, John, *The French Revolution and the Church*. London 1969.

Martin, Georges, *Histoire et Genéalogie de la Maison de la Tour du Pin*. Lyon 2006.

Masson, M., *Arthur Dillon*. Revue de Paris 1910.

Staël, Mme de, *Seize lettres inédites à Gouvernet,* ed. Charles de Portairols. Paris 1913.

Rohan Chabot, Alix de, *Madame de la Tour du Pin: Le talent du Bonheur*. Paris 1997.

Source notes

Chapter 1

5. 'Of all the faubourg's . . .': see Charles Lefeuve, *Histoire de Paris rue par rue, maison par maison* (Paris 1875).
7. 'The Marquis de Bombelles . . .': Journal (Geneva 1977), p. 143.
8. 'On 16 May . . .': see Antonia Fraser, *Marie Antoinette* (London 2001).
9. 'In 1770 . . .': see Richard Cobb, *Paris and Its Provinces* (London 1975); Louis-Sebastien Mercier, *Tableau de Paris* (Paris 1979).
11. 'Unwanted children . . .': see Claude Delasselle, '*Les Enfants abandonnés au XVIIIième siècle*', Annales, Économies, Sociétés, Civilisations, Jan./Feb. 1975.
11. 'Much of the life of the capital . . .': see Isabelle Backouche, *La Trace du fleuve, la Seine et Paris 1750–1850* (Paris 2000).
12. 'To the west . . .': see William Howard Adams, *The French Garden 1500–1800* (London 1979).
12. 'Approaching the city . . .': Stacy Schieff, *A Great Improvisation: Franklin, France and the Birth of America* (New York 2005), p. 45.
14. 'Food was glorious . . .': Philip Mansel, *Prince of Europe: The Life of Charles-Joseph de Ligne: 1735–1814* (London 2003), p. 53.
17. 'the *Encyclopédistes* . . .': see Robert Darnton, *The Great Cat Massacre and Other Episodes in French Cultural History* (New York 1984); Roy Porter and Marie Mulvey Roberts (eds), *Pleasure in the 18th Century* (London 1966).
20. 'In rooms that were . . .': see Benedetta Craveri, *The Art of Conversation* (New York 2004).
21. 'the Duchesse de Mazarin . . .': Mme de Genlis, *Mémoires* (Paris 2004), p.183.
22. 'Rousseau's call . . .': see Robert Forster and Orest Ranum (eds), *Medicine and Society in France: Selection from Annales, Vol. 6* (Baltimore 1980).
24. 'The Archbishop shared . . .': Duc de Lauzun-Biron, *Mémoires* (Paris), p. 117.
25. 'Adèle d'Osmond . . .': *Mémoires de la Comtesse de Boigne* (Paris 1931), p. 43.

Chapter 2

27. 'By the 1770s . . .': see Eva Jacobs *et al.*, *Women and Society in 18th-century France* (London 1979); Mme la Comtesse de Miremont, *Traité de l'éducation des femmes* (Paris 1779); Royer Chartre, *L'éducation en France du XVI au* XVIII *siècles* (Paris 1976); Jean-Jacques Rousseau, *Oeuvres Complètes*, Vol. IV. (Paris 1959).
28. 'Like her mother . . .': see Jean Gallois (ed.,) *Musiques et Musiciens au Faubourg Saint-Germain* (Paris 1996); Duc de Lauzun-Biron, *Mémoires* (Paris).
29. 'There was another . . .': see Durand Echeverria, *Mirage in the West: A History of the French Image of American Society to 1815* (Princeton 1957); Hector St John de Crèvecoeur, *Letter from an American Farmer* (Paris 1782).
32. 'Turgot, the King's . . .': see Thomas E. Crow, *Painters and Public Life in 18th-century Paris* (Newhaven 1985).
32. 'Informed that they lived . . .': see Samuel Breck, *Recollections 1771–1862* (Philadelphia 1877).
33. 'His superior officers . . .': Vicomte de Noailles, *Marins et soldats français en Amérique: 1778–1783* (Paris 1903), p. 169.
35. 'Masses and celebratory . . .': see Jean Chalon (ed.), *Mémoires de Mme de Campan: Première femme de chambre de Marie Antoinette* (Paris 1988).
36. 'The court at Versailles . . .': Jacques Levron, *A la Cour de Versailles aux XVI–XVIII siècles* (Paris 1965), p. 297.
36. 'Clothes, like meals . . .': Philip Mansel, *Dressed to Rule* (New Haven 2005), p. 56.
37. 'botanise in a watered meadow . . .': Arthur Young, *Travels in France during 1787, 1788, 1789* (London 1905), p. 89.
37. 'For Lucie and her mother . . .': see Aileen Ribiero, *Dress in 18th-century Europe: 1715–1789* (New Haven 2002).
38. 'Smell, ever a . . .': Alain Corbin, *The Foul and the Fragrant: Odor and the French Social Imagination* (New York 1986), p. 74.
41. 'Despite the efforts . . .': see Robert Forster and Orest Ranum (eds), *Medicine and Society in France: Selection from Annales, Vol.* 6 (Baltimore 1980); Alessa Johns (ed.), *Dreadful Visitations: Confronting Natural Catastrophes in the Age of Enlightenment* (London 1999); Roy Porter (ed.), 'The Medical History of Water and Spas', *Medical History* Supp. No. 10 (London 1990).

Chapter 3

44. 'With Monsignor Dillon . . .': see Louis Audibert, *Le Dernier Président des Etats Généraux de Languedoc* (Bordeaux 1868); Bernard Plongeron, *La Vie quotidienne du clergé Français au XVIIIième siècle* (Paris 1974); Philippe Ariès et Georges Duby (eds), *Histoire de la vie privée*, Vol. 3 (Paris 1986).

50. '"Monsignor," he told . . .': Nigel Aston, *The End of an Elite: The French Bishops and the Coming of the Revolution 1786–1790* (Oxford 1992), p. 43.
54. 'The Duc de Chartres . . .': René Héron de Villefosse, *L'Anti-Versailles ou le Palais-Royal de Philippe Egalité* (Paris 1974), p. 196.
55. 'The English, estranged . . .': see J. M. Thompson, *English Witnesses of the French Revolution* (Oxford 1938); Constantia Maxwell, *The English Traveller in France 1658–1815* (London 1932).
55. '"We seem to want . . ."': Josephine Grieder, *Anglomania in France: Fact, Fiction and Political Discourse* (Geneva 1985), p. 25.
56. 'Two brothers . . .': Alistair Horne, *Seven Ages of Paris: Portrait of a City* (London 2002), p. 173.
57. 'For a while, Cagliostro . . .': see Stanislas Jean de Boufflers, *Vie* (Lille 1860).
57. 'And then there were the exotic animals . . .': see L. Robbins, *Elephant Slaves and Pampered Parrots: Exotic Animals in 18th Century Paris* (London 2002); Jean-Jacques Marquet de Vasselot, *La Ménagerie de Versailles: Revue de l'Histoire de Versailles* (Paris 1899).
58. 'It was, wrote Thomas Blaikie . . .': *Diary of a Scotch Gardener at the French Court at the End of the 18th Century* (London 1931), p. 142.
60. 'Louis, Prince de Rohan . . .' see Antonia Fraser, *Marie Antoinette* (London, 2001)
62. 'The Queen was now . . .': see Léonard Autie, *Recollections of Léonard, Hairdresser to Marie Antoinette* (London 1912); Emile Langlade, *Rose Bertin: The Creator of Fashion at the Court of Marie Antoinette* (London 1933).
70. 'The Princesse d'Hénin . . .': see *Revue d'Histoire de Versailles et de Seine et Marne* (Paris 1923), p. 83; Vicomtesse de Noailles, *Vie de la Princesse de Poix* (Paris 1855), p. 27.

Chapter 4

82. 'This was the Princesse de Beauvau . . .': Stanislas Jean de Boufflers, *Vie* (Lille 1860), p. 216.
83. 'In portraits . . .': Pierre Pluchon, *Nègres et Juifs au XVIIIième siècle* (Paris 1984), p. 135.
83. 'It was the Duc de Guines . . .': Comte d'Hézecques, *Page à la cour de Louis XVI* (Paris 1987), p. xxx.
83. 'Most of the habitués . . .': Duchesse d'Abrantès, *Histoire des Salons de Paris*, Vol. 1 (Brussels 1837) p. 46.
83. 'In 1786, the Neckers' . . .': see Maria Fairweather, *Madame de Staël* (London 2005).
84. 'In 1784, Jefferson . . .': see William Howard Adams, *The Paris Years of Thomas Jefferson* (New Haven 1997).
85. 'In the exalted . . .': Stacy Schieff, *A Great Improvisation: Franklin, France and the Birth of America* (New York 2005), p. 230.

90. 'Everywhere, people were poor . . .': see Olwen H. Hufton, *The Poor of Eighteenth-Century France* (Oxford 1974).
92. 'Among these aristocratic . . .': Michel Poniatowski, *Talleyrand et l'ancienne France: 1754–1789* (Paris 1988), p. 507.
93. '*Libelle* literature . . .': see Robert Darnton and Daniel Roche (eds), *Revolution in Print: The Press in France 1775–1800* (Berkeley 1989).
93. 'In *Charles IX* . . .': Simon Schama, *Citizens* (London 1989), p. 415.
97. '"Here drops" . . .': see Gouverneur Morris, *A Diary of the French Revolution* (London 1939).
97. 'Most of the Third . . .': see Timothy Tackett, *Par la Volonté du Peuple* (Paris 1997).

Chapter 5

104. 'It was rumoured . . .': J. Forneron, *Histoire générale des émigrés pendant la révolution française* (Paris 1884), p. 124; see also Simon Schama, *Citizens* (London 1989), p. 365.
105. 'At Versailles . . .': see Lucien de Clully, *La Tour du Pin* (Paris 1909).
105. '"The Queen's entourage . . ."': Jules Flammermont, *Les Correspondances des agents diplomatiques étrangers en France avant la Révolution* (Paris 1946), p. 247.
107. 'Even musicians were . . .': Jean Chalon (ed.), *Mémoires de Mme Campan* (Paris 1988), p. 282.
108. 'Pamphlets and news-sheets . . .': Robert Darnton and Daniel Roche, *Revolution in Print: the Press in France 1775–1800* (Berkeley 1989), p. 82.
108. 'They would serve, said Brissot . . .': John Brewer and Roy Porter (eds), *Consumption and the World of Goods* (London 1993), p. 412.
109. 'Like his father . . .': Château de Vêves, Private papers.
110. 'The neat white gowns . . .': Olwen H. Hufton. *Women and the Limits of Citizenship in the French Revolution* (Toronto 1992), p. 8.
115. 'After a first night . . .': Antonia Fraser, *Marie Antoinette* (London 2001), p. 283.
116. 'Paris, as Gouverneur Morris . . .': Gouverneur Morris, *A Diary of the French Revolution* (London 1939), p. 138.
116. 'In the first issue . . .': Alison Ribiero, *Dress in Eighteenth-Century Europe 1715–1789* (New Haven 2002) p. 53.
116. 'Before they finally . . .': see Edna Hindie Lemay, *La Vie quotidienne des députés aux Etats Généraux: 1789* (Paris 1987).
117. 'One of these was . . .': see François Furet and Mona Ozouf, *A Critical Dictionary of the French Revolution* (London 1989).
119. 'Horace Walpole, hearing . . .': Duff Cooper, *Talleyrand* (London 1932), p. 57.
120. 'This obsession with . . .': Harold T. Parker, *The Cult of Antiquity and the French Revolution* (Chicago 1937), p. 35; and see Alfred Copin, *Talma et la Révolution Française* (Paris 1887).

121. 'With them came . . .': see Georges Snyders, *Le Goût Musical en France au XVIIième et XVIIIième siècles* (Paris 1968).
123. '"It is time . . ."': see Paul Carbonel, *Histoire de Narbonne des Origines à l'époque contemporaine* (Narbonne 1956).

Chapter 6

125. 'The first French . . .': Dorette Berthoud, *Le Général et la Romancière* (Neuchâtel 1959), p. 150.
125. 'Mme de Staël herself loved . . .': Lucie Achaud, *Rosalie de Constant, sa famille et ses amis*, Vol. 2 (Paris n.d.), p. 12.
126. 'But among the émigrés . . .': Philippe Godet, *Mme de Charrière et ses Amis*, Vol. 1 (Geneva 1906), p. 400.
127. 'Soon after reaching . . .': M. de Bouillé, *Mémoires sur la révolution française* (London 1797), p. 161.
128. '"The army," he warned . . .': Lucien de Clully, *La Tour du Pin* (Paris 1909), p. 157.
130. 'In the Tuileries . . .': see Comte de Ségur, *Mémoires ou Souvenirs et Anecdotes*, Vol. 3 (Paris 1824), p. 590.
131. 'In caricatures . . .': [Anonymous pamphlet], *La Ménagerie Nationale* (Paris 1790).
132. 'Paris in the winter . . .': J. G. Alger, 'British Colony in Paris 1792–1793', *English Historical Review* 13 (1898), p. 25. And see J. G. Alger, *Englishmen of the Revolution* (London 1889).
133. 'Burke's view . . .': see Jacques Godechot, *Le Directoire vu de Londres: Annales historiques de la Révolution française* (Paris 1950).
133. 'Though he attacked . . .': Robert Forster, 'The Survival of the Nobility during the French Revolution', *Past and Present* 37 (July 1967), p. 186.
135. 'On the eve of . . .': Michel Poniatowski, *Talleyrand et les années occultées: 1789–1792* (Paris 1995), p. 292.
135. '"The people," he warned . . .': Colin Jones, *The Great Nation: France from Louis XV to Napoleon* (London 2002), p. 435.
136. 'On 2 April . . .': Simon Schama, *Citizens* (London 1989), p. 464.
137. 'Among its early members . . .': Lilian Crété, *La Traité des nègres sous l'ancien régime* (Paris 1989), p. 258.
137. 'But the Antilles . . .': see Gabriel Debien, *Les esclaves aux Antilles françaises, XVII–XVIIIième siècles* (Gaudeloupe 1974).
137. 'As Montesquieu . . .': Pierre Pluchon, *Nègres et Juifs au XVIIIième siècle: Le racisme et le siècle des lumières* (Paris 1984), p. 135.
138. 'Women, however . . .': see Vera Lee, *The Reign of Women in 18th Century France* (Cambridge MA 1975); Richard Rand, *Intimate Encounters: Love and Domesticity in 18th-Century France* (Princeton 1997); Carla Hesse, *The Other Enlightenment* (Princeton 2001).
140. 'The city was quiet . . .': see Henry-Paulin Panon Desbassayns, *Voyage à Paris pendant la révolution: 1790–1792* (Paris 1985).
140. '*Mesdames Tantes* . . .': see *Livre Journal de Madame Éloffe,*

marchande de modes, couturière, lingère ordinaire de la reine et des dames de sa Cour. Vols 1 and 2 (Paris 1885).

142. 'Before hearing of their capture . . .': John Keane, *Tom Paine: A Political Life* (London 1995), p. 317.
142. '"People call him . . ."': Antonia Fraser, *Marie Antoinette* (London 2001), p.326.
143. '"To aggravate . . ."': William Augustin Miles, *Correspondence on the French Revolution: 1789–1817* (London 1819), p. 253.

Chapter 7

144. 'When doubts . . .': Alphonse Aulard, *Études et leçons sur la Révolution française*, Vol. 2 (Paris 1914) p. 281.
144. 'Even so, Mme de Staël . . .': Mme de Staël, *Seize lettres inédites de Madame de Staël à Gouvernet*, ed. Charles de Portairols (Paris 1913), 11 March 1791.
146. 'The word "émigré" . . .': see Kirsty Carpenter and Philip Mansel (eds), *The French Émigrés in Europe and the Struggle against Revolution* (London 1999); Jean Vidalenc, *Les Emigrés français, 1789–1825* (Caen 1963).
146. '*L'émigration élégante* . . .': see Vicomte de Broc, *Dix ans d'une femme pendant l'émigration* (Paris 1893).
147. '"The Dutch," he wrote . . .': Ministère des Affaires Etrangères, Holland, letter of 31 January 1792. (Paris)
148. 'Robespierre had about him . . .': see J. G. Millingen, *Recollections of Republican France between 1790–1801* (London 1848).
149. 'Soon, men all over . . .': Simon Schama, *Citizens* (London 1989), p. 508.
150. 'On 20 June, a mob . . .': Antonia Fraser, *Marie Antoinette* (London 2001), p. 343.
154. 'Arthur immediately declared . . .': see Théodore de Lameth, *Notes et Souvenirs* (Paris 1914).
155. 'As a friend observed . . .': General Bertrand, *Lettres à Fanny*, ed. Suzanne de la Vaissière-Orfila (Paris 1978), p. 70.
155. '"We cannot be calm . . ."': see Robert and Isabelle Tombs, *That Sweet Enemy: The French and the British from the Sun King to the Present* (London 2006).
161. 'It was no longer . . .': see Aileen Ribiero, *Fashion in the French Revolution* (London 1988).
163. 'Louis had been . . .': see Comte d'Hézecques, *Page à la cour de Louis XVI* (Paris 1987).
164. 'An account of the trial . . .': see John Moore, *A Journal during a Residence in France from the Beginning of August to the Middle of December 1792* (Boston 1794).
166. 'An English visitor . . .': Frédéric Masson, *Le Département des Affaires Etrangères pendant la révolution: 1787–1804* (Paris 1877), p. 271.

167. 'Lucie was extremely reluctant . . .': Archives Nationales, Paris, Series T 595/281 1–4.

Chapter 8

170. 'On 1 July . . .': Archives Nationales, Paris, W/345.
171. 'Among his few possessions . . .': Archives Nationales, Paris, F/17/1195.
171. 'In the national . . .': Archives Nationales, Paris, T281.
172 'With the murder . . .': see F. W. Blagdon, *Paris as It Was and as It Is: 1801–1802* (London 1803), p. 127; *Thermomètre du Jour*, 2 August 1793.
173. 'M. de la Tour du Pin . . .': Lucien de Clully, *La Tour du Pin* (Paris 1909), p. 366.
174. 'On Marie Antoinette's . . .': Hector Fleischmann, *Behind the Scenes in the Terror* (London 1914), p. 65.
175. 'Every day . . .': see C. A. Dauban, *Les Prisons de Paris sous la Révolution* (Paris 1870); *Le Moniteur*, 4 September 1793.
176. 'In Bordeaux, "as in Paris" . . .': see Raymond Celeste, *Les anciennes sociétés musicales* (Bordeaux 1900); Camille Jullian, *Histoire de Bordeaux* (Bordeaux 1895); Alan Forrest, *Society and Politics in Revolutionary Bordeaux* (Oxford 1975); J. L. Barraud, *Vieux Papiers Bordelais* (Paris 1910).
177. 'Bordeaux would not experience . . .': see Aurélien Lignereux, *Gendarmes et policiers dans la France de Napoléon* (Paris 2002); P. Bécamps, 'Détenus et proscrits pendant la Révolution à Bordeaux', *Revue Historique de Bordeaux* (1958); Anne de Mathau, *Mémoires de Terreur: L'an 11 à Bordeaux* (Bordeaux 2002).
184. 'Jeanne-Marie-Ignace-Thérésia . . .': see Christian Gilles, *Madame Tallien: La reine du Directoire* (Biarritz 1999); Thérèse Charles-Vallin, *Tallien: le mal aimé de la Révolution* (Paris 1997); Comte de Paroy, *Mémoires* (Paris 1895).
184. 'Even Gouverneur . . .': Gouverneur Morris, *A Diary of the French Revolution* (London 1989), p. 138.
190. 'He was to be replaced . . .': see Pierre Gascar, *L'Ombre de Robespierre* (Paris 1979).
193. '"Heads," remarked . . .': see Remy Bijaoni, *Prisonniers et Prisons de la Terreur* (Paris 1996); Jean-Paul Bertrand, *La Vie quotidienne en France au temps de la Révolution* (Paris 1983).
193. 'After Hébert . . .': see Jean-Paul Bertrand, *Camille et Lucile Desmoulins: Un couple dans la tourmente* (Paris 1986).
193. 'Dillon, said his accusers . . .': Archives Nationales, Paris, W/345.
193. '"If," he had written . . .': *Le Vieux Cordelier*, 8 July 1793.

Chapter 9

195. 'Sturdy . . .': see Melvin Maddocks, *The Atlantic Crossing* (Alexandria 1981).

199. 'By 1794 . . .': see Warren S. Tryon, *A Mirror for Americans: Life and Manners in the US 1790–1870* (Chicago 1952); Durand Echeverria (ed.), *Mirage in the West: A History of the French Image of American Society up to 1815* (Princeton 1957); Beatrice F. Hyslop, 'American Press Reports of the French Revolution: 1789–1794', *New York Historical Society Quarterly* (October 1958), p. 329.
200. 'Some of these . . .': see J. G. Rosengarten, *French Colonists and Exiles in the United States* (Philadelphia 1907), p. 126. See also Echeverria, op. cit.; J. P. Brissot de Warville, *New Travels in the United States of America: 1788* (Cambridge, MA 1964); Alexandre Capitaine, *La Situation économique et sociale des États Unis à la fin du XVIIIième siècle d'après les voyageurs français* (Paris 1926); Marquis de Chastellux, *Travels in North America in the Years 1780, 1781, 1782* (New York 1963); Benjamin Franklin, *Information to Those Who Would Remove to America* (London 1784).
203. 'It is Dillon . . .': *Le Moniteur*, 14 April 1794.
203. 'Not long before . . .': Château de Vêves, Private papers. Letter of 29 frimaire.
203. 'At six o'clock . . .': *Independent Chronicle and Universe*, 5 June 1794.
203. 'The frail . . .': see Helen Maria Williams, *Letters containing a Sketch of the Politics of France from 31 May 1793 to 28 July 1794* (Dublin 1795).
204. 'On the sandy . . .': see Anne Grant, *Memoirs of an American Lady* (Albany 1876); Tom Lewis, *The Hudson* (Virginia 2005); Count Paolo Andreani, *Along the Hudson and the Mohawk* (Philadelphia 2006).
205. 'While the 1783 . . .': see Alan Taylor, *The Divided Ground: Indians, Warriors, Settlers and the Northern Borderland of the American Revolution* (New York 2006).
207. 'It depicted Louis . . .': *The Albany Register*, 14 August 1794.
209. 'if I have to stay . . .': Echeverria, op. cit. p. 183.
209. 'To your care . . .': Library of Congress, Washington, DC, Hamilton Papers.
209. 'as old as the world . . .': Michel Poniatowski, *Talleyrand aux États Unis: 1794–1796* (Paris 1967), p. 371.
210. 'Among the last. . .': unpublished journal of Philippe de Noailles. Archives Nationales, Paris.
212. 'With 150 acres . . .': see Ira Berlin, *Generations of Capitivity* (Cambridge, MA 2003); Simon Schama, *Rough Crossings: Britain, the Slaves and the American Revolution* (London 2005); Thomas F. Gossett, *Race: The History of an Idea in America* (New York 1997).
217. 'Later, in his immensely . . .': Duc de la Rochefoucault Liancourt, *Travels through the United States of America, the Country of the Iriquois and Upper Canada in the years 1795, 1796 and 1797* (London 1799), p.383.
217. 'M. du Petit-Thouars . . .': see Rosengarten, op. cit., p. 137; Bergasse du Petit-Thouars (ed.), *Aristide Aubert du Petit-Thouars: héro d'Aboukir 1760–1798: lettres et documents inédits* (Paris 1937).

219. 'the *Albany Register* carried . . .': the *Albany Register*, 14 August 1795.
219. 'Nearly all the larger . . .': see Isaac Weld, *Travels through the States of North America during the Years of 1795, 1796 and 1797* (London 1799).
221. '"a noble temple . . ."': Comte de Ségur, *Mémoires ou Souvenirs et Anecdotes*, Vol. 3 (Paris 1824), p. 389.
221. 'There was the Vicomte . . .': see Comte de Volney, *A View of the Soil and Climate of the United States of America* (New York 1968), p. 364.
221. 'From Paris too . . .': *Le Courrier Français*, Philadelphia, August 1795.
222. 'Jefferson, who was . . .': see Waverley Root and Richard de Rochemond, *Eating in America: A History* (New York 1976).
223. 'As Hamilton observed . . .': *Papers of Alexander Hamilton*, Vol. XVII (New York 1972), p. 587.

Chapter 10

232. 'Though Frédéric's name . . .': Archives Nationales, Paris, F/7/5990.
232. 'Now Le Bouilh . . .': Archives Municipales, Bordeaux, Inventory '*Ci-devant Chateau de Bouilh*'.
233. 'their properties stripped down . . .': Archives Municipales, Bordeaux, Box 1, File 43.
234. 'Up and down the country . . .': see Marcel Marion, *Le Brigandage pendant la Révolution* (Paris 1934).
235. 'Though the streets . . .': see Pierre Chauvet, *Essai sur la Propreté de Paris* (Paris 1798).
235. 'At least 14 . . .': see Frédéric Jean Laurent Meyer, *Fragments sur Paris* (Hamburg 1790).
236. 'a depravation . . .': Helen Maria Williams, *Letters containing a Sketch of the Politics of France from 31 May 1793 to 28 July 1794* (Dublin 1795), p. 29.
236. 'At balls, lit . . .': Madame de Bawr, *Mes Souvenirs* (Paris 1853), p. 166.
237. 'In the Jardin des Plantes . . .': see Paul Lacroix, *Directoire, Consulat et Empire: moeurs et usages, lettres, sciences et arts en France* 1795–1815 (Paris 1884); Édmond and Jules de Goncourt, *Histoire de la société française pendant le Directoire* (Paris 1840); *Au temps des merveilleuses: la société parisienne sous le Directoire et le Consulat*, Musée Carnavalet (Paris 2005).
239. 'the supreme *bon ton* . . .': Jacques Godechot, *La Vie quotidienne en France sous le Directoire* (Paris 1977), p. 102.
240. 'There were melancholy . . .': Goncourt and Goncourt, op. cit., p. 38.
241. 'When she dressed . . .': Christian Gilles, *Madame Tallien: La Reine du Directoire* (Biaritz 1999) p. 267.
244. 'The *Journal de Paris* . . .': *Journal de Paris*, 30 July 1797. See also Maurice Herbette, *Une ambassade turque sous le Directoire* (Paris 1802).

246. 'Within hours . . .': see Victor Pierre, 'Les émigrés et les commissions militaires après fructidor', *Revue des Questions Historiques*, Paris October 1884.

Chapter 11

248. 'By 1797 . . .': see Kirsty Carpenter and Philip Mansel, *The French Émigrés in Europe and the Struggle against Revolution* (London 1999); Jacques Godechot, *Le Directoire vu de Londres: Annales de la Révolution française* (Paris 1950).
248. '"*La patrie* . . ."': see Micheline de Vallée, *Les Emigrés de 1793* (Segueville-en-Bersin 1991).
249. 'Though by 1797 . . .': see Robin Eagles, *Francophilia in English Society: 1748–1815* (Basingstoke 2000).
250. 'The violence and confusion . . .': Diana Donald, *The Age of Caricature: Satirical Prints in the reign of George III* (London 1996), p. 142.
250. '*The Times* warned . . .': *The Times*, 9 March 1797.
250. 'reports of "stout . . ."': National Archives, HO 1/3, Emigré correspondence. GLRO, Kew.
251. 'London at the end of . . .': see Christopher Hibbert, *The English: A Social History: 1066–1945* (London 1987), and Christopher Hibbert, *London: The Biography of a City* (London 1969); Roy Porter, *English Society in the 18th Century* (London 1982); François Crouzet, 'England and France in the 18th Century', in *Social Historians in Contemporary France* (New York 1972); Matthew O. Grenby, 'Révolution française et Littérature Anglaise', *Annales historiques de la Révolution Française* 4 (2005), pp. 101–44.
252. 'where visitors were warned . . .': Peter Thorold, *The London Rich* (London 1999), p. 157.
252. 'The French also remarked . . .': see Mme Vigée-Lebrun, *Souvenirs* (Paris 1984).
253. 'It was the fog . . .': Comte de Montloisier, *Souvenirs d'un émigré: 1791–1798* (Paris 1951), p. 187.
253. 'Just occasionally . . .': Josephine Grieder, *Anglomania in France. Fact, Fiction and Political Discourse* (Geneva 1985), p. 57.
255. 'One of the servants . . .': Julien Sapori, private communication.
255. 'the six elderly bishops . . .': Archives Nationales, Paris, F/1/4336.
257. 'Cossey Hall . . .': see Ernest G. Gage, *Costessey Hall* (Norwich 1991).
258. 'Ever practical . . .': see *The Jerningham Letters*, 1780–1843 (London 1896), Summer 1795.
259. 'Those who had not grown . . .': Johanna Schopenhauer, *A Lady Travels* (London 1988), p. 155; Carpenter and Mansel, op. cit., p. 64.
260. '*The Times* . . .': *The Times*, 9 January 1793.
260. 'Our fortunes . . .': National Archives, Kew, London, Bouillon Papers, PC/1/118A.

260. 'M. de Rodire . . .': National Archives, Kew, London, Bouillon Papers, PC/118AB; T93.9; T93.57.
261. 'Could some of these . . .': *The Times*, 30 August 1796.
261. 'They went, when they had . . .': Porter, op. cit. p. 257.
261. 'The Abbé Tardy . . .': see *Manuel du voyageur à Londres* (London 1800).
262. 'monks entertained . . .': see Pierre Bessand-Massenet, *Les Deux Frances: 1799–1804* (Paris 1949).
262. 'The Comtesse de Guery . . .': M. le Vicomte Walsh, *Souvenirs de Cinquante Ans* (Paris 1845), p. 160.
262. 'when guests left . . .': see Baron Portalis, *Henry Pierre Danloux, peintre de portraits et son journal: 1753–1809* (Paris 1910).
263. 'the Archbishop offered . . .': Private Dillon family papers, 13 September 1797.
264. 'Richmond lay . . .': Thorold, op. cit., p. 66; see also Judith Fitson, *French Refugees in Richmond: 1785–1815* (Richmond 1998).
265. 'Horace Walpole, living . . .': see *Correspondence*, 2 vols (London 1851), 27 September 1791.
265. 'When the Princesse d'Hénin . . .': Linda Kelly, *Juniper Hall* (London 1991), p. XIV.
265. 'Mme de Staël's brilliance . . .': Maria Fairweather, *Madame de Staël* (London 2005), p. 171.
266. 'the fattest . . .': Adèle de Boigne, *Mémoires: Souvenirs d'une tante,* 4 vols (Paris 1908), Vol. 3, p. 8.
266. 'The once famously . . .': see Amanda Foreman, *Georgiana, Duchess of Devonshire* (London 1998) and Amanda Foreman, *Georgiana's World* (London 2001).
269. 'her childhood friend, Amédée . . .': see A. Bardoux, *La Duchesse de Duras* (Paris 1898).
273. 'Napoleon's eye . . .': J. Christopher Herald, *The Age of Napoleon* (London 1963), p. 78.

Chapter 12

278. 'Several of the most unpopular . . .': see Aurélien Lignereux, *Gendarmes et Policiers dans la France de Napoléon* (Paris 2002).
278. '"It has become . . ."': Vicomte de Broc, *Dix ans d'une femme pendant l'émigration* (Paris 1893), p. 289.
279. 'Frédéric had written . . .': Library of Congress, Washington, DC, Hamilton Papers, ALS, 21 February 1798.
280. 'most other émigrés . . .': Catherine Wilmot, *An Irish Peer on the Continent. 1801–1803* (London 1924), p. 72.
280. 'One of the first . . .': Henri d'Alméras, *La Vie parisienne sous le Consulat et l'Empire* (Paris 1909), p. 365; see also Edmond and Jules de Goncourt, *Histoire de la société française pendant la Révolution* (Paris 1889).

280. 'Guests quickly . . .': Duchesse d'Abrantès, *Histoire des salons de Paris*, Vol. 2 (Brussels 1837), p. 9.
280. 'According to her great-niece . . .': Joseph Turquan, *Les femmes de l'émigration: 1789–1814* (Paris 1911), p. 289.
281. 'Musical soirées . . .': see Edmond and Jules de Goncourt, *Histoire de la société française pendant le Directoire* (Paris 1840).
282. 'She had a rival . . .': Adèle de Boigne, *Mémoires: Souvenirs d'une tante*, 4 vols (Paris 1908), Vol. 2 p. 177.
283. 'After dinner . . .': J. F. Reichardt, *Un hiver à Paris sous le Consulat: 1802–1803* (Paris 1850), p. 73.
283. 'These new salons . . .': Sophie Gay, *Salons célèbres* (Paris 1837), p. 22.
284. 'Napoleon preferred . . .': Marie-Blanche d'Arneville, *Parcs et jardins sous le premier Empire* (Paris 1981), p. 31.
285. 'Consular Paris smelt . . .': Alain Corbin, *The Foul and the Fragrant: Odor and the French Social Imagination* (New York 1986), p. 196.
285. 'It made women . . .': Mme de Genlis, *Mémoires* (Paris 2004), p. 324.
286. 'Napoleon let it . . .': see Mme de Rémusat, *Mémoires* (London 1880), Vols 1 and 2.
289. 'Under Napoleon's drive . . .': see Michel Figeac, *Destins de la noblesse bordelaise* (Bordeaux 1996).
290. 'It was from her visitors . . .': see J. G. Lemaistre, *A Rough Sketch of Modern Paris* (London 1803).
290. 'There was also news of Talleyrand . . .': Duff Cooper, *Talleyrand* (London 1932), p. 134.
293. 'With the peace came . . .': see *The Journal of Bertie Greethead: An Englishman in Paris* (London 1853); August von Kotzebue, *Travels from Berlin through Switzerland to Paris in the year 1804* (Paris 1850); John Goldsworth Alger, *Napoleon's British Visitors and Captives 1801–1815* (London 1904).
295. 'There was much jostling . . .': Thomas Thornton, *A Sporting Tour through France in the Summer of 1802* (London 1806), p. 125.
297. 'Around Notre Dame . . .': Marie-Louise Bivet, *Le Paris de Napoléon* (Paris 1963), p. 312.
297. 'When Napoleon entered . . .': José Cabanis, *Le Sacre de Napoléon* (Paris 1970), p. 23.
299. 'According to a malicious . . .': Joseph Turquan, *Mme de Montesson: Douainière d'Orléans: 1738–1806* (Paris 1904), p. 289.
300. 'The Archbishop spent . . .': Adèle de Boigne, op. cit., p. 142.

Chapter 13

308. 'My prefects . . .': see Jean Savant, *Les Préfets de Napoléon* (Paris 1958).
308. 'I have often been told . . .': Archives Nationales, Paris, Dossier Prefects F1b1/166/15 Ministère des Affaires Etrangères.

309. 'Over his prefects . . .': Jacques Regnier, *Les Préfets du Consulat et de l'Empire* (Paris 1907), p. 26.
310. 'Belgium had been . . .': see Jean Cathelin, *La Vie quotidienne en Belgique sous le régime français 1792–1815* (Paris 1966); Felix Maguette, *Les Émigrés français aux Pays Bas* (Brussels 1907); L. de Lanzac de Laborie, *La Domination française en Belgique* (Paris 1895); Janet Polasky, *Revolution in Brussels: 1787–1793* (Brussels 1987).
314. 'The new four-horse . . .': see Jean Robiquet, *La Vie quotidienne au temps de Napoléon* (Paris 1942).
315. '"I am happy for you . . ."': Archives Municipales, Chateauxroux, Bertrand private papers.
316. 'At home, the Senate . . .': Alphonse Aulard, *Etudes et leçons sur la Révolution française*, Vol. 8 (1914), p. 291.
316. 'illustrations showing . . .': Anne-Marie Kleinert, *Le Journal des Dames et des Modes* (Stuttgart 2001), p. 291.
316. 'But the city of Paris . . .': see Jean Tulard, *Le Grand Empire 1804–1815* (Paris 1982); and Jean Tulard, *Napoléon et la noblesse d'Empire* (Paris 1979).
317. '"Adopt neither . . ."': Comtesse de Bradi, *Du Savoir-Vivre en France au XIXième siècle* (Paris 1858), p. 31.
318. '"I am sorry for you . . ."': Henri d'Alméras, *La Vie parisienne sous le Consulat et l'Empire* (Paris) p. 309.
318. 'He was a man': see Mme de Rémusat, *Mémoires* (Paris 1880).
319. 'Grimod de la Reynière . . .': Giles MacDonogh, *A Palate in Revolution* (London 1987), p. 201.
320. 'She continued . . .': Mme de Staël to Talleyrand, see Maria Fairweather, op. cit., 3 April 1808.
320. 'Before returning to Paris . . .': Château de Vêves, Private papers; see A. Bardoux, *La Duchesse de Duras* (Paris 1898); G. Pailhes, *La Duchesse de Duras et Chateaubriand* (Paris 1910).
325. 'As Louis-Antoine Bourrienne . . .': Serje Grandjean, *Inventaire après décès de l'Impératrice Joséphine à Malmaison* (Paris 1965), p. 41.
326. 'The proxy marriage . . .': See Prince Charles de Clary-et-Aldringen, *Trois mois à Paris lors du mariage de l'Empereur Napoléon 1er et de l'Archduchesse Marie-Louise* (Paris 1914).
327. 'On 28 April . . .': see Charlotte de Sor, *Napoléon en Belgique et en Hollande, 1811* (Brussels 1839).
334. 'The Comte de Merode . . .': see *Souvenirs* (Brussels 1872).
335. 'Posters were seen . . .': Tulard, op. cit., p. 158.
336. 'You will laugh at me . . .': Private papers, Lucie to Mme de Duras, 8 May 1811.
336. 'Better still . . .': Archives Nationales, Paris, F/1b1/166/15.
336. 'but not before Lucie . . .': General Bertrand, *Lettres à Fanny*, ed. Suzanne de la Vaissière-Orfila (Paris 1978), p. 289.
336. 'One morning, when Lucie was . . .': Archives Nationales, Paris, Dossiers Personnels: de la Tour du Pin.

Chapter 14

339. 'Frédéric, she wrote . . .': see Angélique de Maussion, *Les Rescapés de Thermidor* (Paris 1975).
341. 'Her one fault . . .': Private papers, letter to Mme de Duras, 25 July 1813.
343. 'His treachery . . .': see Robin Harris, *Betrayer and Saviour of France* (London 2006).
344. 'Wurtembergers had . . .': Archives Nationales, Paris, AF1V 1670.
345. 'As a young man . . .': see Philip Mansel, *Louis XVIII* (London 1981); T. E. B. Howarth, *Citizen King* (London 1961).
346. 'In 1799, his niece . . .': Gilbert Stenger, *Grandes dames du XIXième siècle* (Paris 1911), p. 12.
347. 'Next day, the Senate . . .': see Robert Christopher, *Napoleon on Elba* (London 1964).
348. 'The Cossacks . . .': Mme de Chastenay, *Mémoires: 1771–1815* (Paris 1987), p. 506.
351. 'For the moment, the imperfections . . .': see Philippe Sussel, *La France et la bourgeoisie: 1815–1850* (Paris 1970).
352. '"We must thank . . ."': José Cabanis, *Charles X, roi ultra* (Paris 1972), p. 59.
352. 'Just the same . . .': Adèle de Boigne, *Mémoires: Souvenirs d'une tante*, 4 vols (Paris 1908), Vol. 3. p. 298.
354. 'Talleyrand's own entourage . . .': see Philip Ziegler, *The Duchess of Dino* (London 1985).
354. 'Vienna, in September . . .': Duff Cooper, *Talleyrand* (London 1932), p. 244; see also Philip Mansel, *Prince of Europe: The Life of Charles-Joseph de Ligne 1735–1814* (London 2003).
357. 'Her epitaph on Napoleon . . .': Henri Rossi, *Mémoires aristocratiques feminins 1789–1848* (Paris 1998).
358. 'The reaction of French society . . .': See Anne Martin-Fugier, *La vie élégante* (Paris 1990)
359. 'Not everyone agreed . . .': Anne-Marie Kleinert, *Le Journal des Dames et des Modes* (Stuttgart 2001), p. 220.
359. 'Writing to Castlereagh . . .': Beckles Wilson, *The British Embassy* (London 1927), p. 33.
360. 'Frédéric, declaring . . .': Château de Vêves, Family papers.
361. 'To Mme de Staël . . .': Château de Vêves Family papers, 5 April 1815.
363. 'the King spent hours at table . . .': see Theo Fleischman, *Le Roi de Gand* (Brussels 1953).
363. 'Brussels was immensely . . .': Château de Vêves Private papers, letter of 7 May 1815.
363. '"This is without . . ."': see Lady Caroline Capel and Dowager Countess of Uxbridge, *The Capel Letters* (London 1955).
364. 'Wellington had reached . . .': see Theo Fleischman and Winant Aerts,

Bruxelles pendant la Bataille de Waterloo (Brussels 1956); Richard Holmes, *Wellington: The Iron Duke* (London 2003); Sir William Fraser, *Words on Wellington* (London 1889); Comte d'Haussonville, *Ma jeunesse 1814–1830* (Paris 1885).

Chapter 15

366. 'Alexandre Mercier . . .': *Journal de la Campagne de Waterloo* (Paris 1933), p. 106.
366. 'Chateaubriand . . .': see *Mémoires d'Outre-Tombe* (Paris 1849).
367. '"We have conquered . . ."': Guillaume de Bertier de Sauvigny, *Metternich et la France après le Congrès de Vienne* (Paris 1968), p. 120.
368. '"Mercy," observed . . .': T. E. B. Howarth, *Citizen King* (London 1961), p. 131.
370. 'St Helena . . .': see Betsy Balcombe, *To Befriend an Emperor* (Welwyn Garden City 2005); Frédéric Masson, *Napoléon à Sainte Hélène* (Paris 1912); Barry E. O'Meara, *Napoléon dans l'exil* (Paris 1993).
373. '"What shall I tell you . . ."': Private papers, Château de Vêves letter 19 May 1816.
373. 'Its only drawback . . .': Edmund Boyce, *The Belgian Traveller* (London 1816), p. 29.
373. 'Frédéric's official . . .': Archives Nationales, Paris, Ministère des Affaires Étrangères, Hollande, Vol. 617, p. 380.
374. 'In The Hague . . .': Archives Nationales, Paris, Ministère des Affaires Étrangères, Hollande/Pays Bas, 1817–1818, 618.
375. 'Mme de Staël, too . . .': Adèle de Boigne, *Mémoires: Souvenirs d'une tante* (Paris 1908), Vol. 3, p. 366.
375. 'these were the "green . . ."': José Cabanis, *Charles X, roi ultra* (Paris 1972), p. 122.
376. 'Louis XVIII desired . . .': see Philip Mansel, *The Court of France: 1789–1830* (Cambridge 1988).
378. '"Ah my God!" . . .': G. Pailhes, *La Duchesse de Duras et Chateaubriand* (Paris 1910), p. 128.
378. 'On 13 February . . .': see Duchesse de Maille, *Souvenirs des deux restaurations* (Paris 1984).

Chapter 16

381. 'They came to see . . .': see James Fenimore Cooper, *Excursions in Italy* (London 1838); William Hazlitt, *Notes of a Journey through France and Italy* (London 1826); Jeremy Black, *The British and the Grand Tour* (London 1985).
382. 'A Roman colony . . .': see John Chetwood Eustace, *A Classical Tour through Italy* (London 1841); Marianne Baillie, *First Impressions on a Tour upon the Continent in the Summer of 1818* (London 1819), William M. Johnston, *In Search of Italy* (London 1987).

383. 'For a French ambassador . . .': see Denis Mack Smith, *The Making of Italy: 1796 –1866* (London 1968), and Denis Mack Smith, *Cavour* (London 1985); G. de Bertier de Sauvigny, *Metternich et la France après le Congrès de Vienne* (Paris 1968).
384. 'Frédéric, quickly . . .': Archives Nationales, Paris, Ministère des Affaires Étrangères, Sardaigne 287/1820.
385. '"Will they" . . .': Château de Vêves, Private papers, letter of 14 February 1821.
385. 'After Naples . . .': Archives Nationales, Paris, Ministère des Affaires Étrangères, Sardaigne, 9 February 1821.
385. 'He requested . . .': see Lady Theresa Lewis (ed.), *Journals and Correspondence of Miss Berry from the Years 1782–1852* (London 1866).
389. 'Bertrand, constantly . . .': see *The Jerningham Letters, 1780–1843* (London 1896).
390. 'Not long before, Charlotte . . .': Château de Vêves, Family papers. Private diary.
392. 'By early 1824 . . .': José Cabanis, *Charles X, roi ultra* (Paris 1972), p. 289.
392. 'On the afternoon . . .': Alain Corbin, *The Foul and the Fragrant: Odor and the French Social Imagination* (New York 1986), p. 122.
394. '"I have always . . ."': Château de Vêves, Family papers, letter of 6 January 1823.
395. '"Perhaps I am . . ."': Château de Vêves, Family papers, letter of February 1824.
396. 'For many travellers . . .': Benjamin Colbert, *Shelley's Eye: Travel Writing and the Aesthetic Vision* (London 2005), p. 125.
396. 'The Piazza di Spagna . . .': Maurice Andrieux, *Les Français à Rome* (Paris 1968), p. 354.
397. 'She was still very beautiful . . .': E. J. Delécluze, *Impressions Romaines: Carnet de Route d'Italie: 1823–1824* (Paris 1942), p. 36.
398. '"But I hardly . . ."': Château de Vêves, Family papers, letter of 28 November 1825.
402. 'Frédéric, as outspoken . . .': Henri Contamine, *Diplomatie et diplomates sous la Restauration* (Paris 1970), p. 206.

Chapter 17

404. 'Spring . . .': see Alain Corbin, *The Foul and the Fragrant: Odor and the French Social Imagination* (New York 1986); Anne Martin-Fugier, *La Vie élégante* (Paris 1990); Philip Mansel, *The Court of France: 1789–1830* (Cambridge 1988).
405. 'At court . . .': T. E. B. Howarth, *Citizen King* (London 1961), p. 132.
405. 'In this outfit . . .': see Comtesse d'Agoult, *Souvenirs et journaux* (Paris 1990).
409. 'But the Dauphin . . .': José Cabanis, *Charles X, roi ultra* (Paris 1972), p. 442.

411. '"My one wish . . ."': Château de Vêves, Family papers, letter to Hadelin, n.d.
412. 'Among those . . .': see Thérèse Rouclette, *La Folle Équipée de la Duchesse de Berry* (La Roche sur Yon 2004); Général Dermoncourt, *La Vendée et Madame* (Paris 1834); Gustave Gautherot, *L'héroique Comtesse: Correspondance de la Comtesse Auguste de la Rochejacquelein* (Paris 1922).
414. 'It was at this point that Aymar . . .': Château de Vêves, Family papers, unpublished memoir by Aymar de la Tour du Pin.
418. 'The Fort du Hâ . . .': See Jean-Jacques Déogracias, *Le fabuleux destin du Fort du Hâ* (Bordeaux 2006).
423. 'The walls of Paris . . .': see Fanny Trollope, *Paris and the Parisians* (London 1836); Duchesse de Maillé. *Souvenirs des deux restaurations* (Paris 1984).

Chapter 18

431. 'Their first glimpse . . .': see M. Curreli and A. L. Johnson (eds), *Paradise of Exiles: Shelley and Byron in Pisa* (Salzburg 1988).
432. '"I am busy describing . . ."': Château de Vêves, Family papers, letter to Félicie. n.d.
433. '"Our civilisation . . ."': T. E. B. Howarth, *Citizen King* (London 1961), p. 304.

Time Line

1770	25 February	Birth of Lucie-Henriette Dillon in the rue du Bac, Paris
1774		Louis XVI and Marie Antoinette ascend the throne of France
1778	5 April	Departure of Lucie's father, Arthur Dillon, for the American War of Independence
1782	8 September	Death of Lucie's mother, Thérèse-Lucy Dillon
1787	22 May	Lucie marries Frédéric de Gouvernet and is presented at court
1789	5 May	Estates General meet in Versailles
	14 July	Fall of the Bastille
1790		M. de la Tour du Pin made Minister for War
	19 May	Birth of Humbert
1791	20 June	Flight of royal family to Varennes
	October	Lucie and Frédéric (and Humbert) leave for Holland where he is now ambassador
1792	March	Frédéric dismissed
	20 April	France declares war on Austria
	August	Prussian and Austrian troops invade France
	10 August	The storming of the Tuileries and the massacre of the Swiss Guard
	22 September	French Republic proclaimed
1793	January	Lucie returns to Paris
	21 January	Louis XVI guillotined
	March	Lucie and Frédéric go to Le Bouilh, Bordeaux
	September	Birth of Séraphine
	October	Marie Antoinette guillotined
1794	January	Terror reaches its peak in Bordeaux
	March	Lucie, Frédéric and the children leave for America
	13 April	Arthur Dillon is guillotined
	28 April	Jean-Frédéric de la Tour du Pin is guillotined
	June	Lucie and Frédéric buy a farm near Albany
1795	September	Death of Séraphine

1796	6 May	Lucie, Frédéric and Humbert return to Bordeaux via Spain
	1 November	Birth of Charlotte
	2 November	Directoire set up; it would last until November 1799
1797	July	Lucie, Frédéric, Humbert and Charlotte go to Paris
	4 September	*Coup d'état* of 18 fructidor
	November	Lucie, Frédéric and the children flee to England
1798		Birth of Edward and, three months later, his death
1799		Lucie, Frédéric, Humbert and Charlotte return to Paris via Holland
	November	Consulat set up, with Napoleon as First Consul
1800	13 February	Lucie gives birth to Cécile
	September	Family goes to settle at Le Bouilh
1802	25 March	Treaty of Amiens signed with Britain
	2 August	Napoleon appointed First Consul for Life
1804	28 May	Napoleon becomes Emperor
1805–6		Series of military victories by Napoleon across Europe
1806	18 October	Birth of Aymar
1808	12 May	Frédéric appointed Prefect of the Dyle and family moves to Brussels
1810	1–2 April	Having divorced Josephine, Napoleon marries Marie-Louise of Austria
	End of April	Lucie attends on Marie-Louise in Brussels
1813	May	Marriage of Charlotte to Auguste de Liederkerke
		Frédéric dismissed from Brussels but appointed Prefect of Amiens. Family moves to Amiens.
1812		Disastrous Russian campaign
1814	1 January	The Allies invade France
	20 April	Napoleon sails for Elba
	3 May	Louis XVIII arrives in Paris
	1 November	Frédéric sent to represent France at Congress of Vienna
		Lucie settles in Paris
1815	20 March	Napoleon returns to Paris and Louis XVIII flees to Ghent
		Lucie returns to Brussels
	18 June	Napoleon defeated at Waterloo and sent to be held on St Helena; in his entourage are Lucie's half-sister Fanny and her husband
		Frédéric returns as Ambassador to Holland
1816	28 January	Death of Humbert, aged 25, in a duel
1817	20 March	Death of Cécile, at the age of 17, from tuberculosis

TIME LINE

1820	1 January	Lucie decides to write her memoirs
		Frédéric appointed Ambassador to Turin
1822	1 September	Death of Charlotte, at 25, from tuberculosis. Her 2-year-old daughter, Cécile, comes to live with Lucie
1824	16 September	Death of Louis XVIII; his brother ascends throne as Charles X
1830	February	Lucie and Frédéric visit Paris for the first time in 10 years
	2 August	Abdication of Charles X; Louis-Philippe becomes King
		Frédéric resigns and moves back to Le Bouilh
1832		Aymar implicated in failed coup by the Duchesse de Berri
	December	Frédéric is sent to prison in the Fort du Hâ
1833	20 March	Frédéric is released and they move to Italy
1836		Félicie lends them her house in Lausanne
1837	26 February	Death of Frédéric
1842	November	Lucie and Aymar move first to Lucca and then to Pisa
1848		Year of revolution in Europe
1852		Louis-Napoleon adopts the title of Napoleon III
1853	2 April	Death of Lucie in Pisa at the age of 83

Index

INDEX

INDEX

INDEX